Data Analysis Using Stata

Second Edition

ULRICH KOHLER
Wissenschaftszentrum Berlin für Sozialforschung
Berlin, Germany

FRAUKE KREUTER
Joint Program in Survey Methodology
University of Maryland
College Park, Maryland

A Stata Press Publication
StataCorp LP
College Station, Texas

Second edition 2009

Published by Stata Press, 4905 Lakeway Drive, College Station, Texas 77845
Typeset in LaTeX 2ε
Printed in the United States of America
10 9 8 7 6 5 4 3 2 1

ISBN-10: 1-59718-046-7
ISBN-13: 978-1-59718-046-7

Acknowledgments

This is now the second English edition of our book, which was first published in Germany in 2001. It has been an exciting journey, and we are proud that our book has become stronger and stronger as Stata has evolved. We are extremely grateful to Sophia Rabe-Hesketh for her extensive and very useful comments on several drafts. We thank our students in the Joint Program of Survey Methodology at the University of Maryland, College Park and at the Wissenschaftszentrum Berlin for their many questions pointing to sections that were unclear. We also benefited greatly from several discussions with Richard Valliant about the peculiarities of survey data. Several of our revisions were inspired by Phil Schumm's review in the *Stata Journal*. Participants of Statalist suggested that we include exercises, which we were happy to do with the help of Michael Lemay.

Textbooks, and in particular self-learning guides, improve especially through feedback from readers. Among many others we therefore thank: C. Balsam, J. Boy, J. Breyer, C. Brzinsky-Fay, N. Eger, U. Fingerlos, S. Gundert, A. Hasudungan, A. Hense, B. Hinderlich, C. Jorns, M. Kiesel, G. Koch, V. Mahlberg, A. Marold, S. Mensah, S. Müller, V. Nakavachara, D. Neff, A. Neumeyer, T. Pereira, S. Radyakin, L. Reinhard, C. Schnirch, U. Schuhmacher, K. Sonderhof, and A. Spermann.

Many other factors contributed to creating a usable textbook. Half the message of the book would have been lost without good data. We greatly appreciate the help and data we received from the German Institute for Economic Research (DIW). We also used datasets from the Stata manuals to expand the scope in the exercise sections. Maintaining an environment that allowed us to work on this project was not always easy; we thank Jens Alber and Roger Tourangeau for supporting us at work and thank our families and friends for supporting us at home.

We both take full and equal responsibility for the entire book. We can be reached at kkstata@web.de, and we always welcome notice of any remaining errors and suggestions for improvements.

Berlin and College Park
October 2008

Ulrich Kohler
Frauke Kreuter

Contents

Tables

Figures

Preface

As you may have guessed, this book discusses data analysis, but especially data analysis *using* Stata. We intend for this book to be an introduction to Stata, but at the same time the book also explains, for beginners, the techniques used to analyze data.

Data Analysis Using Stata does not merely discuss Stata commands but demonstrates all the steps of data analysis using practical examples. The examples are related to public issues, such as income differences between men and women, and elections, or to personal issues, such as rent and living conditions. This approach allows us to avoid using social science theory in presenting the examples and to rely on common sense. We want to emphasize that these familiar examples are merely standing in for actual scientific theory, without which data analysis is not possible at all. We have found that this procedure makes it easier to teach the subject and use it across disciplines. Thus this book is equally suitable for biometricians, econometricians, psychometricians, and other "metricians"—in short, for all who are interested in analyzing data.

Our discussion of commands, options, and statistical techniques is in no way exhaustive but is intended to provide a fundamental understanding of Stata. Having read this book and solved the problems in it, the student should be able to solve all further problems to which Stata is applicable.

We strongly recommend to both beginners and advanced readers that they read the preface and the first chapter (*"The first time"*) attentively. Both serve as a guide throughout the book. Beginners should read the chapters in order while sitting in front of their computers and trying to reproduce our examples. More advanced users of Stata may benefit from the extensive index and may discover one or another useful trick when they look up a certain command. They may even throw themselves into programming their own commands. Those who do not (yet) have access to Stata are invited to read the chapters that focus on data analysis, to enjoy them, and maybe to translate one or another hint (e.g., about diagnostics) into the language of the statistical package they do have access to.

Structure

"The first time" (chapter 1) shows what a typical session of analyzing data could look like. To beginners, this chapter conveys a sense of Stata and explains some basic concepts such as variables, observations, and missing values. To advanced users who already have experience in other statistical packages, this chapter offers a quick entry into Stata.

They will find many cross-references within this chapter, which can therefore be viewed as an extended table of contents.

The rest of the book is divided into three parts:

Chapters 2–6 serve as an introduction to the basic tools of Stata. Throughout the subsequent chapters, these tools are used extensively. It is not possible to portray the basic Stata tools, however, without using some of the statistical techniques explained in the second part of the book. The techniques described in chapter 6 may not seem useful until you begin working with your own results, so you may want to skim chapter 6 now and read it more carefully when you need it.

Throughout chapters 7–9, we show examples of data analysis. In chapter 7, we present techniques for describing and comparing distributions. Chapter 8 introduces linear regression using Stata. It explains in general terms the technique itself and shows how to run a regression analysis using an example file. Afterward, we discuss how to test the statistical assumptions of the model. We conclude the chapter with a discussion of sophisticated regression models and a quick overview of further techniques. Chapter 9, in which we describe regression models for categorical dependent variables, is structured in the same way as the previous chapter to emphasize the similarity between these techniques.

Chapters 10–12 deal with more advanced Stata topics that beginners may not need. In chapter 10, we explain how to read and write files that are not in the Stata format. At the beginning of chapter 11, we introduce some special tools to aid in writing do-files. You can use these tools to create your own Stata commands and then store them as ado-files, which are explained in the second part of the chapter. It is easy to write Stata commands, so many users have created a wide range of additional Stata commands that can be downloaded from the Internet. Chapter 12 discusses these and other resources.

Using this book: Materials and hints

The only way to learn how to analyze data is to do it. To help you learn by doing, we have provided data files (available on the Internet) that you can use with the commands we discuss in this book. You can access these files from within Stata or by downloading a zip archive.

Please do not hesitate to contact us if you have any trouble obtaining these data files and do-files.[1]

- If the machine you are using to run Stata is connected to the Internet, you can download the files from within Stata. To do this, type the following commands in the Stata Command window (see the beginning of chapter 1 for information about using Stata commands).

1. The data we provide and all commands we introduce assume that you use Stata 10 or higher. Please contact us if you have an older version of Stata.

```
. mkdir c:\data\kk2
. cd c:\data\kk2
. net from http://www.stata-press.com/data/kk2/
. net get data
. mkdir kksoep
. cd kksoep
. net get kksoep
. cd ..
```

If you are using a Macintosh or Unix system, substitute a suitable directory name in the first two commands.

- The files are also stored as a zip archive, which you can download by pointing your browser to http://www.stata-press.com/data/kk2/kk2.zip.

  To extract the file `kk2.zip`, create a new folder: `c:\data\kk2`. Copy `kk2.zip` into this folder. Unzip the file `kk2.zip` using any program that can unzip zip archives. Most computers have such a program already installed. If not, you can get one for free using the Internet.[2] Make sure to preserve the `kksoep` subdirectory contained in the zip file.

Throughout the book, we assume that your current working directory (folder) is the directory where you have stored our files. This is important if you want to reproduce our examples. At the beginning of chapter 1, we will explain how you can check where your current working directory is. Make sure that you do not replace our files with a modified version of the same file; avoid using the command `save, replace` while working with our files.

We cannot say it too often: the only way to learn how to analyze data is to analyze data yourself. We strongly recommend that you reproduce our examples in Stata as you read this book. A line that is written `in this font` and begins with a period (which itself should not be typed by the user) represents a Stata command, and we encourage you to enter that command in Stata. Typing the commands and seeing the results or graphs will help you better understand the text, since we sometimes omit output to save space.

As you follow along with our examples, you must type all commands shown, as they build on each other within a chapter. Some commands will only work if you have entered the previous commands. If you do not have time to work through a whole chapter at once, you can type the command

```
. save mydata, replace
```

before you exit Stata. When you get back to your work later, type

```
. use mydata
```

and you will be able to continue where you left off.

2. For example "pkzip" is free for private use, developed by the company PKWARE. You can find it at http://www.pkware.com.

Most examples use a slightly modified version of the German Socio-Economic Panel, either on a single year (1997) or on the entire panel dataset from 1984 to 2002. In some sections, we use other data to explain specific tasks. At the beginning of these sections, you will see the command `preserve` and at the end, the command `restore`. If you are working through such a sequence and you need to interrupt your work, type `restore` before you type the command `save mydata, replace`.

The exercises at the end of each chapter use either data from our data package or data used in the Stata manuals. StataCorp provides these datasets online.[3] They can be used within Stata using the command `webuse` *filename*. However, this command assumes that your computer is connected to the Internet. Otherwise, you have to download the respective files manually from a different computer.

This book contains a lot of graphs, only one of which was not generated with Stata. You can reproduce all other graphs from the analysis examples. For the more complicated graphs, we have included do-files in our file package so that you can reproduce these graphs (the name of the do-file needed for each graph is given in a footnote under the graph).

If you do not understand our explanation of a particular Stata command or just want to learn more about it, use the Stata `help` command, which we explain in chapter 1. Or you can see the printed manuals (to which the online help refers) for more details. For example, [R] **summarize** refers to the entry describing the `summarize` command in the *Stata Base Reference Manual*. [U] **18 Programming Stata** refers to chapter 18 of the *Stata User's Guide*. When you see a reference to a keyword in one of the *Reference* manuals, you can use the online help (see section 1.3.17) to get information on that keyword.

Teaching with this manual

We have found this book to be useful for introductory courses in data analysis, as well as for courses on regression and on the analysis of categorical data. We have used it in courses at universities in Germany and the United States. When developing your own course, you might find it helpful to use the following outline of a course of lectures of 90 minutes each, held in a computer lab.

To teach an introductory course in data analysis using Stata, we recommend that you begin with chapter 1, which is designed to be an introductory lecture of roughly 1.5 hours. You can give this first lecture interactively, asking the students substantive questions about the income difference between men and women. You can then answer them by entering Stata commands, explaining the commands as you go. Usually, the students name the independent variables used to examine the stability of the income difference between men and women. Thus you can do a stepwise analysis as a question-and-answer game. At the end of the first session, the students should save their commands in a log file, and as a homework assignment they should produce a commented do-file (it might be helpful to provide them with a template of a do-file).

3. http://www.stata-press.com/data/r10/

The next two lectures should work with chapters 3–5 and can be taught a bit more conventionally than the introduction. It will be clear that your students will need to learn the *language* of a program first. These two lectures need not be taught interactively but can be delivered section by section without interruption. At the end of each section, give the students time to retype the commands and ask questions. If time is limited, you can skip over sections 3.3 and 5.6, whereas you should make time for a detailed discussion of sections 5.1.3 and 5.1.4 and the examples in them. Both sections contain concepts that will be unfamiliar to the student but are very powerful tools for users of Stata.

One additional session should suffice for an overview of the commands and some interactive practice in the graphs chapter.

Two sessions can be scheduled for chapter 7. One example for a set of exercises to go along with this chapter is given by Donald Bentley and is described on the web page http://www.amstat.org/publications/jse/v3n3/datasets.dawson.html. The necessary files are included in our file package.

Three sessions should be scheduled for chapter 8. According to our experience, even with an introductory class, you can cover sections 8.1, 8.2, and 8.3 in one session each. We recommend that you let the students calculate the regressions of the Anscombe data (see page 203) as a homework assignment or an in-class activity before you start the session on regression diagnostics. Also we recommend that toward the end of the course, you spend two sessions on chapter 10 introducing data entry, management, and the like, before you end the class with chapter 12, which will point the students to further Stata resources.

In addition to using this book for a general introduction to data analysis, you can use it to develop a course on regression analysis (chapter 8) or categorical data analysis (chapter 9). As with the introductory courses, it is helpful to begin with chapter 1, which gives a good overview of working with Stata and solving problems using the online help. Chapter 12 makes a good summary for the last session of either course.

1 “The first time”

Welcome! In this chapter, we will show you several typical applications of computer-aided data analysis to illustrate some basic principles of Stata. Advanced users of data analysis software may want to look through the book for answers to specific problems instead of reading straight through. Therefore, we have included many cross-references in this chapter as a sort of distributed table of contents. If you have never worked with statistical software, you may not immediately understand the commands or the statistical techniques behind them. Do not be discouraged. Reproduce our steps anyway. If you do, you will get some training and experience working with Stata. You will also get used to our jargon and get a feel for how we do things. If you have specific questions, the cross-references in this chapter can help you find answers.

Before we begin, you need to know that Stata is command-line oriented, meaning that you type a combination of letters, numbers, and words at a command line to perform most of Stata's functions. With Stata 8 and later versions, you can access most commands through pulldown menus. However, we will focus on the command line throughout the book for several reasons. We think the menu is rather self-explanatory. If you know the commands, you will be able to find the appropriate menu items. The look and feel of the menu depends on the operating system installed on your computer, so using the command line will be more consistent, no matter what system you are using. Also switching between the mouse and the keyboard can be tedious. Finally, once you are used to typing the commands, you will be able to write entire analysis jobs, so you can later replicate your work or easily switch across platforms. At first you may find using the command line bothersome, but as soon as your fingers get used to the keyboard, it becomes fun. Believe us, it is habit forming.

1.1 Starting Stata

We assume that Stata is installed on your computer as described in the *Getting Started* [GS] manual for your operating system. If you work on a PC using the Windows operating system, you can start Stata by selecting **Start** > **All Programs** > **Stata 10**.

On a Macintosh system, you start Stata by double-clicking on the Stata symbol. Unix users type the command `xstata` in a shell. After starting Stata, you should see the default Stata windowing: a Results window; a Command window, which contains the command line; a Review window; and a Variables window, which shows the variable names.

1.2 Setting up your screen

Instead of explaining the different windows right away, we will show you how to change the default windowing. In this chapter, we will focus on the Results window and the command line. You may want to choose another font for the Results window so that it is easier to read. Right-click in the Results window. In the pop-up menu, choose **Font...** and then the font you prefer.[1] If you choose the suggested font size, the Results window may not be large enough to display all the text. You can resize the Results window by dragging the borders of the window with the pointer until you can see the entire text again. If you cannot do this because the Stata background window is too small, you must resize the Stata background window before you can drag the Results window to the size you want.

Make sure that the Command window is still visible. If necessary, move the Command window to the lower edge of the Stata window. To move a window, point to the title of the window and hold the left mouse button as you move the window. For beginners, it is helpful to dock the Command window with a double-click to the background. Stata for Windows has many options for manipulating the window layout; see [GSW] **4 The Stata user interface** for more details.

Your own windowing will be saved as the default when you exit Stata. You can restore the initial windowing by selecting **Edit** > **Preferences** > **Manage Preferences** > **Load Preferences** > **Factory Settings**. You can have multiple sets of saved preferences; see [GS] **18 Setting font and window preferences**.

1.3 Your first analysis

1.3.1 Inputting commands

Now we can begin. Type the character `d` at the command line, and press *Enter* or *Return*. You should see the following text in the Results window:

```
. d
Contains data
  obs:              0
 vars:              0
 size:              0 (100.0% of memory free)
Sorted by:
```

1. In Mac OS X, right-click on the window you want to work with, and from **Font Size**, select the font size you prefer. In Unix, right-click on the window you want to work with, and select **Preferences...** to display a dialog allowing you to choose the fonts for the various windows.

You have now typed your first Stata command. The letter `d` is an abbreviation for the command `describe`, which describes data files in detail. As you are not working with a data file yet, the result is, of course, not very exciting. However, you can see that entering a command in Stata means that you type a letter or several letters (or words) and press *Enter*.

Throughout the book, every time you see a word in `this font` preceded by a period, you should type the word in the command line and press *Enter*. You type the word without the preceding period and you must preserve uppercase and lowercase letters. Stata is case sensitive. In the example below, you type `describe` at the command line:

```
. describe
```

1.3.2 Files and the working memory

The output of the above `describe` command is more interesting than it seems. In general, `describe` provides information about the number of variables and number of observations in your dataset.[2] As we did not load a dataset, `describe` shows zero variables ("vars") and observations ("obs").

`describe` also indicates the percentage of the working memory (RAM) allocated to Stata. Unlike many other statistical software packages, Stata loads the entire data file into the working memory of your computer. Most of the working memory is reserved for data, and some parts of the program are loaded only as needed. This system ensures quick access to the data and is one reason why Stata is much faster than many other conventional statistical packages.

The working memory of your computer gives a physical limit to the size of the datasets with which you can work. Thus you might have to install more memory to load really big data files. But given the usual hardware configurations today, problems with the size of the data file are rare.

Besides buying new memory, there are a few other things you can do if your computer is running out of memory. We will explain what you can do in section 10.6.

1.3.3 Loading data

Let us load a dataset. To make things easier in the long run, change to the directory where the data file is stored. In what follows, we assume that you have copied our datasets into `c:\data\kk2`.

To change to another directory, use the command `cd`, which stands for "change directory", followed by the name of the directory to which you want to change. If the directory name contains blanks (spaces), you must enclose the name in double quotes. To move to the proposed data directory, type

```
. cd c:\data\kk2
```

2. You will find more about the terms *variable* and *observation* on pages 5–6.

With Macintosh you can use colons instead of slashes. Alternatively, slashes can be used on all operating systems.

Depending on your current working directory and operating system, there may be easier ways to change to another directory (see [D] **cd** for details). You will also find more information about folder names in section 3.1.8 on page 53.

Check that your current working directory is the one where you stored the data files by typing `dir`, which shows a list of files that are stored in the current folder:

```
. dir
  <dir>   4/21/08  7:41  .
  <dir>   4/21/08  7:41  .
   1.1k   4/21/08  7:36  an1cmdkk.do
   0.2k   4/21/08 11:31  an1kk.do
   2.2k   4/21/08 11:31  an2kk.do
   0.5k   4/21/08 11:31  an2_0kk.do
   1.6k   4/21/08 11:31  analwe.dta
   1.5k   4/21/08 11:31  anbeta.do
   0.2k   4/21/08  7:36  anchap1.do
  (output omitted)
```

Depending on your operating system, the output may look slightly different. You will not see the line indicating that some of the output is omitted. We use this line throughout the book to save space.

In displaying results, Stata pauses when the Results window fills and it displays `—more—` on the last line if there are more results to display. You can display the next line of results by pressing *Enter* or the next page by pressing any other key except the letter *q*. You can use the scroll bar at the side of the Results window to go back and forth between pages of results. To change the maximum number of lines, you can scroll back to—say, 50,000—by typing

```
. set scrollbufsize 50000
(set scrollbufsize will take effect the next time you launch Stata)
```

on the command line, but this will only take effect when you restart Stata.

When you typed `dir`, you should have seen a file called `data1.dta` among those listed. If there are a lot of files in the directory, it may be hard to find a particular file. To reduce the number of files displayed at a time, you can type

```
. dir *.dta
```

or display only the desired file by typing `dir data1.dta`. Once you know that your current working directory is set to the correct directory, you can load the file `data1.dta` by typing

```
. use data1
```

The command `use` loads Stata files into working memory. The syntax is straightforward: Type `use` and the name of the file you want to use. If you do not type a file extension after the filename, Stata assumes the extension `.dta`.

For more information about loading data, see chapter 10. That chapter may be of interest if your data are not in a Stata file format. Some general hints about filenames are given in section 3.1.8.

1.3.4 Variables and observations

Once you load the data file, you can look at its contents by typing

```
. describe

Contains data from data1.dta
  obs:         3,340                          SOEP'97 (Kohler/Kreuter)
 vars:            47                          19 Jun 2003 16:33
 size:       237,140 (98.9% of memory free)   (_dta has notes)
-------------------------------------------------------------------------------
              storage  display     value
variable name   type   format      label      variable label
-------------------------------------------------------------------------------
persnr          long   %8.0f                  Person ID
intnr           long   %8.0f                * Interviewer ID
state           byte   %8.0f       bul      * State (Bundesland 97)
gender          byte   %8.0f       sex        Gender
ybirth          int    %8.0f                  Year of Birth (YYYY)
ymove           byte   %8.0f                * Year moved in place
ybuild          byte   %8.0f       bauj     * Year house was build
hcond           byte   %23.0f      renov    * House condition (renovation or
                                                not)
sqfeet          float  %8.0f                * Home size in sqft.
rooms           byte   %8.0f                * Number of rooms (bigger than 18
                                                sqft.)
fseval          byte   %14.0f      wgurt    * Home size evaluation
kitchen         byte   %8.0f       janein   * Kitchen yes/no
shower          byte   %8.0f       janein   * Bath/Shower yes/no
wc              byte   %8.0f       janein   * Toilet yes/no
heating         byte   %8.0f       janein   * Central Heating yes/no
cellar          byte   %8.0f       janein   * Cellar yes/no
balcony         byte   %8.0f       janein   * Balcony oder terrace yes/no
garden          byte   %8.0f       janein   * Garden (yard) yes/no
phone           byte   %8.0f       janein   * Phone yes/no
renttype        byte   %11.0f      wohnst   * Status of living
rent            int    %8.0f                * Rent monthly
renteval        byte   %12.0f      mietur   * Evaluation of monthly rent
hhtype          byte   %26.0f      ntyphh1  * Type of household
htype           byte   %21.0f      htyp     * Type of house
area            byte   %16.0f      wum      * Neighborhodd
np11701         byte   %16.0f      zuf        General life satisfaction
np0105          byte   %16.0f      zuf        Satisfaction with living
                                                conditions
np9401          byte   %10.0f      pia        Party identification - yes/no
np9402          byte   %8.0f       pib        Party identification - Party
  (output omitted)
```

The data file `data1.dta` is a subset of the German Socio-Economic Panel (GSOEP), a longitudinal study of a sample of private households in Germany. The same households, individuals, and families have been surveyed yearly since 1984 (GSOEP-West). To protect data privacy, the file used here contains only information on a random subsample of all

GSOEP respondents, with some minor random changes of some information. The data file includes 3,340 respondents, called observations (*obs*). For each respondent, different information is stored in 47 variables (*vars*), most of which contain the respondent's answers to questions of the GSOEP survey questionnaire.

Throughout the book, we use the terms "respondent" and "observations" interchangeably to refer to units for which information has been collected. A detailed explanation of these and other terms is given in section 10.1.

Below the second solid line in the output is a description of the variables. The first variable is `persnr`, which, unlike most of the others does not contain survey data. It is a unique identification number for each person. The remaining variables include information about the interviewer who conducted the interview, the state in which the respondent lives, the respondent's year of birth, and so on. To get an overview of the names and contents of the remaining variables, scroll down. (Remember, you move one line further down by pressing *Enter* and one whole page by pressing any other key except the letter *q*.)

To begin with, we want to focus on a subset of variables. For now we are less interested in the information about housing than we are about information on respondents' political attitudes. Therefore, we want to remove from the working dataset all variables in the list from the variable recording the year the respondent moved into the current place (`ymove`) to the variable describing the respondents' worries about employment stability (`np9507`):

```
. drop ymove-np9507
```

1.3.5 Looking at data

Using the command `list`, we get a closer look at the data. The command lists all the contents of the data file. You can look at each observation by typing

```
. list

     +------------------------------------------------------------------------+
  1. | persnr    intnr  |  state   gender |  ybirth                      hhpos |
     |   2229   145700  |     NW      men |    1955          head of household |
     |------------------------------------------------------------------------|
     |        hhsize    |         marital    |                            edu  |
     |             1    |       unmarried    |                          other  |
     |------------------------------------------------------------------------|
     |                          voc |  yedu | emp | occ | hhinc |    income   |
     |                           no | 10.00 |   1 |  61 |   915 |       832   |
     |------------------------------------------------------------------------|
     |                                  egph                                  |
     |                semi- and unskilled manual workers                      |
     +------------------------------------------------------------------------+
```

```
  2. | persnr    intnr      state   gender   ybirth                hhpos |
     |   3994   256862     Thuer.      men     1971    head of household |
     |--------------------------------------------------------------------|
     |     hhsize            marital                                  edu |
     |          1          unmarried              maturity qualification |
     |--------------------------------------------------------------------|
     |                    voc    yedu   emp   occ   hhinc           income |
     |                     no   13.00     7    42     813                0 |
     |--------------------------------------------------------------------|
     |                                egph                                 |
     |                                   .                                 |

  3. | persnr    intnr      state   gender   ybirth                hhpos |
     |   6326   166979   Schl.Hst    women     1980                child |
     |--------------------------------------------------------------------|
     |     hhsize            marital                                  edu |
     |          4          unmarried                     still in school |
     |--------------------------------------------------------------------|
     |                    voc    yedu   emp   occ   hhinc           income |
     |                     no    7.00     7    53    4830                0 |
     |--------------------------------------------------------------------|
     |                                egph                                 |
     |                     service class 1                                 |

  4. | persnr    intnr      state   gender   ybirth                hhpos |
     |   8660   120826         NW      men     1980                child |
     |--------------------------------------------------------------------|
     |     hhsize            marital                                  edu |
     |          6          unmarried                     still in school |
     |--------------------------------------------------------------------|
     |                    voc    yedu   emp   occ   hhinc           income |
     |                     no    7.00     7    62    2398                0 |
     |--------------------------------------------------------------------|
     |                                egph                                 |
     |              skilled manual workers                                 |
```

(output omitted)

In a moment, you will see how to make `list` show only certain observations and how to reduce the amount of output. The first observation is a man born in 1955; he is single and employed full-time. The second observation is another single man, born in 1971. The period as the entry for the variable `egph` indicates that there is no information recorded in this variable for person number 3,994. There are various possible reasons for this; for example, the interviewer never asked this particular question of the man, or the man refused to answer it. If a period appears as an entry, Stata calls it a "missing value" or just "missing".

In Stata, a period or a period followed by any character `a` to `z` indicates a missing value. Later in this chapter, we will show you how to define missings (see page 12). A detailed discussion on handling missing values in Stata is provided in section 5.4, and some more general information can be found on page 310.

1.3.6 Interrupting a command and repeating a command

Not all the observations in this dataset can fit on one screen, so you may have to scroll through many pages to get a feel for the whole dataset. Before you do so, you may want to read this section.

Scrolling through more than 3,000 observations is tedious, so using the `list` command is not very helpful with a large dataset like this. Even with a small dataset, `list` can display too much information to process easily. However, sometimes you can take a glance at the first few observations to get a first impression or to check on the data. In this case, you would probably rather stop listing and avoid scrolling to the last observation. You can stop the printout by pressing *q*, for quit. Anytime you see —`more`— on the screen, pressing *q* will stop listing results.

Rarely will you need the key combination *Ctrl+Break* (Windows), *command-.* (Macintosh), or *Break* (Unix), which is a more general tool to interrupt Stata.

1.3.7 The variable list

Another way to reduce the amount of information displayed by `list` is to specify a "variable list". When you append a list of variable names to a command, the command is limited to that list. By typing

```
. list gender income
```

you get information on gender and monthly net income for each observation.

To save some typing, you can access a previously typed `list` command by pressing *Page Up*, or you can click once on the command `list` displayed in the Review window. After the command is displayed again in the command line, simply insert the variable list of interest. Another shortcut is to abbreviate the command itself, in this case by typing the character *l* (lowercase letter L). A note on abbreviations: Stata commands are usually short. However, several commands can be shortened even more, as we will explain in section 3.1.1. You can also abbreviate variable names; see "Abbreviation rules" in section 3.1.2.

Scrolling through 3,340 observations might not be the best way to learn how the two variables `gender` and `income` are related. For example, we would not be able to judge whether there are more women or men in the lower-income groups.

1.3.8 The in qualifier

To get an initial impression of the relationship between gender and income, we might examine the gender of the 10 respondents who earn the least. It would be reasonable to first sort the data on the variable `income` and then list the first 10 observations. We can list only the first 10 observations by using the `in` qualifier:

```
. sort income
. list gender income in 1/10

     +-----------------+
     | gender   income |
     |-----------------|
  1. |  women        0 |
  2. |  women        0 |
  3. |    men        0 |
  4. |  women        0 |
  5. |  women        0 |
     |-----------------|
  6. |  women        0 |
  7. |  women        0 |
  8. |  women        0 |
  9. |    men        0 |
 10. |    men        0 |
     +-----------------+
```

The `in` qualifier allows you to restrict the `list` command or almost any other Stata command to certain observations. You can write an `in` qualifier after the variable list, or, if there is no variable list, after the command itself. You can use the `in` qualifier to restrict the command to data that occupy a specific position within the dataset. For example, you can obtain the values of all variables for the first observation by typing

```
. list in 1
```

and you can obtain the values for the second to the fourth observations by typing

```
. list in 2/4
```

The current sort order is crucial for determining each observation's position in the dataset. You can change the sort order by using the `sort` command. We sorted by `income`, so the person with the lowest income is found at the first position in the dataset. Observations with the same value (in this case, income) are sorted randomly. However, you could sort by `gender` among persons with the same income by using two variable names in the sort command, for example, `sort income gender`. Further information regarding the `in` qualifier can be found in section 3.1.4.

1.3.9 Summary statistics

Researchers are not usually interested in the specific answers of each respondent for a certain variable. In our example, looking at every value for the `income` variable did not provide much insight. Instead, most researchers will want to reduce the amount of information and use graphs or summary statistics to describe the content of a variable. Probably the best-known summary statistic is the arithmetic mean, which you can obtain using the `summarize` command. The syntax of `summarize` follows the same principles as the `list` command and most other Stata commands: the command itself is followed by a list of the variables that the command should use.

You can obtain summary statistics for income with

```
. summarize income

    Variable |       Obs        Mean    Std. Dev.       Min        Max
-------------+--------------------------------------------------------
      income |      3034    1349.207    1245.701          0      12438
```

This table contains the arithmetic mean (`Mean`), as well as information on the number of observations (`Obs`) used for this computation, the standard deviation (`Std. Dev.`) of the variable `income`, and the smallest (`Min`) and largest (`Max`) values of income in the dataset.

As you can see, only 3,034 of the 3,340 observations were used to compute the mean because there is no information on income available for the other 306 respondents—306 people have a missing value for income. The average income of those respondents who reported their income is $1,349. The minimum is $0, which means there is at least one person in the dataset who has no personal income. The highest reported income in this dataset is $12,438 a month. The standard deviation of income is approximately $1,246.

As with the `list` command, you can `summarize` a list of variables. If you use `summarize` without specifying a variable, summary statistics for all variables in the dataset are displayed:

```
. summarize

    Variable |       Obs        Mean    Std. Dev.       Min        Max
-------------+--------------------------------------------------------
      persnr |      3340     3628319     2094513       2229    7254426
       intnr |      2779    160384.1    79833.91         19     505404
       state |      3339    8.245583    4.277823          0         16
      gender |      3340    1.522455    .4995703          1          2
      ybirth |      3340     1951.72    18.33337       1902       1981
-------------+--------------------------------------------------------
       hhpos |      3340    .6625749    1.234272          0         11
      hhsize |      3340    2.576946    1.291716          1         11
     marital |      3340    2.142515    1.418671          1          6
         edu |      3312    2.479771    1.667339          1          7
         voc |      3302    3.107208    2.522246          1         10
-------------+--------------------------------------------------------
        yedu |      3292    11.38548    2.408573          7         18
         emp |      3340    3.695509    2.866571          1          7
         occ |      2100    51.71048    12.53318         10         70
       hhinc |      3201    1921.357    1068.343         18      14925
      income |      3034    1349.207    1245.701          0      12438
-------------+--------------------------------------------------------
        egph |      2079    5.246753    3.190941          1         11
```

In chapter 7, we will discuss further statistical methods and graphical techniques for displaying variables and distributions.

1.3.10 The if qualifier

Assume for a moment that you are interested in possible income inequality between men and women. You can determine if the average income is different for men and for women by using the `if` qualifier. The `if` qualifier allows you to process a command, such as the computation of an average, conditional on the values of another variable. However, to use the `if` qualifier, you need to know that in the `gender` variable men are coded as 1 and women are coded as 2. How you discover this will be shown on page 100.

If you know the actual values of the categories in which you are interested, you can use the following commands:

```
. summarize income if gender==1

    Variable |       Obs        Mean    Std. Dev.       Min        Max
-------------+--------------------------------------------------------
      income |      1455    1670.816    1293.905          0      10945

. summarize income if gender==2

    Variable |       Obs        Mean    Std. Dev.       Min        Max
-------------+--------------------------------------------------------
      income |      1579    1052.854    1121.029          0      12438
```

You must type a double equal-sign in the `if` qualifier. Typing a single equal-sign within the `if` qualifier is probably the most common reason for the error message "invalid syntax".

The `if` qualifier restricts a command to those observations where the value of a variable satisfies the `if` condition. Thus you see in the first table the summary statistics for the variable `income` only for those observations that have 1 as a value for `gender`—meaning men. The second table contains the mean income for all observations that have 2 (meaning women) stored in the variable `gender`.

Most Stata commands can be combined with an `if` qualifier. As with the `in` qualifier, the `if` qualifier must appear after the command and after the variable list, if there is one. When you are using an `in` qualifier with an `if` qualifier, the order in which they are listed in the command line does not matter.

Sometimes you may end up with very complicated `if` qualifiers, especially when you are using logical expressions such as "and" or "or". We will discuss these in section 3.1.5.

1.3.11 Define missing values

As you have seen in the table above, men earn on average substantially more than women, $1,671 compared with $1,053. However, you might argue that, for good reason, a higher percentage of women have no personal income. So, you argue, it would make more sense to compare only those people who actually have personal income. To achieve this goal, you can expand the `if` qualifier, for example, by using a logical "and" (see section 3.1.5).

Another way to exclude persons without personal income is to change the content of `income`. That is, you change the `income` variable so that an income of zero is recorded as a missing value, here stored with the missing-value code `.a`. This change automatically omits these cases from the computation. To do this, use the command `mvdecode`:

```
. mvdecode income, mv(0=.a)
      income: 366 missing values generated
```

This command will exclude values of zero in the variable `income` from your analysis.

There is much more to be said about encoding and decoding missing values. In section 5.4, you will learn how to reverse the command you just entered and how you can specify different types of missing values. For general information about using missing values, see page 310 in chapter 10.

1.3.12 The by prefix

Now let us see how you can use the `by` prefix to obtain the last table with a single command. A *prefix* is a command that is written preceding the main Stata command, separated by a colon. The command prefix `by` has two parts: the command itself and a variable list. We call the variable list that appears within the `by` prefix the *bylist*. When you include the `by` prefix, the original Stata command is repeated for all categories of the variables in the *bylist*. The dataset must be sorted by the variables in the *bylist*. Here is one example in which the *bylist* contains only the variable `gender`:

```
. sort gender

. by gender: summarize income

-------------------------------------------------------------------------------
-> gender = men

    Variable |       Obs        Mean    Std. Dev.       Min        Max
-------------+--------------------------------------------------------
      income |      1344    1808.808     1250.11          2      10945

-------------------------------------------------------------------------------
-> gender = women

    Variable |       Obs        Mean    Std. Dev.       Min        Max
-------------+--------------------------------------------------------
      income |      1324    1255.631    1115.402          1      12438
```

The output above is essentially the same as that on page 11, although the values have changed slightly, because we have changed the income variable using the `mvdecode` command. The `by` prefix changed only the table captions. However, compared with the `if` qualifier, the `by` prefix offers some advantages. The most important is that you do not have to know the values of each category. When you use `by`, you need not know whether the different genders are coded with 1 and 2 or with 0 and 1, for example.[3] The `by` prefix saves typing time, especially when the grouping variable has more than two categories or when you use more than one grouping variable. The `by` prefix allows

3. You can learn more about coding variables on page 310.

you to use several variables in the *bylist*. If the *bylist* contains more than one variable, the Stata command is repeated for all possible combinations of the categories of all variables in the *bylist*.

The `by` prefix is one of the most useful features of Stata. Even advanced users of other statistical software packages will be pleasantly surprised by its usefulness, especially when used in combination with commands to generate or change variables. For more on this topic, see sections 3.2.1 and 5.1.3.

1.3.13 Command options

Let us go back to exploring income inequality between genders. You might argue that using the arithmetic mean of `income`, even when combined with its minimum and maximum, is an inadequate way to compare the two subgroups. These values are not sufficient to describe the income distribution, especially if, as you may suspect, the distribution is positively skewed (skewed to the right). You can obtain more statistics from the `summarize` command by specifying options. Options are available for almost all Stata commands.

In contrast with the prefix commands and qualifiers discussed so far, options are command specific. For most commands, a certain set of options is available with a command-specific meaning. You specify options at the end of the command, after a comma.

The `summarize` command has only a few options. An important one is `detail`. Specifying this option returns several types of percentiles, among them the median (the 50th percentile) and the first and third quartiles (the 25th and 75th percentiles, respectively); the already known mean, minimum, and maximum; and the second through fourth "moments": the variance, skewness, and kurtosis.

```
. summarize income, detail

         Personal (work related) income before tax 1997
                              (monthly)
-------------------------------------------------------------
      Percentiles      Smallest
 1%           24              1
 5%          126              2
10%          243              3       Obs                2668
25%        582.5              3       Sum of Wgt.        2668

50%       1315.5                      Mean           1534.293
                        Largest       Std. Dev.      1216.817
75%       2204.5           9950
90%         2959          10945       Variance        1480643
95%         3482          11443       Skewness       1.898076
99%         5475          12438       Kurtosis       11.55445
```

The `detail` option does the same thing, even if you use `if` and `in` qualifiers and the `by` prefix command. You can add any list of variables or any `if` or `in` qualifiers, as well as any prefix; the function of the option is always the same. You can check this yourself by typing

```
. by gender: summarize income if edu==4, detail
```

After entering this command, you will obtain the income difference between men and women for all respondents who have at least the *Abitur* (which is the German qualification for university entrance). Interestingly enough, the income inequality between men and women remains, even if we restrict the analysis to the more highly educated respondents.

More general information about options can be found in section 3.1.3.

1.3.14 Frequency tables

In addition to simple descriptive statistics, frequencies and cross-classified tables (univariate and bivariate frequency tables) are some of the most common tools used for beginning a data analysis. The Stata command for generating frequency tables is `tabulate`. This command must include a variable list, consisting of one or two variables. If you use one variable, you get a one-way frequency table of the variable specified in the command:

```
. tabulate gender
```

Gender	Freq.	Percent	Cum.
men	1,595	47.75	47.75
women	1,745	52.25	100.00
Total	3,340	100.00	

If you specify two variables, you get a two-way frequency table:

```
. tabulate emp gender
```

Employment status 1997	Gender men	women	Total
1	870	691	1,561
2	83	160	243
3	37	37	74
4	17	32	49
6	9	4	13
7	579	821	1,400
Total	1,595	1,745	3,340

The first variable entered in the variable list of the `tabulate` command forms the row variable of the cross-classified table, and the second variable forms the column variable. Absolute frequencies are written in the table cells, the contents of which you can change using appropriate options. The crucial options for this command are `row` and `column`, which return row and column percentages. Other `tabulate` command options return information about the strength of the relationship between the two variables. As we explained for the `summarize` command, you can use options with `if` or `in` qualifiers, as well as command prefixes.

For more information about `tabulate`, see section 7.2.1. See below for an example of column percentages in a cross-classified table.

1.3.15 Variable labels and value labels

Looking at the table you just created, you decide that you need to make it more readable. Right now, the rows of the table show the numerals 1 to 7. Without knowing what these numbers mean, you will not know how to interpret them. Assigning labels to the variables and their values—and even to the entire dataset—can help make their meaning clear. In Stata, you assign labels using the `label` command.

See section 5.5 for a detailed explanation of the `label` command. Here we will simply show you how the `label` command works. First, let us change the label for the variable `emp`, which is currently "Employment status 1997", to "Status of employment in '97":

```
. label variable emp "Status of employment in '97"
```

The command is the same whether a label was previously assigned to the variable. To assign value labels to the values of `emp`, we type the command below. Because the command is too long for one line in this book, we had to break it. We used > to indicate this break, but you should not use the > in the command line; Stata breaks lines automatically.

```
. label define emplb 1 "full-time" 2 "part-time" 3 "retraining" 4 "irregular"
> 5 "not working" 6 "military service" 7 "unemployed", modify
. label values emp emplb
```

You may want to reenter the `tabulate` command used above so you can see the results of the labeling commands. The following command displays column percentages and suppresses the display of frequencies:

```
. tabulate emp gender, column nofreq

   Status of  |
employment in |        Gender
          '97 |       men      women |     Total
--------------+----------------------+----------
    full-time |     54.55      39.60 |     46.74
    part-time |      5.20       9.17 |      7.28
   retraining |      2.32       2.12 |      2.22
    irregular |      1.07       1.83 |      1.47
military service |   0.56       0.23 |      0.39
   unemployed |     36.30      47.05 |     41.92
--------------+----------------------+----------
        Total |    100.00     100.00 |    100.00
```

From this table, you can see that men are more often employed full-time than women (55% versus 40%). Women are more often employed part-time or not employed at all, which may partially explain our finding that women earn on average less than men. We would not be surprised to observe income differences based on employment status. A graph might help us check this hypothesis.

1.3.16 Graphs

Graphs provide a quick and informative way to look at data, especially distributions. Comparative box-and-whisker plots are a nice way to compare a distribution of one variable, such as `income`, across different subgroups, in this case employment status:

```
. graph box income, over(emp)
```

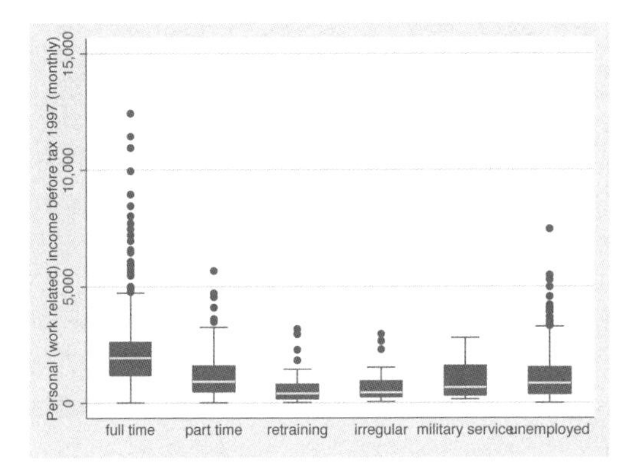

This graph command is composed of the `graph box` command, the variable `income`, and the option `over(emp)`, which specifies the grouping variable. (See chapter 6 for a discussion of other possible options.)

Looking at the six box plots shown in the Graph window, you can see that income is right skewed for all subgroups, with many outliers. The median (indicated by the middle line in the box) for the full-time employees is, as we assumed, higher than that for all the other groups. If there are relatively more part-time working women represented in the dataset, the median gross income must be smaller. Therefore, we might observe that income inequality is due to the division of labor within couples rather than due to gender discrimination at work. When we further analyze income inequality, we should at least control for employment status.

1.3.17 Getting help

From the results we obtained in the previous section, it seems reasonable to take into account the difference in employment status. In part, we took care of that by excluding all persons without income from our analysis. However, that did not affect the issue of part-time employees. One possible way to consider the effects of employment status and gender on income is to use linear regression.

We do not expect you to know how to do linear regression analysis at this point; we discuss it in detail in chapter 8. However, the following paragraphs should be—we hope—easy to understand, even if the techniques are not familiar.

Unfortunately, you do not yet know the Stata command for running a linear regression. This is, in fact, a very common situation: you learn about a statistical technique and would like to know if Stata has a built-in command for it. You can find out by using the `search` command.

The `search` command scans through a database of Stata resources for the expressions you entered and displays any entries that match. `search` is not case sensitive.

Here you could begin searching for one of the following terms:

```
. search Linear Regression
. search Model
. search OLS
```

Typing these commands will provide you with a series of entries, all of them somehow related to the searched-for term. For the first two commands, the list is rather long; the last command is quickest. Ordinary least squares (OLS) is an estimation technique used for linear regression. Some entries refer to articles in the *Stata Journal* (SJ) or its predecessor, the *Stata Technical Bulletin* (STB). Other entries are answers to frequently asked questions (FAQs) that have been given on Stata web pages. All these, as well as other resources and information about Stata, will be described in section 12.1.

Usually the entries refer to commands and their accompanying references to the *online help functions*. For example, among other entries you will find

```
[R]     regress . . . . . . . . . . . . . . . . . . . . . . Linear regression
        (help regress)
```

Now you know that there is a `regress` command for linear regression; you can see a detailed explanation by typing

```
. help regress
```

or by clicking on the blue word "regress".

To display the online help system, type `help` at the command line. You do not need Internet access to use the online help, as all the necessary information is already stored on your machine. The online help contains help text for every Stata command. The help text will be displayed on the screen, and you can scan the pages of the help text as usual. To see the help file for a specific command, type `help` followed by the name of the command.

Help entries are all structured the same way. They begin with the structure of the command—the "syntax diagram", followed by a fairly detailed explanation of the command's function and in turn by a description of available options. At the end of each help entry, you will find examples showing how to use the command, together with cross-references to related commands.

For example, the description of `regress` tells you that it fits a model of *depvar* on *indepvars* using linear regression. To do this, you must type the command itself followed by the dependent (endogenous) variable and then type the list of independent (exogenous) variables. Below we give an example of linear regression, and we will explain regression in chapter 8. For more information about reading Stata syntax, see chapter 3.

1.3.18 Recoding of variables

In our analysis of income inequality, income is the dependent variable while gender and employment status are independent variables. Unfortunately, the two independent variables are problematic:

1. Gender is a dichotomous nominal variable. Conventionally, such variables are included in a linear regression analysis coded 0 and 1 for the two outcomes. However, in our dataset, `gender` is coded as 1 and 2, so we recommend constructing a new variable in the required form:

   ```
   . generate men = 1 if gender == 1
   . replace men = 0 if gender == 2
   ```

   This command says, "Generate the variable `men` with the value 1 if the value of the variable `gender` is equal to 1, or leave it as missing otherwise. Then replace the values in the variable `men` with 0 if the value of the variable `gender` is equal to 2."

2. Employment status is also a nominal variable. But because it is not dichotomous, we cannot use its values the way they appear in the data file. However, right now we are interested only in a brief survey of income inequality, so we can limit our analysis to the dichotomy between full-time and part-time employment:

   ```
   . generate fulltime = 1 if emp == 1
   . replace fulltime = 0 if emp == 2
   ```

   The `fulltime` variable now has a missing value for all respondents that work neither full-time nor part-time.

You will find more examples and a detailed description of the commands for creating and changing variables in chapter 5. Recoding variables is probably the most time-consuming part of data analysis, so we recommend that you spend some time learning these commands.

1.3.19 Linear regression

Now that you have generated all the variables you want to include in the regression model, the remaining task is simple. The command to compute a linear regression is `regress`. Just type `regress` and, after the command itself, the name of the dependent variable, followed by a list of independent variables:

```
. regress income men fulltime
      Source |       SS       df       MS              Number of obs =    1560
-------------+------------------------------           F(  2,  1557) =   74.73
       Model |   225941371     2   112970685           Prob > F      =  0.0000
    Residual |  2.3538e+09  1557  1511751.31           R-squared     =  0.0876
-------------+------------------------------           Adj R-squared =  0.0864
       Total |  2.5797e+09  1559  1654739.04           Root MSE      =  1229.5

------------------------------------------------------------------------------
      income |      Coef.   Std. Err.      t    P>|t|     [95% Conf. Interval]
-------------+----------------------------------------------------------------
         men |   450.6853   63.14927     7.14   0.000     326.8187    574.5518
    fulltime |   805.6375   92.05069     8.75   0.000     625.0811    986.1939
       _cons |    965.281   87.49397    11.03   0.000     793.6626    1136.899
------------------------------------------------------------------------------
```

The average monthly income of respondents with values of zero on all independent variables—that is, the part-time working women—is $965. Their full-time working colleagues earn on average around $806 more (which means roughly $1,771 a month). Independent of the income difference between full-time and part-time workers, men still earn, on average, $451 more than women. Therefore, the income inequality between men and women cannot be explained by the higher proportion of part-time working women in the data file.

The validity of these conclusions depends on many different factors, among which are several statistical aspects that we will address in chapter 8.

1.4 Do-files

Suppose that the regression results computed in the preceding section are interesting enough that we want to keep them. When analyzing data, you want to make sure you can reproduce the results. We will discuss this basic rule of scientific work in chapter 2, but some of its aspects are so important to us that we want to introduce the basic procedure at this point.

To begin, type the command

```
. doedit
```

This command calls the Stata Do-file Editor, which allows you to create and modify Stata do-files and other ASCII text files. You can also use any text editor, such as NotePad, TextPad, UltraEdit, or (X)Emacs.[4]

4. See http://fmwww.bc.edu/repec/bocode/t/textEditors.html for a discussion of different text editors.

The Do-file Editor is in the foreground. It is basically an empty sheet in which you can enter any text—including, of course, Stata commands, which is what we want to do now. To begin, type the following lines:

top: an1.do
```
1: use data1, clear
2: mvdecode inc, mv(0)
3: generate men = 1 if gender == 1
4: replace men = 0 if gender == 2
5: generate fulltime = 1 if emp == 1
6: replace fulltime = 0 if emp == 2
7: regress income men fulltime
```
end: an1.do

Be sure that you type only the plain text. Do not type the solid lines, `an1.do`, or the numbers with the colons at the beginning of each line. We have indicated the text to be entered by placing it between the two solid lines. The line numbers help us refer to specific lines in text files. The word `an1.do` shown at the top is the filename under which you should save your text when you finish.

The lines you typed are the commands that you will need to reproduce the regression analysis. The only new thing here is the option `clear` in the `use` command, which is needed because there are already data in memory that need to be replaced by the dataset you want to use.

Now save the text using the filename `an1.do`. Make sure that you save the file in the same directory that you are using while working with Stata. (If you followed our suggestion, this will be `c:\data\kk2`.) To save the file with the Stata Do-file Editor, select **File > Save As...**, or choose its equivalent if you use another text editor. In many Windows editors, you must specify the file type to be saved; be sure to save it as an ASCII text file with a `.do` extension.

Now switch back to Stata by clicking on any Stata window. Another way to switch between an editor and Stata is to use the key combination *Alt+Tab*. (On a Macintosh keyboard, you use the Apple command key with *Tab*.)

As soon as you are back at the Stata command line, type

```
. do an1.do
```

You should see a list of all the Stata commands you entered in the Do-file Editor. The command `do` causes Stata to execute the do-file that contains all the Stata commands that you want to execute. For example, to run the do-file `an1.do`, you type `do an1.do`. However, you can omit the extension `.do` and simply type `do an1`; Stata will understand that you are running a do-file (see section 3.1.8). If you are still in the Do-file Editor, you can select **Tools > Do**.

If you saw an error message, you most likely made a typing error in your do-file. As soon as Stata detects an error, it stops executing the do-file and displays the corresponding error message. In this case, switch back to your editor and carefully check the commands you entered. Correct the error, save the file, and switch back to Stata. Try the do-file again. Remember that you can repeat a command by pressing *Page Up*. It might be a good idea to repeat the sequence of switching to an editor, saving, switching back to Stata, and executing the do-file a few times, to make it a habit.

If you received the error message

```
file an1.do not found
r(601)
```

you may not have saved `an1.do` in the correct directory. You can check whether the file is in your current working directory by typing

```
. dir *.do
```

If `an1.do` does not appear in the list, type

```
. pwd
```

to determine the location of your current working directory. Then switch back to your editor and save the file in the correct directory.

1.5 Exiting Stata

Now you are at the end of your first session. Before you stop, save all the commands you typed in this session. You save the commands by right-clicking anywhere in the Review window, selecting **Save All...** from the pulldown menu, and saving the contents to the file `an1cmd.do`. Make sure that this file is stored in the same working directory you have been using during this session! We will come back to this file in the next chapter.

Now you can finish your first Stata session. Type the command `exit`.

```
. exit
no; data in memory would be lost
r(4);
```

Well, that is obviously not the way to do it. The reason is that commands like `generate` and `replace` have changed the data. Exiting Stata without saving the changes in the dataset would cause those changes to be lost forever. To do this, you must explicitly request that Stata exit without saving your changes by using the `clear` option. If you want to save the data, you can use the `save` command. You should specify a filename after `save`, and it is fine to use a new filename, such as `mydata`:

```
. save mydata
```

The file will be saved in the current working directory with the name `mydata.dta`. If you pick a name that is already in use for another file, you will get an error message. Stata tries to ensure that you do not accidentally overwrite data. The only way to overwrite a file is to use the option `replace`, e.g., `save mydata, replace`.

In most cases, you should avoid using the command `save, replace`, which will use the name of the file you currently have loaded and cause an irreversible loss of the original version of the data.

Now that you have saved the file, you can exit Stata with `exit`.

You do not really need to save the changes to the data, as you have already saved the commands that created the changes. You can reproduce the current state of your work by running the do-file again on the original data. If you do not want to save the file—and we recommend that you not save it (see chapter 2)—you can simply exit Stata by typing

```
. exit, clear
```

1.6 Exercises

1. The command `cd` changes the working directory to the specified drive and directory. Explore the command by changing Stata's working directory to the following directories:

   For Windows user

   a. `c:\Documents and Settings\All Users\Desktop`
   b. Starting from the desktop go up one directory in the directory tree.
   c. `c:\Program Files`
   d. `c:\data\kk2\`

   For Linux user

   a. `~/Desktop`
   b. Starting from the desktop go up one directory in the directory tree.
   c. `/usr/local`
   d. `~/data/kk2/`

2. Try the following commands. Write a short description of the task they perform:

   a. `copy`
   b. `mkdir`
   c. `type`
   d. `winexec`

3. Load `data1.dta` into memory and try the following commands. Again, write a short description of the task they perform:

    a. `codebook`
    b. `inspect`
    c. `browse`
    d. `edit`
    e. `lookfor`

4. Find the Stata commands to execute the tasks listed below. Give an example with `data1.dta` for each of these commands.

    a. Delete variables
    b. Chi-squared test for bivariate tables
    c. Correlation matrix
    d. Cronbach's alpha
    e. Factor analysis
    f. Kernel density estimation
    g. Locally weighted scatterplot smoother (LOWESS)

2 Working with do-files

Science claims to be objective. A central criterion for objectivity is *intersubjective traceability* (Popper 1994, 18); in other words, other people should be able to confirm the results, using the same methods, or to criticize them on the grounds of problematic details. This requires that you diligently document every decision you make during your research.

Few areas of scientific work are as easy to document as statistical data evaluation, yet every now and then, some students cannot show how they obtained a particular result. This problem is not restricted to students. When trying to reproduce 62 empirical economic studies from the renowned *Journal of Money, Credit and Banking*, Dewald, Thursby, and Anderson found that only 22 of the addressed authors provided their data and programs. Twenty of them did not reply, and for 20 others, the data did not exist. Only one of the 22 articles for which the data and programs were available was well documented.[1]

Poor organization is likely the cause of the cases of nonexistent data and the 21 analyses that were badly, or not at all, documented. This chapter will show you how to prevent such problems by using do-files. As discussed in section 1.4, do-files are simple text files that contain Stata commands, which are executed one after the other. Using do-files is the best way to guarantee reproducibility, but you must take care to organize your do-files properly.

2.1 From interactive work to working with a do-file

Even though do-files are important, we usually begin our analyses interactively. We try different models, delete outliers, transform variables, construct indices, and so on. As you work on an analysis, you should, however, try to document the essential steps you take in a do-file so you can reproduce them later. Stata provides two ways to record your steps.

2.1.1 Alternative 1

At the end of chapter 1, we asked you to right-click on the Review window and save the review contents to the file `an1cmd.do`. Once you save the review contents to a file,

1. Quoted from Diekmann (1998).

you can produce a do-file from those saved commands. To do this, type the following command:[2]

```
. doedit an1cmd.do
```

The doedit command opens the Stata Do-file Editor. If you specify a filename with the command, that file is opened in the Do-file Editor. If you did not follow our example in chapter 1, you can open the file an1cmdkk.do, which you previously downloaded; see the *Preface*. Your file an1cmd.do should look something like our an1cmdkk.do, the first several lines of which contain

```
                                              top: an1cmdkk.do
1: d
2: describe
3: cd c:\data\kk2
4: dir
5: set scrollbufsize 50000
6: dir *.dta
7: use data1
8: describe
9: drop  ymove - np9507
                                              end: an1cmdkk.do
```

Whatever your file looks like, it reproduces the commands you typed while following the examples in chapter 1. It is a list of Stata commands, and hence nearly a complete do-file. But you will want to remove any commands that you do not need. For example, you might see that a command has been entered several times in a row, so usually only the last one will be needed.

Some commands are unnecessary in a do-file, such as describe, list, and variations of summarize. The only really essential commands in our example are those that are directly related to the regression analysis: those needed to reproduce the results or that deliver important extra information. Decide for yourself which commands you want to delete. This is what your file might look like after you have deleted some commands

```
                                              top: an2.do
 1: use data1
 2: drop  ymove - np9507
 3: mvdecode income, mv(0=.a)
 4: sort gender
 5: by gender: summarize  income
 6: summarize  income, detail
 7: label variable emp "Status of employment in '97"
 8: label define emplb 1 "full time" 2 "part time" 3 "retraining" ///
 9:   4 "irregular" 5 "not working" 6  "military service" 7 "unemployed", modify
10: label values emp emplb
11: generate men = 1 if gender == 1
12: replace men = 0 if gender == 2
13: generate fulltime = 1 if emp == 1
14: replace fulltime = 0 if emp == 2
15: regress income men fulltime
                                              end: an2.do
```

2. Remember that your working directory is c:\data\kk2. For details, see page 3.

This do-file could now be run in Stata without error. Unfortunately, the command for defining the label is too long to print within the margins of this book, so we broke it into two lines. We will show you better ways to deal with long command lines in section 2.2.2.

Before we discuss a second way to save your work as a do-file, please save your current file as a do-file. We recommend that you name it `an2.do`. Make sure that you save the file in the directory (e.g., `c:\data\kk2`) in which you are working in Stata with this book.

2.1.2 Alternative 2

In the previous section, you preserved your interactive work by saving the contents of the Review window. Another way to do this is to use the `cmdlog` command. We will show you an example of how to use this command, but be aware that the example uses advanced statistical techniques. Do not worry about the statistics; just concentrate on creating the do-file.

Our example extends the analysis from chapter 1, where we found that women generally earn less than men. A multiple regression model showed that this inequality is due only partly to the higher rate of part-time employment among women.

You, however, still have doubts about your results. You argue:

- The income of the working population increases with age. At the same time, women are still more likely than men to give up working in favor of family responsibilities. The group of working women therefore is proportionally younger and therefore earns less than the group of working men.
- The income inequality between men and women is dying out. For many years, women ranked the objective of a career lower than that of starting a family, resulting in generally lower professional ambition and correspondingly lower incomes for working women. The situation is different today, as young women pursue their careers as ambitiously as men, so the income inequality is found only among older women.

To verify these hypotheses, you must first determine the necessary steps of the analysis. You can start by doing some interactive trials. Change to the Stata Command window. Before you begin, you should reproduce your original findings. To save some typing, we have made creating the two new variables slightly more elegant. You can find out more on page 76.

```
. use data1, clear
. mvdecode income, mv(0)
. generate men = gender == 1 if gender < .
. generate fulltime = emp == 1 if emp <= 2
. regress income men fulltime
```

Now you can begin to turn your hypotheses into analyses. Since you already know that your trials should result in a do-file, you should record your interactive session. Type the following command:

```
. cmdlog using an2.do
file an2.do already exists
r(602)
```

Explanation: Typing `cmdlog using` instructs Stata to create a file in which all subsequent commands will be recorded. Along with the command you must specify the name of the file to be created, here, `an2.do`. If you do not type the filename extension (`.do`), Stata will use the extension `.txt`.

An error message then informs you that the file `an2.do` already exists, which is true, because you saved such a file above (page 27). As always, you cannot lose data in Stata unless you explicitly request to lose it. You can use the `replace` option to overwrite the previous do-file with a new one. However, that would not be good advice in this case. The file `an2.do` contains all your analyses up to now, which you would also like to keep. You need to use a different name, or better still, add the subsequent analysis directly to the previous analyses. This is what the option `append` is for:

```
. cmdlog using an2.do, append
```

You think the results are biased because young women with low incomes make up much of the group of working women. You should therefore control for age by creating an `age` variable from the year of birth:

```
. generate age = 1997 - ybirth
```

`age` is a continuous variable. You should center this variable, i.e., subtract the mean from every value; otherwise, the constant in the multiple regression gives you the estimated average income of zero-year-olds, which has no use.

When centering your `age` variable, you should modify only the cases for which you run the regression analysis. These are the working persons with valid values for all the variables used in the regression model. There are many ways to do this, which we encourage you to find for yourself. If you like, you can type the command

```
. cmdlog off
```

You can then try out commands without saving them in `an2.do`. When you have found a solution, type

```
. cmdlog on
```

and retype the command you want to use. After you type `cmdlog on`, your entries are once again saved in `an2.do`.

To center age based on the observations in the model, start by creating a variable named miss, which for every observation contains the number of missing values of the model variables:

```
. egen miss = rowmiss(income fulltime men age)
```

We have not explained the command used here, but be assured that it works; you will just have to restrain your curiosity until page 85.

We use the newly created variable miss to focus on those observations that have no missing values for the specified variables:

```
. summarize age if miss==0
    Variable |       Obs        Mean    Std. Dev.       Min        Max
-------------+--------------------------------------------------------
         age |      1560    41.01346    14.18531         16         91
```

Finally, we create another variable named age_z, in which the mean age (41) is subtracted from the age of every person:[3]

```
. generate age_z = age - 41 if miss==0
```

Now for the second hypothesis. You believe that income inequality is not an issue among younger persons. This can be modeled by what is called an "interaction effect".[4] To model the interaction effect, create the variable agemen by typing

```
. generate agemen = age_z * men
```

Now you can better evaluate your hypotheses:

```
. regress income men fulltime age_z agemen
      Source |       SS       df       MS              Number of obs =    1560
-------------+------------------------------           F(  4,  1555) =   38.13
       Model |   230403251     4  57600812.7           Prob > F      =  0.0000
    Residual |  2.3493e+09  1555  1510826.31           R-squared     =  0.0893
-------------+------------------------------           Adj R-squared =  0.0870
       Total |  2.5797e+09  1559  1654739.04           Root MSE      =  1229.2

------------------------------------------------------------------------------
      income |      Coef.   Std. Err.      t    P>|t|     [95% Conf. Interval]
-------------+----------------------------------------------------------------
         men |   448.7167   63.14638     7.11   0.000     324.8557    572.5778
    fulltime |   813.8009   92.17844     8.83   0.000     632.9938    994.6081
       age_z |  -4.939898   3.081272    -1.60   0.109    -10.98378    1.103989
      agemen |    6.88735   4.398057     1.57   0.118    -1.739399     15.5141
       _cons |   960.3107   87.54728    10.97   0.000     788.5875    1132.034
------------------------------------------------------------------------------
```

The model fits the average income of middle-aged, part-time working women as \$960. Contrary to our expectations, women's income does not seem to increase with age, instead it decreases. The coefficient of −4.94 for the variable age_z implies that the

3. A much more elegant way of centering variables can be found in chapter 4.

4. For more about interaction effects, see section 8.4.2.

mean income of women decreases by $4.94 for each one-year increase in age. For men, the net effect of a one-year increase in age is equal to the coefficient on `age_z` (−4.94) plus the coefficient on the interaction term `agemen` (6.89), which equals 1.95. Hence the mean income of men increases by $1.95 for each one-year increase in age. Therefore, the income inequality between men and women, which at the mean age amounts to $448.72, is higher among the older interviewees and lower among the younger interviewees. Even so, for the 20-year-olds, there is still a considerable income gap.

You now know that your initial estimate (page 19) for the income inequality between men and women, $451, is slightly overestimated for younger people. However, these hypotheses do not sufficiently explain the income difference between men and women. The general findings of income inequalities persist even for younger people.

This result also seems interesting enough to be preserved in a reproducible way. Because you have recorded your commands, that is no longer a problem. Close the log file, and change back to the Do-file Editor:

```
. cmdlog close
. doedit an2.do
```

Now you find the commands you entered last at the end of the file `an2.do`. The commands are already formatted for a do-file. Only faulty commands need to be deleted.

You might need to make some cosmetic changes, which we will explain.

2.2 Designing do-files

The do-file below and on the following page should be very similar to your `an2.do`, except that it will include some modifications: comments, line breaks, and some Stata commands that are useful for do-files. The do-file also includes line numbers, which are not part of the do-file but are there to orient you when we talk about specific parts of the file. In the following paragraphs, we will walk you through these modifications.

top: an2.do

```
 1: * Income inequality between men and women in Germany (GSOEP-data)
 2: * ---------------------------------------------------------------
 3:
 4: version 10
 5: set more off
 6: capture log close
 7: log using an2, replace
 8:
 9: * Data: GSOEP 1997, Sample
10: use data1, clear                                // -> Note 1
11: drop ymove - np9507
12:
13: * Descriptive statistic of income
14: summarize income
15: mvdecode income, mv(0=.a)                       // -> Note 2
16: sort gender
17: by gender: summarize income
18: summarize income, detail
```

```
19:
20: * Employment by gender
21: label define emplb 1 "full time" 2 "part time" 3 "retraining" ///
22:  4 "irregular" 5 "not working" 6 "military service" 7 "unemployed", modify
23: label values emp emplb
24: tabulate emp gender, colum nofreq            // -> Note 3
25:
26: * Preparation for the regression analysis
27: generate men = gender == 1                   // gender
28: generate fulltime = emp == 1                 // fulltime vs. part-time
29: replace fulltime = . if emp >  2             // -> Note 4
30:
31: * Regression analysis I
32: regress income men fulltime
33:
34: * Preparation regression analysis II
35: generate age = 1997 - ybirth                 // age
36: egen miss = rowmiss(income fulltime men age)
37: summarize age if miss==0                     // -> Note 5
38: generate age_z = age - r(mean) if miss == 0
39: gen agemen = age_z * men                     // age x gender
40:
41: * Regression analysis II
42: regress income men fulltime age_z agemen
43:
44: log close
45: exit
46:
47: Description
48: -----------
49:
50: This is an analysis of income inequality between men and women in Germany.
51: Hypothesis see Kohler/Kreuter (2009, chapters 1 and 2). The higher amount
52: of part time working women is not a sufficient explanation for the inequality
53: in average income between men and women. And in addition, even though there
54: is a higher income inequality among older people, younger women
55: are still affected.
56:
57: Notes:
58: ------
59:
60: 1) SOEP - Education-Sample and random selection of one person per household
61:    (Created with crdata1.do).
62: 2) Respondents with income equal to zero are excluded from further analysis
63: 3) Women are more often part time employed than men. It is reasonable to
64:    control for employment status.
65: 4) This command excludes all respondents that are not employed.
66: 5) Centering the age variable -- see Aiken/West (1991)
```

end: an2.do

2.2.1 Comments

One change in an2.do is that we included comments to make it easier for you to understand the do-file. Using comments, you can insert titles, explain why an analysis was carried out, or attach keywords to the results.

There are three ways to add comments. You can add a line of comments by putting an asterisk (*) at the beginning of a line; see lines 1, 2, and 9 of `an2.do` (page 30) for an example.

You can also add comments within a line by adding //. Everything that is written on a line after // is ignored; see lines 27–29 for examples.

A third way to include comments is with the /* */ syntax. Stata ignores everything, including line breaks, between the /* and */.

2.2.2 Line breaks

We also added a command to label the variable for employment status. In the first version of our do-file (see page 26), that command is too long to fit on one page. This is not a problem for Stata; it can deal with such long lines. But avoiding long lines makes your do-file more readable. Moreover, some other text editors might split long lines into several lines, causing Stata to produce an error when running your do-file. So it is a good idea to restrict lines to no more than 75–80 characters.

For long commands, you may have to continue the command over several lines. In Stata, commands end with a line break. If you put a line break at the end of a line and write part of a command on the next line, Stata thinks that you have already entered the entire command and interprets the second line as a new command.

There are two ways to continue commands over more than one line. First is the `#delimit` command.

`#delimit` defines the character that indicates the end of a command. You can use either a line break (the default) or a semicolon. You can change delimiters within do-files. For example, you could type

```
#delimit ;
label define emplb
 1 "Full-time"
 2 "Part-time"
 3 "Retraining"
 4 "Irregular"
 5 "Unemployed"
 6 "Milit. serv."
 7 "N. working";
#delimit cr
```

Typing `#delimit ;` changes the character marking the end of the command to the semicolon. The `label define` command now spans several lines. Typing `#delimit cr` changes the delimiter back to the line break (carriage return).

Some Stata users always use the semicolon to mark the end of commands in do-files. We do not do this because we often run only extracts of our do-file, which is much more cumbersome when using the semicolon as the command delimiter.

The second method of continuing commands across lines is to add three slashes, as shown in lines 21 and 22 of our example do-file (page 30). You already know that after //, everything up to the line break is interpreted as a comment—but the line break itself is interpreted as a line break. With /// the line break is part of the comment, so Stata does not notice the line break and the command continues on the next line.

2.2.3 Some crucial commands

Some commands are recommended in all do-files. Every do-file should begin with a series of similar commands:[5]

```
top: an2.do
✂
4: version 10
5: set more off
6: capture log close
7: log using an2, replace
end: an2.do
```

Explanation: With version, you specify the version of Stata for which the do-file was written. By doing so, you ensure that your do-file will still run without errors in future versions of Stata. When you include the command version, Stata works the same way as the specified version for the duration of the do-file. Thus version should always be the first command in a do-file.

After the version command, the sequence can vary. In our example, we have first deactivated the partitioned display of output. With set more off, the do-file runs without interruption. However useful it may be to break up the output into screen pages in interactive sessions, it is useless when you are running a do-file—at least if you are saving the results of the do-file in a file. You do not need to undo this setting at the end of the do-file. The specification is valid *locally*, i.e., only for the do-file.

Now you should ensure that the results of your do-file are actually saved in a file. Generally, you will use the log using command for this, which is why we call the files with the Stata results "log files". The command works very much like cmdlog using. It instructs Stata to create a file in which the output of all subsequent commands will be written. If you specify a filename, the log file will have that name: here it is called an2. If you do not specify a filename extension, Stata uses the extension .smcl. After you enter log using an2, everything that appears on the screen is saved in the file an2.smcl until you enter log close.

Stata stores the results in a format called the Stata Markup and Control Language (SMCL). With SMCL, log files stored on the hard disk display in the same format as output in the Results window. You can view the log files by typing the view command, which opens a new window, the Stata Viewer, containing the contents of the specified file. For example, to view the file named an2.smcl you type

```
. view an2.smcl
```

5. The recommendations presented here are taken from the Stata NetCourse 151. For information on the Stata Internet courses, see section 12.1.

SMCL is also useful for translating the log files to HTML or LaTeX. If you want to insert Stata output into a word processor, such as Microsoft Word, you may want to translate the log file to plain ASCII with `translate`:

```
. translate an2.smcl an2.log
```

You can even save your log files as plain ASCII from the beginning by specifying the filename after `log using` with the extension `.log`.

Now let us return to our example do-file. Here we use `log using` without the extension because we want Stata to save the log file in SMCL. We always use the same filename for the log file as for the do-file; that is, the do-file `an2.do` stores its results in `an2.smcl`. Because the file `an2.smcl` may already exist—maybe created by an earlier version of `an2.do`—you should specify the `replace` option to overwrite the obsolete log file.

The `log using` command does not directly follow the `set more off` command. Instead we use the `log close` command to ensure that no log file is already open. However, we placed `capture` before the `log close` command. If a log file had not been open when we ran the do-file, entering `log close` would have resulted in an error message—and aborted the do-file. So, we included the `capture` command before the `log close` command. `capture` forces Stata to ignore the error message and continue running the do-file.

You can place `capture` before any Stata command so that Stata ignores any error messages from that command and continues running the command. In this case, it ensured that the do-file would continue to run if there were no open log file to be closed. Usually, you will not want to `capture` error messages; if a command does not work properly, you will want to know about it. However, using `capture` with `log close` within a do-file is an exception because `log close` leads to an error message only if no log file is open. Here you do not need `log close`, and the do-file can simply continue with the next command, here `log using an2, replace`.

If you had omitted `capture log close`, an error message would occur every time you ran this do-file with a log file already open. This can happen if you have previously written some interactively entered commands in a log file and have forgotten to close the log file.

The end of every do-file should contain the following commands:

```
top: an2.do
✂
44: log close
45: exit
end: an2.do
```

`log close`, as we explained, is necessary for closing the previously opened log file. `log close` is executed only if the do-file runs without errors. If the do-file does not run completely because of an error, the log file remains open.

exit is more interesting. Although chapter 1 presented exit as a command for exiting Stata, in do-files exit ends the do-file and returns you to the interactive command entry mode. In fact, typing exit in a do-file is not necessary, but there are two good reasons for doing so. First, the last command in a do-file must include a line break. Had we not pressed *Enter* after typing log close, the log file would not have been closed, and everything we did after the do-file was finished would also be logged. By including the exit command in our do-file, we ensure that the previous command ends in a line break. Second, Stata stops executing the do-files as soon as the exit command is reached, which allows us to include notes at the end of the file without having to use the comment syntax mentioned previously. Notice our description of the file and implementation notes at the end of the do-file.

2.3 Organizing your work

Creating concise and easy-to-read do-files is only the first step toward achieving reproducibility of analyses. The second step is a suitable work plan to ensure that

- important files are not lost,
- you can find your do-files for a particular result again without problems,
- all steps are clearly documented, and
- all analyses can be reproduced easily.

To fulfill these objectives, we suggest the following procedure, which you can tailor to your needs, based on the distinction between two types of do-files:[6] creation do-files and analysis do-files.

It is impossible to distinguish completely between these two types of files. In an analysis do-file, the datasets may still need to be edited, but we try to avoid this and do as much editing as possible in the creation do-files. We will never use a creation do-file to carry out an analysis.

This distinction is useful only if it can be inferred from the filenames. For this reason, the names of all of our creation do-files begin with the letters cr, whereas the names of analysis do-files begin with the letters an. Even if we restrict our filenames to eight characters,[7] we have six characters left to describe the contents of the do-file. This is usually sufficient.

The results of the analysis do-files are always recorded in a log file, which has the same name as the do-file, but with the extension log or smcl—whichever you prefer.

6. We have taken this suggestion from the Stata NetCourse 151. For details on the Stata Internet courses, see section 12.1.

7. Although modern operating systems allow longer names, it makes sense to restrict yourself to short filenames. Eight characters offer maximum compatibility and save the time needed to type long filenames in the commands. For more information about filenames, see section 3.1.8

The analysis do-file we created in this chapter is called `an2.do`. It creates the log file `an2.smcl`.

You use creation do-files to create new datasets. This means that we read in a certain dataset, create new variables, recode existing variables, or delete observations and variables, etc. We then save the newly created dataset as a data file having the name of the create do-file without the letters `cr`. For example, we are using for most of our examples in this book a dataset from a creation do-file called `crdata1.do`.[8] The data file associated with the do-file therefore has the name `data1.dta`.

The separation of creation and analysis do-files makes sense only if you are going to use a particular dataset for several analyses. Modifications of a dataset that apply only to one specific analysis are more likely to belong in the analysis do-file. Modifications that apply to several analyses are better stored in a creation do-file because you need to modify the file only once, saving a lot of time in some cases. You need to decide carefully which dataset modifications you want to save in a creation do-file.

Apart from the creation and analysis do-files, our work plan comprises a further type of do-file: the master do-file, or *master file* for short, which contains a list of `do` commands.

When we begin a new project, we create a do-file with the name `master.do`. In this do-file, we at first write only a title representing the name of our project. Because our master file is a do-file, this title must be marked as a comment (and the file must end with `exit`). The first few entries in `master.do` could, for example, look like this:

```
                                                  top: master.do
 1: * Example analyses for Kohler/Kreuter, Data analysis using Stata
✂
17: exit
                                                  end: master.do
```

Then the actual work begins. Suppose that we were conducting an analysis to use in a fictitious book.

For the first chapter of our book, we wanted to present a small example analysis. We first needed to clarify the nature of the problem, so we did some interactive trials and then finally created a dataset containing data from the social sciences. To create the final version of this dataset, we wrote a do-file called `crdata1a.do`. Of course, it took several attempts to create a do-file that ran absolutely error-free. Then we once again accessed the master file, added the line `do crdata1a`, and commented this entry:

```
                                                  top: master.do
 1: * Example analyses for Kohler/Kreuter, Data analysis using Stata
 2: do crdata1a              // creation extract of GSOEP'97
✂
17: exit
                                                  end: master.do
```

8. For documentation, we have included `crdata1.do` among the files you installed.

After creating the dataset data1a.dta by crdata1a.do, we carried out a plausibility analysis; that is, we intensively searched for errors in the dataset. Such error checking is important to avoid recoding errors. We decided to document our error checking in a do-file so that we can later reconstruct which error sources we checked (and which we did not). We called this ancheck1.do. When we created this do-file, it took several attempts to get it to run error-free. We recorded each attempt in a log file called ancheck1.smcl, which therefore always contained the latest results. When we could run the do-file without error messages, we extended our master.do file as follows:

```
                                                    top: master.do
 1: * Example analyses for Kohler/Kreuter, Data analysis using Stata
 2: do crdata1a                // creation extract of GSOEP'97
 3: do ancheck1                // error checks in data1a.dta
✂
17: exit
                                                    end: master.do
```

Inspecting the results of our error checking, as recorded in ancheck1.smcl, we noticed a small error in the variable for occupational status. We corrected this in a new creation do-file (crdata1b.do), which in turn now creates the dataset data1b.dta. After completing crdata1b.do, we again checked for errors and found no more. We did this error check with a do-file (ancheck2.do), and we included both do-files in the master file and added comments:

```
                                                    top: master.do
 1: * Example analyses for Kohler/Kreuter, Data analysis using Stata
 2: do crdata1a                // creation extract of GSOEP'97
 3: do ancheck1                // error checks in data1a.dta
 4: do crdata1b                // correction of errors from ancheck1.do
 5: do ancheck2                // error checks in data1b.dta
✂
17: exit
                                                    end: master.do
```

We then began the actual analyses. The first analysis dealt with the income inequality between women and men and is contained in an1.do. We also tested this do-file several times before including the line do an1 in our file master.do.

These examples can be extended further. Whenever you complete an analysis, you can add the corresponding do-file to master.do, thus reproducing the sequence of all analyses done in the course of a project. From the comments in master.do, you can quickly determine the rough contents of the respective analyses and reconstruct the analysis from the beginning. You can repeat all the analyses from the beginning by running the master do-file, although this will usually not be necessary.

Once you have added a do-file to master.do, you should not modify it again. If you discover an error in a do-file in the master file, correct the file and add it, *under a new name*, to the end of master.do. This applies especially if you have already carried out many analyses using a dataset that was created by a faulty creation do-file. For example, in our example above, we found in the do-file ancheck1.do that data1.dta contained an error. Instead of correcting the error in crdata1a.do and running the do-file again, we corrected this error in another do-file (crdata1b.do). We would do the same if we had already done several analyses between the time we created the dataset and the time we discovered the error. Assume that your master file contained the following sequence:

```
                                                        top: master.do
 1: * Example analyses for Kohler/Kreuter, Data analysis using Stata
 2: do crdata1a              // creation extract of GSOEP'97
 3: do ancheck1              // error checks in data1a.dta
 4: do crdata1b              // correction of errors from ancheck1.do
 5: do ancheck2              // error checks in data1b.dta
✂
 7: do an1                   // income inequality men/women
 8: do anrent                // description of rents
 9: do anpi                  // part.-identif. and ownership
✂
17: exit
                                                        end: master.do
```

During your current analysis, you realize that the dataset data1b.dta contains a previously undetected error resulting from a faulty recoding in crdata1a.do. Instead of correcting the error in crdata1b.do, you should document the error in a do-file (e.g., anerror.do), correct it in a further do-file (e.g., crdata1c.do), and then repeat the analyses from an1.do to anpi.do. Because you now refer to the new dataset data1c.dta, you should save the corresponding modifications in an1.do to anpi.do under new filenames and document them accordingly in the master file. The result might then resemble the following:

```
                                                        top: master.do
✂
 7: do an1                   // income inequality men/women
 8: do anrent                // description of rents
 9: do anpi                  // part.-identif. and ownership
10:
11: * Error in data1b, -> correction and repetition in an1 - anpi
12: do anerror               // discovery of error in data1b.do
13: do crdata1c              // correction of error
14: do an1c                  // corrected results of an1.do
15: do anrentc               // corrected results of anrent.do
16: do anpic                 // corrected results of anpi.do
17: exit
                                                        end: master.do
```

If you follow this procedure, you can reproduce the entire process, even the mistake itself, at any time. You might think that this will not be useful, but a long time can pass between the error and its discovery, and you may have cited faulty figures in your writing (even some you have published). With the procedure we have outlined, it will be relatively easy, even after a long time, to detect the cause of erroneous figures. There are few things as frustrating as chasing after the determinants of a figure that no longer correspond to the results of a current analysis.

To conserve space on your hard disk, you may not want to create many permanent datasets. Delete every dataset that you have created with a creation do-file as soon as you no longer need it for your current analysis. Examples are the datasets data1a.dta and data1b.dta in the above do-file, which both had problems and were no longer needed after we corrected them. You can delete these datasets using the Stata command erase and document this in the master file:

```
                                                   top: master.do
✂
5: do ancheck2                // error checks in data1b.dta
6: erase data1a.dta
7: do an1                     // income inequality men/women
                                                   end: master.do
```

If you later want to rerun an older analysis do-file, you may see the error message "file data1b.dta not found". You can, however, quickly reproduce the dataset with the corresponding creation do-file.

Likewise, you can also delete log files and then reproduce them accurately later. Finally, you need to make backup copies only of your do-files and the original data file. You can reproduce all the datasets and log files you have created during your project at any time by using the command do master.

2.4 Exercises

1. Open the Stata Do-file Editor.
2. Write a do-file that performs the following tasks:
   a. opens a log file
   b. opens the dataset data1.dta
   c. generates a new variable with the name satisfaction that is equal to the sum of the existing variables np11701 and np0105
   d. drops all variables except satisfaction and income
   e. summarizes satisfaction and income
   f. saves the new dataset with the name 2erase.dta
   g. closes the log file
3. Run your do-file in the following manner:
   a. Use the button in the Do-file Editor two times in a row.
   b. Save the do-file with the name cr2erase.do and run it from the Command window.
   c. Exit and relaunch Stata, and then start the do-file from the command line without opening the Do-file Editor.
   d. Exit Stata, copy the do-file to another directory. Relaunch Stata and start the do-file from the command line.
4. Think of strategies to create do-files that run under all the above conditions.
5. Open and print out the log file created by cr2erase.do using the Stata Viewer.
6. Create a do-file that starts cr2erase.do, and run it.

3 The grammar of Stata

3.1 The elements of Stata commands

As you have seen, Stata commands are made up of specific elements. This section describes these elements of the Stata language and the general rules governing them. These rules apply to all Stata commands, making it easy for you to learn quickly how to use new Stata commands.

Each element of a Stata command can be *required*, *permitted*, or *prohibited*. Obviously, you can use an element of a Stata command only when it is at least permitted. The online help provides information about these elements. For example, if you look at the online help for `summarize` you will find the following syntax diagram:

`summarize` [*varlist*] [*if*] [*in*] [*weight*] [, *options*]

Here you can find most of the elements of the Stata language. For example, *varlist* is an abbreviation for "variable list", which is a language element used in almost every Stata command. *if* stands for the language elements "`if` qualifier" and "expression". You can use any element that is displayed in the syntax diagram. Elements displayed in square brackets are optional; those displayed without them are required.

Now let us examine the elements one by one. We will first describe the command itself and the variable list used in the command. We will then describe options, which further specify the command, and qualifiers, which restrict the command processing to subsets of the data. We will discuss *weight* at the end of the chapter, after we introduce other commonly used commands.

To follow our examples, you will need to use our example dataset:[1]

```
. use data1
```

3.1.1 Stata commands

Each Stata command has a name that specifies the task Stata should do. Commands written by StataCorp are called official commands, and those that users write are called user-written commands. You might want to do something in Stata and find that there is no official Stata command for that. You may find that there is a user-written command

1. Be sure that your working directory is `c:\data\kk2` (see page 3).

written by someone else. Chapter 12 of this book shows you how to find and install such additional commands; chapter 11 shows how to program user-written commands.

Some official commands may be abbreviated.

The shortest possible abbreviation for each command is shown in the syntax diagram of the online help with an underline beneath the smallest part of the command that needs to be typed.

However, it is not always a good idea to use the shortest possible abbreviation for each command, especially in do-files, as overusing abbreviations can make your do-file hard to read. You can use every possible abbreviation between the shortest allowed up to the entire command. For example, from the syntax diagram on page 41, you learn that you can use each of the following abbreviations for `summarize`:

```
. su
. sum
. summ
. summa
. summar
. summari
. summariz
```

Table 3.1 shows some frequently used commands and their abbreviations with the shortest possible abbreviation underlined. We also show a recommended abbreviation, but this is not always the shortest possible abbreviation and sometimes is not an abbreviation at all. We have followed international usage drawn from postings to Statalist (see page 363), but these recommendations are only our opinion.

Table 3.1. Abbreviations of frequently used commands

Command	Recommended abbreviation	Usage
`describe`	`d`	Describe data in memory
`generate`	`gen`	Create new variables
`graph`	`graph`	Graph data
`help`	`h`	Call online help
`list`	`l`	List data
`regress`	`reg`	Linear regression
`summarize`	`sum`	Means, etc.
`save`	`save`	Save data in memory
`sort`	`sort`	Sort data
`tabulate`	`tab`	Tables of frequencies
`use`	`use`	Load data into memory

In this book, we sometimes use abbreviations and sometimes we do not—just as most Stata users do in everyday data analysis. This will help you get used to the different ways of saying, for example, `summarize`.

3.1.2 The variable list

In most commands, you can use a variable list. Within the online help, this is indicated by the term *varlist*. A variable list is a list of variable names, separated by spaces.

List of variables: Required or optional

Some commands allow a variable list but do not require one, whereas others require a variable list. You can learn whether a command requires a variable list by looking at the online help. If the term *varlist* appears in square brackets, the variable list is optional. If *varlist* appears without square brackets, it is required.

Many commands that do not require a variable list will use all existing variables unless you provide a list. For example, specifying `summarize` without a variable list returns the means and standard deviations of all variables in the dataset. Other commands redisplay their previous results if you do not provide a list of variables (e.g., `regress`).

Some commands require a list of variables. This is the case if you cannot repeat the command, or if it is not possible or not useful to apply the command to all variables. If you want to force Stata to apply these commands to all variables, you can specify `_all` instead of a variable list. The command `drop`, for example, deletes specified variables from the dataset. Thus the command

```
. drop ymove ybuild
```

deletes the variables `ymove` and `ybuild`. You cannot reverse this command. Once you drop a variable, the only way to get the variables back is to reload the data file. Therefore, using `drop` without a *varlist* will not apply the `drop` command to all variables, as this would mean dropping the entire dataset. If you really wanted to delete all variables, you would use `drop _all`.[2]

Abbreviation rules

You can abbreviate the variable names in a list of variables to as few characters for a variable as you need to identify it uniquely. For example, the variable `kitchen` in the dataset `data1.dta` is uniquely identified by using the character `k`; no other variable in this dataset begins with *k*. The variable `ybirth`, on the other hand, cannot be distinguished by its first character from the variables `ymove`, `ybuild`, and `yedu`. You must type at least `ybi`.

2. If you have typed `drop _all`, be sure to reload the file before you continue reading.

In addition, you can use the tilde (~) to omit one or more characters of a variable name. For example, you can type

```
. summarize y~h
```

to summarize `ybirth`, as there is only one variable that begins with `y` and ends with `h`.

Both types of abbreviations must match a single variable. If the abbreviation does not uniquely identify a variable, Stata displays an error message. To specify more than one variable, you must do so explicitly. This leads us to the second type of shortcuts for variable lists: specifying more than one variable at once. There are three ways to do so:

1. You can use a question mark to specify variables that have the same names except for one character. For example,

   ```
   . summarize np940?
   ```

   summarizes all variables having names beginning with `np940` followed by a single character. Question marks may be used anywhere in the variable name—at the beginning, in the middle, or at the end.

2. You can use the asterisk (*) as a wildcard character to specify variables that share parts of their names but differ in one or more characters. For example,

   ```
   . summarize np* e*
   ```

   summarizes all variables that begin with `np` or with `e`, regardless of how many characters follow. Just like question marks, wildcards may be used anywhere in a variable name.

3. You can use a hyphen to specify a range of variables that come one after another in the dataset. The order of the variables in the dataset can be found in the output of the `describe` command. From this output,

   ```
   . describe
   ```

   you can see that there are some variables describing the appliances in the dwellings in the dataset. The first one is `kitchen` and the last one is `phone`. To summarize those variables, you can type

   ```
   . summarize kitchen-phone
   ```

It is easy to overuse shortcuts for lists of variables. Instead of typing

```
. summarize rooms kitchen shower wc heating cellar balcony garden phone
```

you could also have typed

```
. su r~s k-ph
```

This is nice if you are working interactively. However, if you need to preserve your work for later, the first version is easier to decipher. We recommend not using abbreviations for variable names in do-files.

Special listings

For some commands, you will see terms like *varname* or *depvar* in the syntax diagram. These terms are used for variable lists that consist of one variable. These single-variable lists are sometimes combined with general variable lists when the order of the variables matters. For example, in the command for linear regression, you must specify the dependent or endogenous variable by putting its name *before* the list of independent variables. Here the order of the variable list matters, so the syntax diagram of the command looks like this:

`regress` *depvar* [*indepvars*] ...

The term *depvar* stands for the dependent variable, whereas *indepvars* stands for the independent variables. As always, the order of the variables within the *indepvars* does not matter.

Likewise, for every Stata command that fits a statistical model, you must specify the dependent variable before a list of independent variables.

3.1.3 Options

Options are used to change the default behavior of a command and are provided for almost every Stata command. In the syntax diagram, you can see that options are allowed if you find the word "*options*" after a comma. Below the syntax diagram you find a list of options that are available for the specific command. The syntax diagram of `summarize`, for example, allows `detail`, `meanonly`, `format`, and `separator()`:

`summarize` ... [, *options*]

options	description
Main	
`detail`	display additional statistics
`meanonly`	suppress the display; calculate only the mean; programmer's option
`format`	use variable's display format
`separator(#)`	draw separator line after every # variables; default is `separator(5)`

Sometimes options are required. Then you will not find a square bracket before the comma in the syntax diagram.

You specify options by placing a *comma* at the end of the command and then listing the options with spaces between them. The comma begins a list of options, but you should not have more than one comma no matter how many options you have. The order of the options does not matter. Here is an example for the `tabulate` command. In the `tabulate` syntax diagram that you can display by typing `help tabulate twoway`, you find among many others the options `row` and `missing`.

`tabulate` ... [, *options*]

options	description
Main	
...	
`row`	report relative frequency within its row of each cell
...	
`missing`	treat missing values like other values
...	

Using the `row` option with the `tabulate` command will give you the percentages within each row of the table (row percentages) in addition to the frequencies. If you add the `missing` option, the table includes respondents who have missing values on the variables specified in the `tabulate` command. In the example below, 1,409 respondents are added to the table although they did not respond to the question on their concerns about crime. For more information about frequency tables, see section 7.2.1.

```
. tabulate gender np9506, row missing

+-------------------+
| Key               |
|-------------------|
|     frequency     |
|  row percentage   |
+-------------------+

           |                 Worry: crime
    Gender |    strong   moderate         no          . |     Total
-----------+--------------------------------------------+----------
       men |       188        462        389        556 |     1,595
           |     11.79      28.97      24.39      34.86 |    100.00
-----------+--------------------------------------------+----------
     women |       136        369        387        853 |     1,745
           |      7.79      21.15      22.18      48.88 |    100.00
-----------+--------------------------------------------+----------
     Total |       324        831        776      1,409 |     3,340
           |      9.70      24.88      23.23      42.19 |    100.00
```

Most options can be abbreviated, and the shortest abbreviation is underlined in the online help. Some commands have a long list of options, so you should carefully review the command-specific online help.

3.1.4 The in qualifier

The `in` qualifier, like the `if` qualifier discussed in the next section, is a command element that limits the execution of the command to a subset of observations. The `in` qualifier is allowed for every Stata command that displays [*in*] in its syntax diagram. The `in` qualifier is composed of the word `in` and a *range* of observations specified in terms of their position in the dataset. If you think about the dataset as a huge table with rows and columns, the number of an observation in the dataset refers to the number of a row in that table.

The range can be a single observation or a range from a certain observation (row) to a second observation below the former observation. You type a slash (/) between the first position and the final position in the range.

For example, you can look at the person ID, the gender, and the year of birth of the 10th observation in the dataset by restricting the command `list` to that observation:

```
. list persnr gender ybirth in 10

     +---------------------------+
     | persnr   gender   ybirth  |
     |---------------------------|
 10. |  21501    women     1924  |
     +---------------------------+
```

And you can get the same information for observations 10–14 by using

```
. list persnr gender ybirth in 10/14

     +---------------------------+
     | persnr   gender   ybirth  |
     |---------------------------|
 10. |  21501    women     1924  |
 11. |  24140      men     1938  |
 12. |  26437      men     1956  |
 13. |  28265      men     1953  |
 14. |  30324    women     1922  |
     +---------------------------+
```

There is no other way to specify a range other than those shown in these two examples. You cannot specify lists of observations or combinations of ranges. Commands such as `list gender in 1 15 26` or `list gender in 1/10 16` would result in an error message. However, you can use a minus sign to specify a range:

```
. list persnr gender ybirth in -5/-1

       +-----------------------------+
       |  persnr   gender   ybirth   |
       |-----------------------------|
3336.  | 7245788    women     1944   |
3337.  | 7247862      men     1979   |
3338.  | 7249881      men     1976   |
3339.  | 7252233      men     1956   |
3340.  | 7254426      men     1973   |
       +-----------------------------+
```

Here `-1` means the observation on the last line of the dataset, i.e., the last observation. Correspondingly, `-5` means the fifth observation from the last, so the command above shows the last five observations of the dataset. If the data have been sorted by year of birth, these will be for the youngest five persons.

When you use the minus sign, do not to confuse the order within the range. The fifth-to-last observation precedes the last observation in the dataset, so you need to insert `-5` *before* `-1`. The command `list persnr gender ybirth in -1/-5` would be invalid. As long as you account for the order of the observations, you can even use combinations of the count order:

```
. list persnr gender ybirth in 3330/-5
```

is valid, as the fifth-to-last observation is 3,336th and therefore is after 3,330th. Instead of `1` and `-1` you can use `f` and `l`, indicating the first and last observations in the dataset.

3.1.5 The if qualifier

The `if` qualifier is allowed for every Stata command that displays [*if*] in its syntax diagram. Like the `in` qualifier, the `if` qualifier is used to run a command only with data from certain observations. The `if` qualifier is composed of the word `if` and an *expression*. To take full advantage of the `if` qualifier you need to be familiar with expressions; see section 3.1.6.

A command with an `if` qualifier uses only observations for which the expression is true, or more precisely, for which the expression is not zero.

In chapter 1, we showed a simple example of using the `if` qualifier:

```
. summarize income if gender == 1

    Variable |       Obs        Mean    Std. Dev.       Min        Max
-------------+--------------------------------------------------------
      income |      1455    1670.816    1293.905          0      10945
```

In this example, the `if` qualifier is `if gender==1`. Note the double equal-sign. With this `if` qualifier, we tell Stata to evaluate the expression `gender==1` for each observation in the data and to use only observations for which the expression is true (not zero).

We have been speaking generally, but let us look at this example in more detail. First, let us look at the income and gender of the first five observations of our dataset:

```
. list income gender in 1/5, nolabel

     +-----------------+
     | income   gender |
     |-----------------|
  1. |    832        1 |
  2. |      0        1 |
  3. |      0        2 |
  4. |      0        1 |
  5. |   4717        1 |
     +-----------------+
```

In the first row of the output, we find an observation with an income of $832 and a gender of 1. We wonder what the expression `gender==1` evaluates to for this observation. We would probably say that the expression is true for this observation, but Stata says that the expression evaluates to 1 for this observation. Stata would make the same evaluation for the second observation. For the third observation, you would say that the expression `gender==1` is false, whereas Stata says that the expression evaluates to 0.

If you add an `if` qualifier to a command, Stata evaluates the expression for every observation in the dataset the same way we just did. Stata then executes the command for all the observations for which the expression is nonzero.

As the `if` qualifier consists of the word `if` and an expression, the general rules for expressions apply (see section 3.1.6). In practice, however, you will most often build expressions that can be true (1) or false (0). Such expressions usually will contain *relational* operators. Here are just three examples. First, the command

```
. summarize income if ybirth < 1979
```

shows the mean income for respondents who were born before 1979. All these observations have a value for year of birth that is smaller than 1979. We can extend the first command to

```
. summarize income if ybirth <= 1979
```

showing the mean income for respondents born in 1979 and before. Finally,

```
. summarize income if ybirth != 1979
```

shows the mean income of all respondents except those born in 1979.

There is one trap associated with the `if` qualifier that you should be aware of—and you might want to mark this paragraph with the brightest color you have. The difficulty arises in expressions that contain the name of a variable with missing values, which are set to $+\infty$ in Stata. Expressions having the relational operator $>$ or $\geq$ therefore evaluate to 1 for observations with missing values.

From the command

```
. tabulate edu, missing nolabel
  Education
   (school)
       1997 |      Freq.     Percent        Cum.
------------+-----------------------------------
          1 |      1,280       38.32       38.32
          2 |        961       28.77       67.10
          3 |         89        2.66       69.76
          4 |        425       12.72       82.49
          5 |        334       10.00       92.49
          6 |        187        5.60       98.08
          7 |         36        1.08       99.16
          . |         28        0.84      100.00
------------+-----------------------------------
      Total |      3,340      100.00
```

we see that the variable `edu`—representing education—contains $187+36 = 223$ nonmissing observations with an education greater than or equal to 6 (whatever that means). There are 28 more observations for which the education is listed as a dot. Stata uses the dot to show that we have no information about the education of those persons. And we can see the dot only because we used the `missing` option with the `tabulate` command.

Now let us summarize the year of birth for those observations having an educational level of 6 or higher:

```
. summarize ybirth if edu >= 6

    Variable |       Obs        Mean    Std. Dev.       Min        Max
-------------+--------------------------------------------------------
      ybirth |       251     1957.92    18.56647       1904       1981
```

Here we get a table based on 251 observations instead of 223. The reason for this is that the 28 observations with missing values are set to $+\infty$ inside Stata, and $+\infty$ clearly is higher than 6. To summarize the observations without including the respondents with unknown education, we need to exclude them explicitly. We can do this using any of the following `if` qualifiers:

```
. summarize ybirth if edu == 6 | edu == 7
. summarize ybirth if edu >= 6 & edu <= 7
. summarize ybirth if edu >= 6 & edu < .
```

These `if` qualifiers use several operators, which we can use to form more complicated expressions. Now we will explain expressions in more detail.

3.1.6 Expressions

Expressions are allowed or required wherever the syntax diagram of a Stata command displays the term *exp*, so expressions can be used at different places within a Stata command. A short version of the syntax diagram of the `generate` command, for example, looks like this:

`generate` *newvar* = *exp* [*if*]

This command requires an expression after the command name. Expressions can also be found at several other places in the Stata language.

Think of an expression as an arithmetic problem, such as $2 + 3$. With Stata this problem is easy to solve, because Stata ships with its own pocket calculator: the `display` command:

```
. display 2+3
5
```

That was a snap! Just like every pocket calculator, the `display` command calculates expressions and displays the result. And as with most pocket calculators, we can calculate somewhat more complicated expressions by combining “operators” and “functions”.

Operators

For an overview of the operators used in expressions, see the online help:

```
. help operators
```

From this list, we see that we can easily calculate the following expressions:

```
. display 2-3
-1
. display 2*3
6
. display 2/3
.66666667
. display 2^3
8
```

We can combine more than one expression, and we can use parentheses to change the order of the calculations:

```
. display 2*3 + 2/3 - 2^3
-1.3333333
. display 2*(3 + 2)/(3 - 2)^3
10
```

Using expressions with logical and relational operators may seem complicated, but it is easy: expressions with relational operators can be true or false, corresponding to 1 and 0, respectively, in Stata.

Stata uses a double equal-sign to test for equality.[3] That is, the expression 2==3 makes Stata determine if 2 and 3 are equal:

```
. display 2==3
0
. display 2==2
1
```

As you can see, Stata says that the expression 2==3 is 0 because this expression is not true: two does not equal three. The expression 2==2 on the other hand is true and therefore evaluates to 1.

The logical operators are most often used to combine different expressions containing relational operators, just like the expressions 2==3 and 2==2. Suppose that you wanted to find out if both expressions are true. You would have Stata evaluate if expression 1 *and* expression 2 are true:

```
. display 2==3 & 2==2
0
```

3. A single equal-sign instructs Stata to make two things equal. Single equal-signs cannot be used in expressions.

To find out if at least one of the expressions is true, you would have Stata evaluate if expression 1 *or* expression 2 is true:

```
. display 2==3 | 2==2
1
```

You can use many operators in an expression. If you build complicated expressions, you should use parentheses to indicate the order to perform the calculations—even if they are not strictly needed.

Finally, Stata expressions can also deal with *strings* (words and letters) if they are enclosed in quotation marks. Let us give an example by asking Stata the following crucial question:

```
. display ("SPSS"=="Stata")
0
```

Now we know that this is wrong.

Functions

Other important tools for building expressions are Stata's *functions*, which are rules that assign *one* specific value to each of a range of given values. A simple function can be expressed as $f(x) = 1$. This function assigns the value 1 to every number x. A slightly more complicated function is $f(x) = \sqrt{x}$, which assigns the square root of x to every number x. Here we will deal only with such nontrivial functions.

Stata has many predefined functions, including the square root, the logarithm, and the exponential. All these functions consist of a function name followed by an argument within parentheses: *function_name*(*argument*).

The function name describes its purpose. `sqrt()` calculates the square root, `ln()` calculates the natural logarithm, and so on.

The argument specifies the values for which the function should be calculated. The argument is itself an expression; that is, you can use simple numbers or complicated expressions of functions and operators inside the parentheses.

Let us try some examples of expression with functions. The following commands calculate $\sqrt{2}$, $\sqrt{2+(3/5)}$, and $e^{\sqrt{|-1|}}$, respectively:

```
. display sqrt(2)
1.4142136

. display sqrt(2 + 3/5)
1.6124515

. display exp(sqrt(abs(-1)))
2.7182818
```

You can find out about all of Stata functions by typing

```
. help functions
```

Depending on your statistical knowledge, you may find this list too large to comprehend. At least, try to remember the functions you already know and take a look at this help file from time to time. As you learn more about Stata, you will find it more and more useful.

3.1.7 Lists of numbers

In the syntax diagrams of some commands, you will find the term *numlist*, which stands for a list of numbers that Stata understands as numbers. As Stata is quite smart about understanding numbers, you can abbreviate lists of numbers, as shown in table 3.2. Examples for useful applications of lists of numbers can be found in sections 3.2.2 and 6.3.4.

Table 3.2. List of numbers and their meanings

Input	Meaning
`1,2,3,4`	1, 2, 3, 4
`1 2 3 4`	1, 2, 3, 4
`1/4`	1, 2, 3, 4
`2 4 to 8`	2, 4, 6, 8
`8 6 to 2`	8, 6, 4, 2
`2 4: 8`	2, 4, 6, 8
`8 6: 2`	8, 6, 4, 2
`2(2)8`	2, 4, 6, 8
`8(-2)2`	8, 6, 4, 2
`8/10 15 to 30 32 to 36`	8, 9, 10, 15, 20, 25, 30, 32, 34, 36
`8/10(5)30(2)36`	8, 9, 10, 15, 20, 25, 30, 32, 34, 36

3.1.8 Using filenames

In the syntax diagrams of some Stata commands, you will find the element `using` *filename*, indicating that these are commands that read or write a file. You refer to the file by inserting the term `using` and the complete name of the file. Sometimes you can omit the term `using`.

Generally, a complete filename consists of a directory, a name, and an extension. The directory is where the file can be found. The name is the name of the file itself, and the extension is usually a file type indicator. The file most often used in this book, `c:\data\kk2\data1.dta`, is in the directory `c:\data\kk2`, has the name `data1`, and has the extension `.dta`, which is the extension for Stata datasets.

How you specify file addresses depends on your operating system: you use colons to separate folder names under Macintosh, slashes under Unix, and backslashes under Windows. Also Stata lets you use the slash to separate folder names in *all* operating systems, which is why we generally use it.

You specify the filename the same way in all operating systems: just type the filename. If the filename (or address) contains blanks or other special characters, you will need to place quotation marks around the complete name. For example, to describe the (fictitious) file `AT & T.dta` in the folder `My Data` you would type `describe using "c:\My Data\AT & T.dta"`.

Whenever you specify a filename, Stata must have the complete name, but this does not mean that you need to type the complete name. Here is what happens if you omit one part or another:

- If you type a filename without the directory, Stata looks in the working (current) directory. You can find the address of the working directory by typing the `pwd` (print working directory) command:

```
. pwd
```

- If you type a filename without the extension, Stata looks for a file with an extension that is appropriate for the specified command. Table 3.3 below shows commonly used commands that look for or save files with the given extension.

Table 3.3. Names of commands and their associated file extensions

Extension	Command
`.dta`	`use`; `save`; `append`; `merge`; `joinby`; `describe`
`.raw`	`infile` (with *varlist*); `infix` (with *varlist*); `insheet`; `outsheet`; `outfile`
`.dct`	`infile` (with *varlist*); `infix` (with *varlist*)
`.smcl`	`log` (`.smcl` or `.log` depending on whether file format is SMCL or ASCII
`.txt`	`cmdlog`
`.do`	`do`; `run`
`.gph`	`graph using`; `graph, saving()`

Finally, the filename does not necessarily refer to a file on your computer but can also mean a file somewhere on the Internet. If your computer has a connection to the Internet, you can load a Stata dataset directly from an Internet location, for example:

```
. use http://www.stata-press.com/data/kk2/data1
```

The same applies for all commands that load a file. However, you cannot write to the web.

3.2 Repeating similar commands

In everyday data analysis, you will often need to retype similar commands over and over again. For example, our data file contains six variables that refer to people's concerns about different aspects of their lives. To compare the answers of men and women for these six variables, you would repeat a `tabulate` command six times:

```
. tabulate np9501 gender
. tabulate np9502 gender
. tabulate np9503 gender
. tabulate np9504 gender
. tabulate np9506 gender
. tabulate np9507 gender
```

Retyping similar commands is boring and, for complex commands, error prone. To avoid this, you can benefit from the tools we describe here.

Stata has several tools for repeating similar commands. We will discuss three of them: the `by` prefix, the `foreach` loop, and the `forvalues` loop (also see Cox 2002a,b). The simplest one is the `by` prefix, which is used to repeat a single command for different observations. The other two are used to loop over elements of lists or number sequences.

Another loop command, `for`, was very popular in older versions of Stata, so you may come across it at some point. However, it is no longer documented because `foreach` and `forvalues` are more powerful and easier to use.

3.2.1 The by prefix

The `by` prefix repeats a command for every group of observations for which the values in the variable list are the same. Most Stata commands accept the `by` prefix; this is indicated immediately following the syntax diagram and options table, for example,

> `by` may be used with `summarize`; see [D] **by**.

The `by` prefix consists of the word `by` or `bysort`, a variable list (which we will call the *bylist*), and a colon. Using the `by` prefix generally requires that the dataset be sorted by the variables of the *bylist*, but there are ways to sort the data on the fly.

The best way to understand the `by` prefix is to use it. Let us summarize the income for each group of the variable `gender`. For this, we need to sort the data by gender and then issue `summarize` with the `by` prefix:

```
. sort gender
. by gender: summarize income

-------------------------------------------------------------------------------
-> gender = men
    Variable |       Obs        Mean    Std. Dev.       Min        Max
-------------+--------------------------------------------------------
      income |      1455    1670.816    1293.905          0      10945

-------------------------------------------------------------------------------
-> gender = women
    Variable |       Obs        Mean    Std. Dev.       Min        Max
-------------+--------------------------------------------------------
      income |      1579    1052.854    1121.029          0      12438
```

As you can see, the `by` prefix forces Stata to summarize the income for every group or category of the *bylist*. First, Stata summarizes the income of the first group (the men), and then it summarizes the income of the second group (the women).

This works equally well if we use a *bylist* with more than just two groups:

```
. by edu, sort: summarize income
```

We have used the option `sort` of the `by` prefix to sort the data by education (`edu`). `sort` is part of the `by` prefix and not part of the command after the colon. Finally, you could have typed

```
. bysort edu: summarize income
```

because `bysort` is just `by` with the `sort` option. Choose whichever syntax you like—and stick with it.

If you provide more than one variable in the *bylist*, Stata does what you have told it to do for each possible combination formed by the values of the *bylist* variables. Therefore, the command

```
. by gender edu, sort: summarize income
```

first has Stata summarize the income of all men with a minimum of 9 years of education ("elementary"). Then it does the same for all men with 10 years of education ("intermediate"). After summarizing the incomes of all educational groups of males, it does the same for each educational group of females.

Some Stata commands allow the `by()` option, which is easy to mistake for the `by` prefix. You need to know the difference: a `by()` option is a part of a specific Stata command, and its function is defined in that command. In other words, the `by()` option works differently with different commands. The `by` prefix, on the other hand, does the same thing with *every* Stata command that allows it: `by` repeats the command for each group indicated by the variable list.

3.2.2 The foreach loop

The `foreach` loop is used to repeat a specific task for each element of a list. This task can be a single, consistently applied command—as with the `by` prefix—or a command that varies slightly within each replication, or even an entire series of Stata commands. Because the `foreach` loop is so flexible, its syntax is slightly more complicated.

A simplified syntax diagram of `foreach` is as follows:

`foreach` *lname listtype list* `{`

commands

`}`

A `foreach` loop always has at least three lines: the first line, which begins the loop, ends with an opening brace. The second line is a Stata command (you can include more lines containing Stata commands). The last line contains a closing brace. You cannot place anything else in the line after the opening brace or in the line containing the closing brace (except for comments).

In the syntax diagram above, the first element of the first line is the command name: `foreach`. Following that are elements you must type: the element name (*lname*), the list type (*listtype*), and the `foreach` list (*list*). Next comes the opening brace.

The `foreach` list is a list of parameters, such as a variable list, a list of numbers, or a list of arbitrary parameters. But you need to tell Stata the list type. For example, if you want to specify a variable list, you use `of varlist`. Finally, you must specify a name for each element of the `foreach` list. The element name is used in the Stata commands between the braces to refer to the elements of the `foreach` list.

There is more to a `foreach` loop, but it is best explained within an example. In what follows, you will type some commands, but make sure you read the fine points carefully before you begin typing.

```
. foreach X of varlist np9501-np9507 {
  2. tabulate ‘X’ gender
  3. }
```

Here are the fine points. The first line begins the loop. After you press *Enter*, you will see the number 2 on your screen. This is just to remind you that everything you are going to type is being processed by `foreach`. You do not need to do anything about that. Just type the `tabulate` command as usual, and then type the third "command", the closing brace. If you make an error somewhere, just press *Enter*, type the closing brace, and press *Enter* again; you will need to start over from the first line again. But remember that you can access the commands you previously entered in the Review window by clicking on the command or by pressing *Page Up*.

Note the symbols before and after the `X` in this command. The symbol before the `X` is a single open quotation mark, and the symbol after the `X` is a single close quotation mark. The two symbols are *not* the same even though they look alike in some fonts. The open quote is a backtick or *accent grave*, whereas the closing right quote is a forward quote or apostrophe. On many American keyboards, the opening quote is found at the top left (near the *Esc* key), whereas the closing quote is found on the right (near the *Enter* key). On European keyboards, the position of both characters vary from country to country. The opening quote is often used to produce the French *accent grave*, which forces you to press the *Spacebar* before the symbol appears on the screen.

Now you may begin entering the commands. If everything works, you will get the output for six `tabulate` commands.

The `of varlist` declares the list type to be a variable list, meaning that the `foreach` list is a list of variable names. The element name is declared to be "X", so we use `X` to represent the current element of the list. Stata requires that you put single quotation marks around the element name in the part within the braces. After the closing brace, Stata begins the loop, successively replacing the `` `X' `` in the second command with the name of each variable in the `foreach` list. Thus `` tabulate `X' gender `` becomes `tabulate np9501 gender` in the first round of the loop, `tabulate np9502 gender` in the second round, and so on.

The types of foreach lists

As we said, you can specify different types of `foreach` lists:

- `of varlist` for lists of variables
- `of newlist` for lists of new variables
- `of numlist` for lists of numbers
- `in` for arbitrary lists of letters, words, and numbers separated by spaces

You have already seen an example of a `foreach` loop with a variable list. The following examples, which we have saved in `foreachkk.do`, show the other list types. You may want to open the file in the Do-file Editor

```
. doedit foreachkk.do
```

to play with these examples. You might want to save the edited version of `foreachkk.do` to, say, `myforeach.do` and run that do-file, or you can type the commands interactively. You will find that in our example do-file, we have indented the part between the opening and the closing brace. Doing so is good style and lets you more easily see what code the `foreach` loop affects. It is also good style to align the closing brace with the first letter of `foreach`. For more about style issues with Stata commands, type

```
. view http://fmwww.bc.edu/repec/bocode/s/stylerules.hlp
```

Our first example uses the list type `of newlist`, which you will find in lines 11–13 of `foreachkk.do`. Here we generate 10 variables containing uniformly distributed random numbers between 0 and 1 by using the Stata random-number generator `runiform()`:

top: foreachkk.do

```
11: foreach var of newlist r1-r10 {
12:         gen `var' = runiform()
13: }
```

end: foreachkk.do

Instead of using the list type `of newlist`, we could have used the list type `of numlist` in this example (see the next example). However, with `of newlist` Stata checks the validity of the variable names to be generated before beginning the loop.

Next we use `of numlist` to replace the variables with newly generated random numbers. As the variables `r1` to `r10` already exist, we need to use `replace` instead of `generate` within the loop:

top: foreachkk.do

```
17: foreach num of numlist 1/10 {
18:         replace r`num' = runiform()
19: }
```

end: foreachkk.do

Finally, here is an example with an arbitrary list:

top: foreachkk.do

```
22: foreach piece in This list has 5 pieces {
23:         display "`piece'"
24: }
```

end: foreachkk.do

Several commands within a foreach loop

You can put more than one command within a `foreach` loop, as shown in the next example, in which we generate a centered version of the variables for income and year of birth. To produce a centered version of a variable, we need to subtract its arithmetic mean from each of its values:

top: foreachkk.do

```
27: foreach var of varlist ybirth income {
28:         summarize `var', meanonly
29:         generate `var'_c = `var' - r(mean)
30:         label variable `var'_c "`var' (centered)"
31: }
```

end: foreachkk.do

We begin by calculating the arithmetic mean (line 28), and then we generate a new variable (line 29) having the name of the old variable with `_c` appended. The term `r(mean)` refers to a "saved result" of the `summarize` command, which contains the value of the mean we just calculated. We will discuss saved results in chapter 4. Finally, we define a variable label for each of the new variables.

3.2.3 The forvalues loop

Once we understand the `foreach` loop, the `forvalues` loop is easy: it is just a shorter way to set up a `foreach` loop with `of numlist` as the *listtype*.

The simplified syntax diagram of `forvalues` is as follows:

```
forvalues lname = range {
        commands
}
```

This looks very much like the syntax diagram of `foreach`. Again there are three lines: the first line begins the loop. The second line is a Stata command (and may be followed by more lines with Stata commands). The last line contains only a closing brace. You cannot place anything in the line after the opening brace or in the same line as the closing brace other than comments.

In the first line of the syntax diagram, you find the command itself—`forvalues`—followed by an element name (*lname*), an equals-sign, a *range*, and then the opening brace. The *range* is similar to a numlist. You can specify a range of numbers by using the rules you have learned from section 3.1.7, except that you cannot list single numbers or more than one range. For example, you can specify `1(1)10` for all integer numbers from 1 to 10, but you cannot specify `1(1)10 15 19` or `1(1)10 12(2)20`.

The element name is an arbitrary name for each number of the *range*. It is used in the Stata commands between the braces to refer to the specified numbers.

Let us try an example. Type the following lines:

```
. forvalues num=1/10 {
  2. replace r`num' = runiform()
  3. }
```

`num` is enclosed in single quotes just like the element name in the `foreach` loop. After you type the third line, Stata should begin the loop, successively replacing `` `num' `` with each number of the specified *range*: Stata will replace once more the contents of the variables `r1` to `r10`.

The `forvalues` loop may seem unnecessary, given that it is the same as a `foreach` loop with the list type `of numlist`. However, `foreach` with a numlist has some limitations; it cannot have more than 1,600 numbers in the list. There is no such limitation with `forvalues`. Moreover, `forvalues` is more efficient for Stata to process.

3.3 Weights

This section is slightly more difficult than the previous sections. We will explain weights as simply as possible, but you may get lost anyway. If so, just skip this section and return to it later.

Weights are allowed for every Stata command that has [*weight*] in its syntax diagram. Think of weights as a way for Stata to treat some observations as more important than others when calculating statistics.

You specify weights through the following general syntax:

`[`*weighttype* `=` *varname*`]`

You must always enclose weights within square brackets. These square brackets have nothing to do with the ones in the syntax diagram that identify optional arguments but are part of the notation of the weights themselves. Inside the square brackets you specify the weight type and the name of the variable that contains the weights (the weight variable). You must specify the right weight type. Make sure you really understand the different weight types before using them (and be somewhat suspicious of results of software packages that do not have different weight types).

There are three main weight types:

- `fweight` "frequency weights",
- `aweight` "analytic weights", and
- `pweight` "probability weights".

If you simply specify `weight` as the weight type, Stata chooses a type for you; different commands have different default weight types. Some commands also use the weight type `iweight`, or *importance weights*. These weights do not have a formal statistical definition, so you should carefully read the description of commands allowing importance weights to learn how they are used. Importance weights are used in programming contexts and are not needed by regular users. The three main weight types have clear statistical definitions, which we will describe.

Frequency weights

Frequency weights are used for weight variables that contain the number of equal observations in the dataset. As this may sound a bit puzzling, we will explain it with an example. To begin, type the following command:

```
. summarize ybirth
    Variable |       Obs        Mean    Std. Dev.       Min        Max
-------------+--------------------------------------------------------
      ybirth |      3340     1951.72    18.33337       1902       1981
```

This command displays a table summarizing data on the year of birth. The calculation is based on 3,340 observations, each of which is a respondent that occurs in the dataset only once.

Now let us save the data for later use with

```
. preserve
```

`preserve` stores the dataset as it is, in a temporary file. After preserving the dataset, you can safely try out whatever you want and then simply type `restore` to get your preserved data back. We will use `preserve` and `restore` quite often in this book.

After preserving the data, you should load and describe `freqwe.dta` from our file package:

```
. use freqwe, clear
. describe
Contains data from freqwe.dta
  obs:            78
 vars:             2                          14 Jun 2005 08:38
 size:           546 (99.9% of memory free)   (_dta has notes)
-------------------------------------------------------------------------------
              storage  display     value
variable name   type   format      label      variable label
-------------------------------------------------------------------------------
ybirth          int    %8.0f                  Year of Birth (YYYY)
n               byte   %12.0g                 Frequency
-------------------------------------------------------------------------------
Sorted by:  ybirth
```

You can see that these data contain only 78 observations and 2 variables—year of birth (`ybirth`) and `n`.

Now `summarize` again the year of birth, but use the weights this time:

```
. summarize ybirth [fweight = n]
    Variable |       Obs        Mean    Std. Dev.       Min        Max
-------------+--------------------------------------------------------
      ybirth |      3340     1951.72    18.33337       1902       1981
```

Comparing these results with those above, we see that they are the same. Again we have 3,340 observations with a mean of 1951.72 and a standard deviation of 18.33.

As our dataset only contains 78 observations, it may come as a surprise that Stata reports 3,340 observations. We should take a closer look at our dataset. As the dataset is sorted by year of birth, we might profit from simply listing some observations one by one:

```
. list in 1/5

     +-------------+
     | ybirth    n |
     |-------------|
  1. |   1902    2 |
  2. |   1904    2 |
  3. |   1906    5 |
  4. |   1907    4 |
  5. |   1908    2 |
     +-------------+
```

In the dataset, there is only one observation for each year of birth; that is, there is only one observation for year 1902, only one observation for 1904, and so on. But each observation is weighted according to the variable `n`. The observations for years 1902 and 1904 are weighted with the factor 2; the observation for 1906 is weighted with the factor 5; and, if you skip to other observations, you will find that the observation for 1927 is weighted with 39.

If you use frequency weights, Stata interprets the weighting variable as if each observation existed in as many copies as are specified in the weighting variable. The `summarize` command above therefore sees not only one but two observations for 1902, and so on.

The dataset `freqwe.dta` contains the same information about the year of birth as the entire `data1.dta` dataset. But rather than listing people with the same year of birth one by one—two observations for 1902, etc.—`freqwe.dta` lists identical observations only once, indicating how many such observations exist. Using frequency-weighted data is therefore a more parsimonious way to store the same data and may be more useful for handling datasets that are too big for the working memory of your computer.

To change a dataset from being unweighted to being frequency weighted, you can use the command `contract`, and the command `expand` is used for the reverse operation. `expand` is useful if a specific command does not allow frequency weights.

Analytic weights

We will explain the idea of analytic weights with yet another example. Load `analwe.dta` into Stata, and summarize the variable `ybirth` once with frequency weights and then a second time with analytic weights:

```
. use analwe

. summarize ybirth [fweight = n]

    Variable |       Obs        Mean    Std. Dev.       Min        Max
-------------+--------------------------------------------------------
      ybirth |      3340     1951.72    1.455569   1943.962       1973

. summarize ybirth [aweight = n]

    Variable |     Obs      Weight        Mean   Std. Dev.       Min        Max
-------------+-----------------------------------------------------------------
      ybirth |      17        3340     1951.72   1.500141   1943.962       1973
```

You get the same mean as before with both weight types. But take a look at the standard deviation (Std. Dev.), which is about 1.46 with frequency weights and 1.50 with analytic weights. Not only are the two values different, but they both differ considerably from the standard deviations calculated in the last section. Which one is correct?

The answer depends on how the data were created. If the weights reflect the number of respondents (observations) born at the respective year of birth, the first result would be correct. But our example dataset was created differently. Let us take a look at some of the data:

```
. list in 1/5

     +---------------------------+
     |    state   ybirth       n |
     |---------------------------|
  1. |   Berlin     1953      79 |
  2. | Schl.Hst     1954      60 |
  3. |       HH     1948      31 |
  4. | Nieders.     1951     293 |
  5. |       HB     1944      26 |
     +---------------------------+
```

The first observation is from Berlin, a German city that is at the same time a German state. The second observation is from Schleswig-Holstein, another state in the far north of Germany. The complete dataset contains 17 observations, one for each state of Germany.[4] For each state, there is a variable for the average year of birth and a weighting variable. The year of birth is the mean of all respondents from the same state, and the weighting variable is the number of respondents used to calculate that mean. Such data sometimes are called *aggregate data*.

Clearly, not all respondents used to calculate the mean for each state have the same year of birth. We do not have 3,340 observations, but rather we have 17 means made from different numbers of observations.[5] You need to use *analytic weights* for such data. With analytically weighted data, each observation on a variable is itself a mean, and the weight represents the sample size used to compute the mean.

Probability weights

The third weight type is probably the most interesting—and may be one of your reasons for preferring Stata over other statistical packages.

Often you analyze data using a sample from a population and use the sample to infer something about the larger population. Many statistical tools assume that the sample is chosen by simple random sampling with replacement, meaning that each element of the population has the same sampling probability. In practice, you rarely have simple

4. Actually, this is not quite true. There are two states—the Rheinland-Pfalz and Saarland—that we put into one category, whereas the western and eastern parts of Berlin are differentiated. Also there is an observation for one respondent with unknown state.
5. See the file `cranalwe.do` for how we created `analwe.dta` from `data1.dta`.

random samples. Complex samples—that is, samples with observations of different sampling probabilities—are much more common.

If you use statistical tools that assume simple random samples, but your dataset comes from complex samples, you can make two mistakes. First, you get biased point estimators; that is, the means, medians, regression coefficients, etc., do not reflect the true values in the population. Second, you incorrectly calculate the sampling distribution of your point estimators, leading you to evaluate confidence intervals erroneously, and so on (Kish 1965).

To avoid the first mistake, you could weight your data with the reciprocal of the sampling probability. Then it would not matter if you use frequency weights or analytic weights, as both lead to the same results. With the second mistake, the issue is more complicated. Both frequency and analytic weights lead to incorrect results. An observation cannot be regarded as referring to more than one respondent, nor can one observation be regarded as an aggregate measure of more than one respondent. Each observation is just one simple observation. Stata's probability weights give correct standard errors in this case.

Stata's tools for dealing with probability weights are part of a broader class of commands for dealing with complex samples. These are called survey commands, which we will introduce in section 8.5.2.

Before you read on, you should restore your data:

```
. restore
```

3.4 Exercises

1. Compute a mean and a standard deviation for each of the following variables in our data file `data1.dta`
   - All variables that begin with "np".
   - All variables with information on the quality of dwellings.
   - All variables related to "satisfaction".
2. Abbreviate the following commands as much as you can (and so that Stata will still recognize them):
   ```
   . summarize ymove ybuild hcond kitchen area
   . regress np0105 ybirth income
   ```
3. Use the command `gen inc = hhinc` to create a new variable `inc` that is equal to `hhinc`, and rerun your last command. Explain why the same command now produces a different result.
4. Read `help set varabbrev` and consider the advantages and disadvantages of `set varabbrev off`.

5. List the person ID, the interviewer number, year of birth, household income, and life satisfaction of the 10 observations with the lowest household income. Then rerun the command without the observation numbers and add a line for the means of the variables in your list.
6. Create a table with gender as a column variable and life satisfaction as a row variable. Include only respondents from West Germany (`state<10`) in the table. Have Stata display only column percentages.
7. Create a table with gender as a column variable and life satisfaction as a row variable, separate for each state.
8. The following sequence of commands creates a standardized variable for year of birth:

```
. summarize ybirth
. gen ybirth_s = (ybirth - r(mean))/r(std)
```

   Create a loop to standardize the following variables in a similar fashion: `ymove`, `sqfeet`, `rooms`, `np11701`, `rent`, `hhsize`, `edu`, `hhinc`, and `income`.
9. Include a command in your loop that labels the variables above (`ymove`, `sqfeet`, etc.) as “standardized”.

4 General comments on the statistical commands

In this book, we refer to all commands that do some statistical calculation as statistical commands. In chapter 1, for example, we used the statistical commands `summarize`, `tabulate`, and `regress`. Despite their different purposes, statistical commands have one feature in common: they store their results internally. This chapter describes these saved results and what you can do with them. You can do further calculations with them, save them as variables, export them to other programs, or just display them on the screen.

To follow our examples, load `data1.dta`:[1]

```
. use data1
```

With saved results, Stata distinguishes between regular statistical commands (r-class) and estimation commands (e-class). Stata stores the results of the regular statistical commands in `r()` and stores the results of the estimation commands in `e()`.

Both `r()` and `e()` can be regarded as *repositories* into which Stata puts the various results of a command. The command `summarize`, for example, is an r-class command that stores its results in the following macros:

`r(N)`	Number of observations	`r(sd)`	Standard deviation
`r(sum_w)`	Sum of the weights	`r(min)`	Minimum
`r(mean)`	Arithmetic mean	`r(max)`	Maximum
`r(Var)`	Variance	`r(sum)`	Sum of variable

Specifying the `detail` option with the `summarize` command saves more statistics.

Why do we need saved results? There are many uses. Generally, whenever you need the result of a command to set up another command, you should use the saved results. Never transcribe the results of commands by hand into other commands. Use saved results instead.

1. You may want to check if your current working directory is `c:\data\kk2`; see page 3.

Here are some examples with the saved results of `summarize`. You can use the saved results to center variables, that is, to subtract the mean of a variable from each of its values:

```
. summarize income
. generate inc_c = income -  r(mean)
```

You can also use the saved results to standardize variables. You first center your variable and divide it by its standard deviation:

```
. summarize income
. generate inc_s = (income - r(mean))/(r(sd))
```

You can use the saved results to make the high values of your variable low and vice versa (mirroring):

```
. summarize income
. generate inc_m = (r(max) + 1) - income
```

You can also use the saved results to calculate the lower and upper bounds of the 95% confidence interval around the mean. Under standard assumptions, the boundaries of the 95% confidence interval are given by $\text{CI} = \overline{x} \pm 1.96 \times \sqrt{s^2/n}$, where $\overline{x}$ is the mean, s^2 is the variance, and n is the number of observations. These figures can be found in the saved results of `summarize`:

```
. summarize income
. display r(mean) + 1.96 *  sqrt(r(Var)/r(N))
. display r(mean) - 1.96 * sqrt(r(Var)/r(N))
```

Every statistical command of Stata saves its main results. To work with a specific result, you must know where Stata stores that result. You can find where saved results are stored in any of the following sources:

- The *Reference* manuals and help files: The entry for each statistical command has a section called "Saved Results", which states the names and meanings of the saved results.
- The command `return list`, which shows the names and contents of the saved results of the last issued r-class command:

  ```
  . summarize income
  . return list
  ```

- The command `ereturn list`, which shows the names and contents or meanings of the last issued e-class command:

  ```
  . regress income yedu
  . ereturn list
  ```

Each new r-class command deletes the saved results of the previous r-class command, and each new e-class command deletes the saved results of the previous e-class command. For example, the commands

```
. summarize income
. tabulate edu gender
. generate inco_c = income - r(mean)
```

do *not* generate a centered version of `income` because `tabulate` deletes all saved results of `summarize` and saves its own results. Some results from `tabulate` have the same names as those of `summarize`, and others do not. `tabulate` does not save `r(mean)`, for example, so the new variable `inco_c` contains missing values.

The transitory nature of the saved results often forces us to save those results more permanently. You can do this using local macros, which unlike saved results are not overwritten until you explicitly do so.[2]

You define local macros using the command `local`. For example, typing

```
. summarize income
    Variable |       Obs        Mean    Std. Dev.       Min        Max
-------------+--------------------------------------------------------
      income |      3034    1349.207    1245.701          0      12438
. local x = r(mean)
```

stores the contents of the saved result, `r(mean)`, in a local macro called `x`. You can use another name—as long as the name is not longer than 31 characters. After defining the local macro, you can use the name of the local macro whenever you need to refer to the result of the `summarize` command above. The contents of the local macro will not be overwritten unless you explicitly redefine the macro.

To use the local macro, you must explicitly tell Stata that `x` is the name of a local macro by putting the name between single quotes ‘ ’. To simply display the content of the local macro `x` you would type

```
. display ‘x’
1349.207
```

If this command does not work for you, you probably incorrectly typed the opening and closing quotes. The opening quote is a backtick or *accent grave*, whereas the closing right quote is a forward quote or apostrophe. On many American keyboards, the opening quote is found at the top left (near the *Esc* key), whereas the closing quote is found at the right, near the *Enter* key. On European keyboards, the position of both characters varies from country to country. Often the opening quote is used to produce the French accent grave, which forces you to press the *Spacebar* before the sign appears on the screen.

Now let us turn to a practical application. You can calculate the difference between the means of a variable for two groups as follows:

---

2. At the end of a do-file or program, any local macros created by that do-file or program are dropped; see chapter 11.

```
. summarize income if gender == 1
    Variable |       Obs        Mean    Std. Dev.       Min        Max
-------------+--------------------------------------------------------
      income |      1455    1670.816    1293.905          0      10945
. local mean0 = r(mean)
. summarize income if gender == 2
    Variable |       Obs        Mean    Std. Dev.       Min        Max
-------------+--------------------------------------------------------
      income |      1579    1052.854    1121.029          0      12438
. display r(mean) - `mean0'
-617.96279
```

Unfortunately you cannot store all saved results of statistical commands in local macros because sometimes results are stored as matrices—the e-class commands store some of their results as matrices. To permanently store the results in matrices, you need to use the Stata matrix commands, which are mainly programming tools, so we will not deal with them here. For more information, see [U] **14 Matrix expressions** or type

```
. help matrix
```

4.1 Exercises

1. Load a subset from the National Longitudinal Survey (NLS) into Stata memory by typing

   ```
   . webuse labor
   ```

2. Request a list of the stored results that are available after running the following commands:

   ```
   . sum whrs
   . sum whrs hhrs
   . sum hhrs, detail
   . tab kl6
   . tab cit kl6
   . count if cit
   . corr whrs ww
   . corr whrs ww faminc
   ```

3. Find out whether the following commands are e-class or r-class:

   ```
   . sum whrs
   . mean whrs
   . ci whrs
   . regress ww whrs we
   . logit cit faminc
   ```

4. Generate a variable containing a centered version of the variable `faminc`; do not input the statistics by hand.

5. Generate a variable containing a z-standardized version of the variable `faminc`; do not input the statistics by hand. Note: The z-standardized values of the variable x are calculated with $z_i = (x_i - \overline{x})/s_x$, where $\overline{x}$ is the mean of variable x and s_x is the standard deviation of x.
6. Create a local macro containing the number of cases in the dataset and display the content of this macro.
7. Display the range of `faminc`.

5 Creating and changing variables

In everyday data analysis, creating and changing variables takes most of the time. Stata has two general commands for these tasks: `generate` and `replace`, which are the bread and butter of Stata users. `egen` and `recode` are also often used; they allow some shortcuts for tasks that would be tedious using `generate` and `replace`.

When we work with data that have not been prepared and preprocessed nicely, we sometimes run into complex data manipulation problems. There is no single technique for solving these problems, which can challenge experienced users as well as newcomers, though `generate` and `replace` are useful. Although we cannot describe all possible data manipulation challenges you might encounter, this chapter will give you some general advice for generating variables for use in Stata. This chapter will give you the tools to solve even the most difficult problems.[1]

To use our examples, load `data1.dta`:[2]

```
. use data1
```

5.1 The commands generate and replace

`generate` creates a new variable, whereas `replace` changes the contents of an existing variable. To ensure that you do not accidentally lose data, you cannot overwrite an existing variable with `generate` and you cannot generate a new variable with `replace`. The two commands have the same command syntax: you specify the name of the command, followed by the name of the variable to be created or replaced. Then you place an equal-sign after the variable name and specify an expression to be created or replaced. The `by` prefix is allowed as well as the `if` and `in` qualifiers.

You can use `generate` to create new variables; for example,

```
. generate newvar = 1
```

generates the new variable called `newvar` with the value of 1 for each observation:

1. Difficult recoding problems appear almost weekly in Statalist. For practice, you should try solving some of them and read the solutions suggested by other readers. For more about Statalist, see chapter 12.
2. You might want to check that your current working directory is `c:\data\kk2`; see page 3.

```
. tabulate newvar

     newvar |      Freq.     Percent        Cum.
------------+-----------------------------------
          1 |      3,340      100.00      100.00
------------+-----------------------------------
      Total |      3,340      100.00
```

You can also create a new variable out of existing variables, for example

```
. generate pchinc = hhinc/hhsize
```

where you divide the household income (**hhinc**) by the number of persons in the household (**hhsize**). **generate** automatically does this for every observation in the data, and the results of the calculations are written to the variable for household income per capita (**pchinc**).

```
. list hhinc hhsize pchinc in 1/4

     +---------------------------+
     | hhinc   hhsize     pchinc |
     |---------------------------|
  1. |   915        1        915 |
  2. |   813        1        813 |
  3. |  4830        4     1207.5 |
  4. |  2398        6   399.6667 |
     +---------------------------+
```

replace changes the content of a variable. Below we change the content of the variable **newvar** to zero.

```
. replace newvar = 0
```

The variable now contains the value 0 for all observations instead of 1:

```
. tabulate newvar

     newvar |      Freq.     Percent        Cum.
------------+-----------------------------------
          0 |      3,340      100.00      100.00
------------+-----------------------------------
      Total |      3,340      100.00
```

5.1.1 Variable names

When working with **generate**, remember some rules about variable names. The names of variables can be up to 32 characters long. However, it is a good idea to keep the names concise to save time when you have to type them repeatedly. To ensure that your data work with earlier versions of Stata or other statistical software packages, you may want to restrict yourself to names up to eight characters long.

You can build your names with letters (A–Z and a–z), numbers (0–9), and underscores (_), but you cannot begin the variable name with a number. The following names are not allowed:

`_all`	`double`	`long`	`_rc`
`_b`	`float`	`_n`	`_skip`
`byte`	`if`	`_N`	`str#`
`_coef`	`in`	`_pi`	`using`
`_cons`	`int`	`_pred`	`with`

Some variables are allowed but not recommended. Avoid using the single letter `e` to prevent confusion with the "e" in scientific notation. Avoid names beginning with an underscore, as such names are used by Stata for internal system variables. Future enhancements may lead to conflicts with such names, even if they are not on this list.

5.1.2 Some examples

Variables created or modified with `generate` or `replace` are assigned the value of the expression after the equal-sign. The general rules for expressions given in section 3.1.6 also apply here.

When generating new variables, you can perform simple calculations, such as

```
. generate rawinc = hhinc - rent
```

where you determine how much of the net income (`hhinc`) is left after subtracting the rent (`rent`). Or

```
. generate age = 1997-ybirth
```

where you determine the age of each respondent at the time the survey was conducted. You can also use mathematical functions to generate new variables; for example,

```
. generate loginc = log(income)
```

calculates the natural logarithm of personal income for each observation and stores the result in the variable `loginc`. The command

```
. replace loginc = log(income)/log(2)
```

produces the base-2 logarithm instead of the natural logarithm. The result of the calculation overwrites the variable `loginc`. Or you can use statistical functions, such as

```
. generate r = runiform()
```

which generates a random variable from a uniform distribution with values ranging from 0 to nearly 1. A random variable with a standard normal distribution (a variable with a mean of 0 and a standard deviation of 1) is generated with

```
. generate r01 = rnormal()
```

You can use every function introduced in section 3.1.6 with **generate** or **replace** and combine them with the operators described in section 3.1.6. Using **generate** and **replace** with the algebraic operators addition, subtraction, multiplication, division, and power is quite straightforward. You might be surprised, however, that you can also use relational operators with **generate** and **replace**. Let us walk through some examples of using **generate** and **replace** with relational operators.

Suppose that we need to generate the variable **minor** to indicate respondents who had not reached the age of legal adulthood in 1997, the time the survey was conducted. This means the respondent will have a value greater than 1979 on the variable for year of birth (**ybirth**). The new variable **minor** must contain the value 1 for all respondents younger than 18 and 0 for all other persons. One way to create this variable is to generate the variable **minor** having a value equal to 0 for all respondents, and then to replace the values of this newly generated variable with 1 for all respondents younger than 18.

```
. generate minor = 0
. replace minor = 1 if ybirth > 1979 & ybirth < .
```

Another way to create the variable **minor** is based on what we explained in section 3.1.5: expressions with relational operators can be true (1) or false (0). Knowing this, we can create a variable indicating minors in a single line; we will call the variable minor2 this time.

```
. generate minor2 = ybirth > 1979
```

The expression ybirth > 1979 is false (0 in Stata) for all interviewees born before 1979, so all those interviewees get the value 0 for the new variable minor2. For those born in 1979 and later, the expression ybirth > 1979 is true (1 in Stata); therefore, the new variable becomes a 1 for these observations:

```
. tabulate minor2
```

minor2	Freq.	Percent	Cum.
0	3,286	98.38	98.38
1	54	1.62	100.00
Total	3,340	100.00	

Note that minor2 will also equal 1 for any observations for which ybirth is missing, because Stata treats missing values as very large numbers. The command

```
. gen minorshh = ybirth > 1979 & hhsize == 1 if ybirth < .
```

generates the variable minorssh that equals 1 for all respondents younger than 18, who at the time of the interview, lived alone (hhsize == 1) and that is zero elsewhere.

You can see that you can easily generate variables with the values 0 and 1. Such variables are called *dummy variables*, or sometimes just *dummies*. You will use them quite often, but be careful if there are missing values among the variables from which you build your dummy. We recommend that you restrict the command to the valid cases by using an `if` qualifier. For more about handling missing values, see section 5.4.

You can mix relational and algebraic operators in your expressions. For example, you can construct an additive index for quality of a dwelling by summing expressions for all the characteristics of a dwelling. Using `generate` and what you have learned about expressions with relational operators that are either true or not true, you can create the additive index for dwelling quality like this:

```
. generate quality = (kitchen==1) + (shower==1) + (wc==1) + (heating==1) +
> (cellar==1) + (balcony==1) + (garden==1)

. tabulate quality
```

quality	Freq.	Percent	Cum.
0	12	0.36	0.36
1	20	0.60	0.96
2	29	0.87	1.83
3	52	1.56	3.38
4	227	6.80	10.18
5	714	21.38	31.56
6	1,334	39.94	71.50
7	952	28.50	100.00
Total	3,340	100.00	

But note that

```
. generate quality2 = kitchen==1 + shower==1 + wc==1 + heating==1 + cellar==1 +
> balcony==1 + garden==1

. tabulate quality2
```

quality2	Freq.	Percent	Cum.
0	3,340	100.00	100.00
Total	3,340	100.00	

would not have led to the result desired.

You do need parentheses around the relational operators in this example. Addition has priority over equality testing, so in Stata the expression `kitchen == 1 + shower == 1 + ···` reads as `kitchen == (1 + (shower == 1 + ···)`. So be careful with complicated expressions! Use parentheses to make complicated expressions clear, even in cases when the parentheses are not required by Stata.

Of course, you can use `generate` and `replace` not only to create dummy variables but for all kinds of recoding. Say, for example, that you would like to reduce our quality index to three categories: poor, medium, and high, where poor indicates two or fewer amenities, medium is three to five, and high is more than five.

It is easy to forget to look for missing values when you generate a new variable from another. Therefore, we recommend generating a new variable with all zeros to start with. After that, you can do all your codings with `replace`.

```
. generate quality3 = 0
. replace quality3 = 1 if quality <= 2
. replace quality3 = 2 if quality > 2 & quality <= 5
. replace quality3 = 3 if quality == 6 | quality == 7
```

Use `tabulate` to check your work. For the example above, this would be `tabulate quality quality3`. If there still are any cases with 0 in the new variable, you may have forgotten to specify a value for some cases. You can set the remaining cases to missing by replacing the zero with a dot.

You can solve most problems using such simple commands. Of course, you can use more complicated `if` qualifiers and more complicated expressions for the values you want to generate.

Many data manipulation problems can be solved using shortcuts or more elegant solutions. You can create the `quality3` variable easily using the `recode` command shown in section 5.2.1. You could have created the additive index `quality` using `egen` (see section 5.2.2). Either way, being fluent in the use of `generate` and `replace` will help you find a solution even when there are no such shortcuts. This is especially true if you know the concepts we explain below. They are a bit more difficult, but it is worthwhile to know them.

5.1.3 Changing codes with by, _n, and _N

Let us intrigue you with a small example. Suppose that you want to generate a variable that represents the number of persons interviewed by each interviewer. At the moment, your data look as follows:[3]

```
. sort intnr
. list persnr intnr in 1/12
```

	persnr	intnr
1.	4489018	19
2.	3477336	19
3.	3058053	19
4.	6416098	19
5.	4400355	19
6.	4773914	19
7.	2871857	19
8.	7187154	19
9.	5019385	19
10.	4680834	287
11.	2332548	287
12.	1650333	1025

3. The order in which the respondents were interviewed by each interviewer may differ in your dataset.

For each subject, you can find the number of the person who conducted the interview. Thus you can easily find that interviewer 19 has conducted 9 interviews and that interviewer 287 has done only 2. You can write these results into your data, and then your data will look like this:

```
       persnr     intnr   intcount
  1.  4489018        19          9
  2.  3477336        19          9
  3.  3058053        19          9
  4.  6416098        19          9
  5.  4400355        19          9
  6.  4773914        19          9
  7.  2871857        19          9
  8.  7187154        19          9
  9.  5019385        19          9
 10.  4680834       287          2
 11.  2332548       287          2
 12.  1650333      1025          1
```

But how do you generate this variable with Stata? Think about it before reading on.

If you have an idea, most likely you have already worked with software for data analysis before. You probably thought of somehow combining the observations for each interviewer into what many software programs call an *aggregate*. If you want to do it that way with Stata, you can use `collapse`. But Stata provides a smarter solution:

```
. by intnr, sort: generate intcount = _N
```

You like this because it is short and fast, keeps the original data file, and requires you to understand only basic concepts, most of which you already know. You know `generate`, and you know the `by` prefix. What you do not know is `_N`. So, let us explain.

To understand `_N`, you need to understand the system variable `_n`, which contains the position of the current observation in the data. You can use `_n` to generate a running counter of the observations:

```
. generate index = _n
. list persnr intnr index in 1/12
```

	persnr	intnr	index
1.	4489018	19	1
2.	3477336	19	2
3.	3058053	19	3
4.	6416098	19	4
5.	4400355	19	5
6.	4773914	19	6
7.	2871857	19	7
8.	7187154	19	8
9.	5019385	19	9
10.	4680834	287	10
11.	2332548	287	11
12.	1650333	1025	12

For the first observation, the new variable `index` contains the number 1, for the second the number 2, for the 10th the number 10, and so on. Used with `by`, the system variable `_n` is the position within each by-group. This way you can generate a running counter for the number of interviews by each interviewer:

```
. by intnr: generate intIndex = _n
. list persnr intnr index intIndex in 1/12
```

	persnr	intnr	index	intIndex
1.	4489018	19	1	1
2.	3477336	19	2	2
3.	3058053	19	3	3
4.	6416098	19	4	4
5.	4400355	19	5	5
6.	4773914	19	6	6
7.	2871857	19	7	7
8.	7187154	19	8	8
9.	5019385	19	9	9
10.	4680834	287	10	1
11.	2332548	287	11	2
12.	1650333	1025	12	1

After learning `_n`, `_N` is easy: it indicates the highest value of `_n`. Without the prefix `by`, `_N` has the same value as `_n` for the last observation, which is just the number of observations in the dataset:

```
. display _N
3340
```

After the `by` prefix, `_N` contains the value of `_n` for the last observation within each by-group. The `_n` for the last observation in the by-group of interviewer 19 in our example is 9. Therefore, `_N` equals 9 in this by-group, and `by intnr: generate intcount = _N` is equivalent to `generate intcount = 9`. In the by-group of interviewer 287, the value of `_n` in the last observation is 2. `by intnr: generate intcount = _N` is now equivalent to `generate intcount = 2`, and so forth. This is what we want to have:

```
. list index persnr intnr intIndex intcount in 1/12
```

	index	persnr	intnr	intIndex	intcount
1.	1	4489018	19	1	9
2.	2	3477336	19	2	9
3.	3	3058053	19	3	9
4.	4	6416098	19	4	9
5.	5	4400355	19	5	9
6.	6	4773914	19	6	9
7.	7	2871857	19	7	9
8.	8	7187154	19	8	9
9.	9	5019385	19	9	9
10.	10	4680834	287	1	2
11.	11	2332548	287	2	2
12.	12	1650333	1025	1	1

Recoding with `by`, `_n`, and `_N` differs slightly from the process of recoding described earlier. It takes some time to get used to it, but once you do, you will find it very useful. Here is one more example.

Suppose that you need a variable that contains a unique value for each combination of marital status and education. The variable should be 1 for married interviewees (`marital==1`) with the lowest educational level (`edu==1`), 2 for married interviewees with secondary education (`edu==2`), etc. The standard way to generate such a variable is

```
. generate maredu = 1 if marital == 1 & edu == 1
. replace maredu = 2 if marital == 1 & edu == 2
```

and so on. After typing 42 commands, you will get what you want. Using the concepts explained above, however, you need only two lines:

```
. by marital edu, sort: generate maredu2 = 1 if _n == 1
. replace maredu2 = sum(maredu2)
```

Why? Consider the following fictitious data:

	marital	edu	step1	step2
1.	1	1	1	1
2.	1	1	.	1
3.	1	1	.	1
4.	1	1	.	1
5.	1	1	.	1
6.	1	2	1	2
7.	2	1	1	3
8.	2	1	.	3
9.	2	1	.	3
10.	2	2	1	4
11.	2	2	.	4
12.	2	2	.	4

The data file contains the variables representing marital status and education, each of which has two categories. The data file is sorted by marital status and education. If you type `by marital edu: generate maredu2 = 1 if _n == 1` you get the variable of step 1. The command assigns 1 to the first observation (`if _n == 1`) of each by-group and missing otherwise. As there are two `by` variables in the command, a new by-group begins when one or both of the variables change. The second command calculates the running sum of the variable generated in the first step. Missing values are treated as zero in this calculation.

5.1.4 Subscripts

Subscripts are commonly used in statistical formulas to specify which observation of a variable you refer to. For example, in the term x_3 the number 3 is a subscript; used in a statistical formula, the term would refer to the third observation of the variable x. Likewise we use subscripts in Stata to refer to a specific observation of a variable. You specify a subscript by putting brackets after a variable name. Inside the bracket, you use an expression to specify the observation in question. For example, `ybirth[1]` would refer to the first observation of the variable `ybirth` or `income[_N]` to the last observation of the variable `income`.

Subscripts can be useful for generating variables and are especially helpful for copying the values from one observation into another. We will show you how this works by using two fictitious datasets.

Typically, you need subscripts in hierarchical datasets like the following:

```
. preserve
. use hierarch, clear
. list
```

	hhnr	income	hhpos	occ
1.	1	0	2	.
2.	1	1100	3	51
3.	1	2600	1	52
4.	2	0	2	.
5.	2	6000	1	15
6.	3	3300	1	62
7.	4	620	3	61
8.	4	1100	2	62
9.	4	7000	1	53
10.	4	0	3	.

Here we have asked every adult in four households (`hhnr`) about their personal income (`income`). Suppose that you would like to have a variable equal to the total income for each household. How can you get this? Try to find a solution yourself before you continue reading.

Our solution is

```
. by hhnr, sort: generate hhinc=sum(income)
. by hhnr: replace hhinc = hhinc[_N]
```

You already know the command `generate hhinc = sum(income)`, which calculates the running sum of income. Here you do not sum from the first observation to the last but sum only from the first observation of each household to the last observation of that household. You do this by typing `by hhnr:` before the command. Then you can find the sum of personal incomes for all adults of the household in the last observation (`_N`) of each household. This number is copied to all other observations of the same household using the command `by hhnr: replace hhinc = hhinc[_N]`.

Similarly, you can find the occupational status (`occ`) of the head of the household (`hhpos==1`) and assign this value to all members of the same household.

```
. sort hhnr hhpos
. by hhnr: generate occ_h = occ[1] if hhpos[1] == 1
```

You can even combine these two tasks in one line using the `by` construction shown below, where the sorting takes place for all variables in the `by` statement. However, when Stata forms the groups to which `generate` will apply, it will ignore the variables in parentheses.

```
. by hhnr (hhpos), sort: generate occ_h2 = occ[1] if hhpos[1] == 1
```

This approach does not copy the contents of the last observation of each household to the other observations but copies the contents of the *first* observation. Using the `if` qualifier ensures that the occupational status of the first observation is indeed the occupational status of the head of the household. This is important because some heads of households may not have been interviewed.

Generating variables with subscripts is often useful when you are working with panel or time-series data. However, for many panel or time-series transformations, Stata offers specific tools for managing your data (see `help xtset` and `help tsset`).

From our examples, you may have gotten the impression that you need subscripts only for special datasets. This is not true. Most datasets contain hierarchical structures in some respect, but even without such structures, you will sometimes need subscripts.

This section and the previous section were harder to understand than the other sections in this chapter. You may ask yourself if there is an easier solution, and the answer is often yes. Highly specialized commands for specific tasks exist. The following section shows you some of them. Unfortunately, these commands often do not help you to better understand the basic concepts of generating variables with Stata. If you cannot find a special command for a specific task, do not give up, but remember there might be a solution in Stata (also see Cox [2002b]).

Before reading on, please restore the original dataset:

```
. restore
```

5.2 Specialized recoding commands

5.2.1 The recode command

You will often need to combine several values of one variable into a single value. Earlier we created the dummy variable minor, indicating respondents that have not reached the legal age of adulthood by the time the survey was conducted. Instead of using the procedure described on page 76, you could also use the command recode. Then you would type

```
. recode ybirth (min/1979 = 0) (1980/max = 1), gen(minor3)
```

With recode you assign new values to certain observations of a new variable according to a coding rule. Using the generate() option stores the results in a new variable instead of overwriting ybirth. For example, above we assigned the value 0 to respondents born before or in 1979, and the value 1 to those born in 1980 or after. The result is stored in the new variable minor3. Here min refers to the lowest value of the variable ybirth, and max refers to the highest value.[4] Missing values in this case do not count as the highest value.

If you wanted to create another variable with three categories, separating out respondents between age 18 and age 21, the recode command would look like this:

```
. recode ybirth (min/1975 = 0) (1976/1979 = 1) (1980/max = 2), gen(minor4)
```

Values specified in the recode command do not have to be in consecutive order.

The recode command also allows you to specify a variable list to be recoded, so you can recode several variables in one step. The following command generates copies of the variables on the appliances in dwellings, but with codes 1 and 2 reversed:

```
. recode kitchen-phone (1=2) (2=1), gen(d1 d2 d3 d4 d5 d6 d7 d8)
```

5.2.2 The egen command

egen provides a large and constantly growing number of extensions to the generate command. These extensions are based on nothing more than one or more generate and replace commands. egen can be seen as a feature designed to spare you from having to think about more complicated command sequences.

The structure of egen is similar to that of generate. The command is followed by a variable name (which is the name of the variable that should be generated), an

4. To categorize variables into quartiles or other percentiles, use xtile and pctile (see help pctile).

equal-sign, and finally the egen function. Unlike the general Stata functions discussed in section 3.1.6, the egen functions are available only within the egen command.

A useful egen function is anycount(*varlist*), which creates a variable containing the count of variables whose values are equal to any of the integer values in the values() option. Therefore, this egen function provides an easy way to form an index from several other variables. Look at the following example, which contains information on how concerned respondents are about several items:

```
. tab1 np950*
```

You may wish to form an index of respondents' concern about the future. One way to do this is to count the number of items that respondents said they were strongly concerned about, where "strongly" is coded as 1. You can do this easily by using the egen function anycount():

```
. egen worried = anycount(np950*), values(1)
. tabulate worried
     np9501 |
     np9502 |
     np9503 |
     np9504 |
     np9506 |
np9507 == 1 |      Freq.     Percent        Cum.
------------+-----------------------------------
          0 |        955       28.59       28.59
          1 |        890       26.65       55.24
          2 |        735       22.01       77.25
          3 |        461       13.80       91.05
          4 |        224        6.71       97.75
          5 |         58        1.74       99.49
          6 |         17        0.51      100.00
------------+-----------------------------------
      Total |      3,340      100.00
```

You can compute the number of missing values for each respondent with the egen function rowmiss(*varlist*):

```
. egen worried_m = rowmiss(np950*)
```

You can use this new variable worried_m in different ways, such as to display the table above only for those observations that have no missing values for any of the variables used in the egen command:

(*Continued on next page*)

```
. tabulate worried if worried_m == 0

     np9501 |
     np9502 |
     np9503 |
     np9504 |
     np9506 |
np9507 == 1 |      Freq.     Percent        Cum.
------------+-----------------------------------
          0 |        187       28.21       28.21
          1 |        159       23.98       52.19
          2 |        143       21.57       73.76
          3 |         92       13.88       87.63
          4 |         45        6.79       94.42
          5 |         20        3.02       97.44
          6 |         17        2.56      100.00
------------+-----------------------------------
      Total |        663      100.00
```

There are many other egen functions, a growing number of which users have created.[5] If you have a specific problem, see the list of egen functions (help egen). Keep in mind that several egen functions are simple applications of generate—simple if you have learned how to look at problems from a Stata perspective.

5.3 More tools for recoding data

The data manipulation tools you have learned so far can help you solve many problems. However, sometimes you will be confronted with variables that contain strings (letters or words) or dates, which require more data-management tools.

Because our previous datasets did not contain strings or dates, we will use a new dataset to provide examples. The dataset mdb.dta contains information about all German politicians who were members of the parliament (*Bundestag*) between 1949 and 1998.[6] For each politician, the dataset contains his or her name, party affiliation, and the duration of his or her membership in parliament.

```
. preserve
. use mdb, clear
```

5.3.1 String functions

String variables can be easily identified in the output of describe, where an entry in the column "storage type" beginning with "str" refers to a string variable:

5. Chapter 12 explains where you can find those functions.
6. See http://www.bundestag.de/htdocs_e/index.html for details.

```
. describe

Contains data from mdb.dta
  obs:         7,918                          MoP, Germany 1949-1998
 vars:            14                          1 Apr 2004 16:15
 size:       878,898 (97.2% of memory free)   (_dta has notes)
-------------------------------------------------------------------------------
              storage  display     value
variable name   type   format      label      variable label
-------------------------------------------------------------------------------
index           int    %8.0g                  Index-Number for Parlamentarian
name            str63  %63s                   Name of Parlamentarian
bmonth          byte   %9.0g                  Month of Beginning
bday            byte   %9.0g                  Day of Beginning
byear           int    %9.0g                  Year of Beginning
enddate         str13  %13s                   Date of End
party           str10  %10s                   Fraction-Membership
constituency    str4   %9s                    Voted in Constituency/Country
                                                Party Ticket
endtyp          byte   %8.0g       endtyp     Reason for Leaving the
                                                Parliament
pstart          int    %d                     Start of Legislative Period
pend            int    %d                     End of Legislative Period
bdate           int    %d                     Date of Birth
ddate           int    %d                     Date of Death
period          str2   %9s
-------------------------------------------------------------------------------
Sorted by:
```

Our dataset contains the following string variables: name, enddate, party, constituency, and period. String variables can be tabulated just like the already familiar numeric variables but cannot be summarized:

```
. summarize party

    Variable |       Obs        Mean    Std. Dev.       Min        Max
-------------+--------------------------------------------------------
       party |         0
```

Because you cannot do calculations with string variables, you may want to change them to numeric variables. You can do this by using either of two methods. First, you can use encode to generate a numeric variable with value labels according to the contents of a string variable. After typing the command, type the name of the string variable to be converted and then type the name of the numeric variable you want to create within the parentheses of the generate() option:

```
. encode party, generate(party_n)
```

encode attaches the value 1 to the category that comes first alphabetically, which is not always desirable. Specifically, you should not use encode for strings containing numbers that merely happen to be stored as strings. This is the case for the variable period, which contains numerals from 1 to 13; however, because they are stored in a string variable, Stata does not know that these numerals represent numbers. String variables like period should be converted to values with the command destring, which mirrors the syntax of encode:

```
. destring period, generate(period_r)
```

You may not always want to convert strings to numeric variables, as many data-management tasks can be done much easier with string variables. You can use many of the tools you have learned about with string variables. For example, to construct a variable that is 1 for all politicians who are members of the Christian Democratic Union and zero for all others, type

```
. generate cdu = party == "CDU"
```

That is, you can use *strings* in expressions if you enclose them in double quotes. However, be careful. Unlike numbers, strings can be lowercase and uppercase. And although we humans tend to be sloppy about whether we use uppercase or lowercase letters, Stata is not. If your strings contain the same information but mix lowercase and uppercase, you should harmonize them by using either the `lower()` or `upper()` functions. The following example uses `upper()` to convert the contents of `party` on the fly, when constructing a variable that is 1 for all politicians who are members of the Social Democratic Party (SPD):

```
. generate spd = upper(party) == "SPD"
```

An important string function is `strpos(`$s1$`,`$s2$`)`, which checks whether one string, $s2$, is part of another string, $s1$, and returns the position where $s2$ is first found in $s1$. For example, using `strpos()` you can find out that the string "Stata" contains the string "a" for the first time at the third position:

```
. display strpos("Stata","a")
3
```

Now take a look at the variable `name`:

```
. list name in 1/5
```

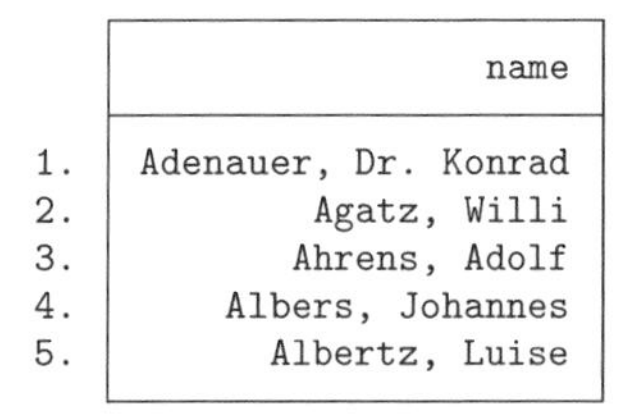

	name
1.	Adenauer, Dr. Konrad
2.	Agatz, Willi
3.	Ahrens, Adolf
4.	Albers, Johannes
5.	Albertz, Luise

As expected, the variable contains the names of all the politicians. If you go through the list of names, you will notice that some politicians have a doctoral degree. You can extract this small piece of information about the politicians' education easily using the `strpos()` function: just check whether the name of a politician contains the string "Dr.":

```
. generate dr = strpos(name,"Dr.") > 0
. tabulate dr

         dr |      Freq.     Percent        Cum.
------------+-----------------------------------
          0 |      5,598       70.70       70.70
          1 |      2,320       29.30      100.00
------------+-----------------------------------
      Total |      7,918      100.00
```

The `strpos()` function is also an important tool if you need to extract a specific part of a string. Suppose that you want to construct a variable that contains only the last names of the parliamentarians. In the variable `name`, first and last names are separated by a comma. The last name of each parliamentarian begins with the first letter and ends one character before the comma. Therefore, the first step in extracting the last name is to know the position of the comma. By typing

```
. gen comma = strpos(name,",")
```

you store this information in the variable `comma`. Now you need to extract the substring of the variable name from the first position up to the position in the variable `comma`. You can do this using the `substr()` function. Inside the parentheses of this function are three arguments, separated by commas: the string from which the substring should be extracted, the position where the substring begins, and the length of the substring. For example, to extract a string of length 4 beginning at the 5th position from the string "Hello, world", type

```
. display substr("Hello world",5,4)
o wo
```

To extract the last two characters of a string, you can count backward by using a negative second argument:

```
. display substr("Hello world",-2,.)
ld
```

That is, negative values for the second argument are used to indicate that the beginning position is counted from the end of the string. A dot in the third argument is used to extract everything from the beginning position to the end of the string.

To extract the last names, you need to extract a substring of the string in the variable `name` from the first position to the position of the comma, which is stored in the variable `comma`. The length of that substring must therefore be the position of the comma minus 1. Hence,

```
. gen str famname = substr(name,1,comma-1)
. list famname in 1/5

     +-----------+
     |   famname |
     |-----------|
  1. |  Adenauer |
  2. |     Agatz |
  3. |    Ahrens |
  4. |    Albers |
  5. |   Albertz |
     +-----------+
```

returns the required result.

Clearly, you can use the same approach to extract the part of the name after the comma:

```
. gen str firstname = substr(name,comma+1,.)
. list firstname in 1/5
```

However, to extract just the first names of the politicians requires more work. In addition to the first names, you will find doctoral degrees and titles of nobility after the comma. To get rid of these suffixes, you can use the **subinstr()** function to replace a substring of an arbitrary string with another string. In the following example, we use this function to replace the string "Dr." in the string variable **firstname** with an empty string—that is, we delete the string "Dr." from the strings in **firstname**.

```
. replace firstname = subinstr(firstname,"Dr.","",.)
```

The first argument within the parentheses is the string or variable that contains the substring to be replaced, the second argument is the substring to be replaced, and the third argument is the replacement string. The fourth argument indicates the number of occurrences to replace; missing means all occurrences.

5.3.2 Date and time functions

Our little dataset contains not only the names of all German politicians who were members of parliament (*Bundestag*) between 1949 and 1998, but also their date of birth, the start and end time of their parliamentary membership, as well as their date of death.[7] We talk about "dates" when there is information on the day, month, and year of the respective event.

Some datasets also include exact information on times of event (e.g., hour, minute, second, or even millisecond).[8] We talk about "times" when there is, at minimum, information on the hour and the minute of an event.

7. For those who are already deceased and for whom this information was put forward to the Parliament's Administration.
8. Similar to the date information, available time data are of varying precision.

Dealing with dates and times require some steps that are quite different from what we discussed so far. We will explain both, starting with date functions.

Dates

Date functions are used to deal with variables containing dates. By date we do not mean, say, a year, but an exact day, such as "January 1, 1960", or "November 9, 1989".

Some datasets, such as our dataset on parliamentarians, use information on dates heavily. Along with each person's name, our dataset contains information on their date of birth (and death), the date they became a member of the parliament, and the date they left the parliament.

In this file, the beginning and end dates are stored in different ways. The beginning dates are stored in three variables: `bday` is short for "begin day" and contains the day the politician joined the parliament, `bmonth` contains the month, and `byear` contains the year:

```
. list name bday bmonth byear in 1/4

     +-------------------------------------------------+
     |                 name   bday   bmonth    byear   |
     |-------------------------------------------------|
  1. | Adenauer, Dr. Konrad      7        9     1949   |
  2. |         Agatz, Willi      7        9     1949   |
  3. |        Ahrens, Adolf      7        9     1949   |
  4. |     Albers, Johannes      7        9     1949   |
     +-------------------------------------------------+
```

The date each member left parliament is stored in a single string variable (`enddate`):

```
. list name enddate in 1/4

     +------------------------------------------+
     |                 name        enddate      |
     |------------------------------------------|
  1. | Adenauer, Dr. Konrad   Sep. 7. 1953      |
  2. |         Agatz, Willi   Sep. 7. 1953      |
  3. |        Ahrens, Adolf   Sep. 7. 1953      |
  4. |     Albers, Johannes   Sep. 7. 1953      |
     +------------------------------------------+
```

Suppose that you want to know the time span between joining and leaving for each politician. Generally, you would want to subtract the beginning date from the end date. However, this is not as easy as it would seem, as you cannot subtract anything from a string variable such as `enddate`. You might therefore decide to store only the year of the exit date and subtract the beginning year from that variable. But this would not be very accurate, because it makes a difference whether the dates are at the beginning or at the end of the beginning and ending years.

When dealing with dates, you need to use "elapsed dates", which are integer numbers counting the days from a standard date. If, for example, January 1, 1960, is used as day 0, then January 2, 1960, is day 1, and November 9, 1989, is day 10,905.

Elapsed dates are clearly preferable to the other ways of storing dates, so Stata has special functions to convert dates from other formats into elapsed dates. To convert dates stored in separate variables into elapsed dates, the function `mdy()` is used. Inside the parentheses you list the names of the variables containing the month, the day, and the year. List the names in the order month, day, and year, and separate the names by commas. This function returns the elapsed date for the accession date as follows:

```
. gen begin = mdy(bmonth,bday,byear)
. list name bday bmonth byear begin in 1/4
```

	name	bday	bmonth	byear	begin
1.	Adenauer, Dr. Konrad	7	9	1949	-3768
2.	Agatz, Willi	7	9	1949	-3768
3.	Ahrens, Adolf	7	9	1949	-3768
4.	Albers, Johannes	7	9	1949	-3768

In Stata, elapsed dates are counted from January 1, 1960, so all dates before then are negative numbers.

To convert dates stored in string variables, you can use the `date()` function. Inside the parentheses, you first state the name of the string variable containing the dates and then, separated by a comma, a "pattern string", which describes the format of the dates in the string variable. If the string variable contains dates like "1989 November 9th" the pattern string will be `ymd` because the order of the date is year, month, and day. In `enddate` the dates have the order month, day, and year, so you need the pattern string `mdy`.

```
. gen end = date(enddate,"MDY")
. list name enddate end in 1/4
```

	name	enddate	end
1.	Adenauer, Dr. Konrad	Sep. 7. 1953	-2307
2.	Agatz, Willi	Sep. 7. 1953	-2307
3.	Ahrens, Adolf	Sep. 7. 1953	-2307
4.	Albers, Johannes	Sep. 7. 1953	-2307

`date()` is quite tolerant about how the dates are stored in your string variable. The `date()` function understands the proper English words and usual abbreviations for the months, does not care about the characters between the different parts of the dates, and can even deal with two-digit years. Here are three further examples, and you might try some more:

```
. display date("November 9 1989","MDY")
10905

. display date("Nov9/1989","MDY")
10905

. display date("11. 9/89","MD19Y")
10905
```

After you convert the entry and exit dates of each politician into elapsed dates, calculating the time span between the two dates is easy:

```
. gen time = end - begin
```

In our dataset, there is one complication. Some politicians are listed more than once if they served in more than one legislative period. If we want to get the entire duration of their membership, we need to sum the durations of all legislative periods. Also we might prefer a dataset with one observation per parliamentarian. Yet another job for recoding with `by`, `_n`, and `_N` (see sections 3.2.1 and 5.1.3).

```
. by name (period_r), sort: replace time = sum(time)
. by name: replace begin = begin[1]
. by name: keep if _n == _N
```

A second complication is more interesting. You may well be convinced that in Stata, it is good to have elapsed dates. However, you find elapsed dates to be cryptic. You might know something about November 9, 1989—the date when the Berlin Wall fell—but the number 10905 is inscrutable. Therefore, it would be helpful for Stata to store dates as elapsed dates but show them in a way that humans can understand. You can do this by setting the display format of a variable containing elapsed dates to the Stata standard date format

```
. format begin %td
```

or build even fancier display formats using the user-specified elapsed date formats described in `help dates and times`:

```
. format end %tdM_d._CY
```

After you have set the display format, commands that respect display formats will show elapsed dates in a way that users can understand. Among the commands that respect display formats is `list`. Therefore, we can provide the following list of those five politicians who were seated for the longest time in the German parliament:

```
. sort time
. list name begin end in -5/l
```

	name	begin	end
2812.	Czaja, Dr. Herbert	06oct1953	December 20. 1990
2813.	Renger, Annemarie	06oct1953	December 20. 1990
2814.	Dollinger, Dr. Werner	06oct1953	December 20. 1990
2815.	Schmidt, Dr. R. Martin	07sep1949	February 18. 1987
2816.	Stuecklen, Richard	07sep1949	December 20. 1990

Time

In Stata, times are handled similarly to dates. To create elapsed times, milliseconds are counted since January 1, 1960 (this is Stata's standard see page 92). A value of one on an elapsed time variable means one millisecond into January 1, 1960. November 9, 1980 at 6:53 PM—which is the time Günter Schabowski declared the opening of the border between East and West Germany—happened 942 432 780 000 milliseconds after January 1 1960, 12:00 AM.

Time diaries are a common example of datasets that include times. In time diaries, respondents are asked to note the start and end time of each activity within a day. For the purpose of illustration, we use the time diary available in the file `diary.dta`. This dataset contains a (fictitious) schedule for one day in the life of a child.[9]

```
. use diary, clear
. list in 1/10
```

	activity	bhour	bmin	estring	location
1.	Sleeping	0	0	7:30	at home
2.	Getting up	7	30	7:40	at home
3.	Using the bathroom	7	40	7:45	at home
4.	Eating	7	45	8:15	at home
5.	On the way	8	45	9:05	in car
6.	In school	9	5	15:15	school
7.	Playing basketball	13	15	17:00	YMCA
8.	On the way	17	0	17:30	in car
9.	Watching TV	17	30	18:00	at home
10.	Eating	18	0	18:25	at home

The dataset contains two variables for the beginning of each activity: one variable captures the hour (`bhour`) and the other variable the minute (`bmin`). The end of an activity is stored in another variable as string. We will show how to convert both start and end data into an elapsed time format.

Knowing that our fictitious child got up at 7:30 AM is not enough. To calculate the elapsed time (the number of milliseconds that have passed since midnight on January 1st in 1960), we need one more piece of information: the day at which we recorded the activities of this child. Let us assume our data recorded activities on November 9, 1989. We can add a variable containing that information to the dataset using one of the date functions we discussed in the previous section:[10]

```
. gen date = date("9.11.1989","DMY")
```

9. The fictitious day followed the questionnaire for children used in the Panel Study of Income Dynamics.
10. This is an arbitrary day we only picked for this example.

Once the date is known, it is fairly easy to create elapsed times. There are a couple of functions that allow you to do this (see `help dates`). For our example, we will use `dhms()`. Here you specify the day as an elapsed date (`d`), then the hour of the event (`h`), the minute (`m`), and finally the seconds (`s`) if you know them. The elapsed time variable `begin` for the start of each activity will be created as follows:

```
. gen double begin = dhms(date,bhour,bmin,0)
```

Let us explain since this command might not be intuitive. First, the keyword "double" ensures that enough storage space is reserved for the elapsed time variable. Remember that times are stored in milliseconds, thus, they often contain large numbers. To avoid rounding errors, you must save those time variables with a special storage type (double). You can find more on storage types in section 5.6. Second, you might notice that we added a zero at the very end of the command. Our data did not contain a variable with information on the second the activity started; thus, we chose to enter a zero instead. Finally, we were fortunate to have the date already stored as an elapsed date variable. However, in the absence of an elapsed date you could use the function `mdyhms()`. Within the brackets you would enter month, day, year, hour, minute, and seconds.

To convert the end time of the activities into an elapsed time variable we need to use a different function. As mentioned above, in our example dataset the end time is stored as a string variable. To convert string variables into elapsed time you would use the function `clock()`. Its functionality is similar to `date()`. Within the brackets, you first list the string variable, which contains the entire time point (that is date and time). Then you give information on the structure of this string variable.

The string variable in our example dataset does not yet contain a variable with all information for the time point. Before we can use `clock()`, we need to add a date to our time information.

```
. gen efull = "9.11.1989" + estring
. gen double end = clock(efull,"DMYhm")
```

The key `DMYhm` indicates that the newly created string variable contains day, month, year, and finally hour and minutes, respectively. Other options can be found in `help dates`.

It is reasonable to define a display format for your time variable, just like you did earlier for dates. The standard format for times is `tc`,

```
. format begin end %tc
```

Once elapsed times are created, we can find out which activity in the life of our fictitious child took the most time. We calculate the difference between the start and end times. The function `minutes()` allows users to save the difference between the start and end times.

```
. gen time = minutes(end-begin)
```

Afterward, you proceed as usual:

```
. by activity, sort: replace time = sum(time)
(5 real changes made)
. by activity: keep if _n == _N
(5 observations deleted)
. sort time
. list activity time
```

	activity	time
1.	Getting up	10
2.	Listening to bedtime story	20
3.	Playing computer games	30
4.	Reading book from library	35
5.	On the way	50
6.	Eating	55
7.	Watching TV	60
8.	Using the bathroom	65
9.	Playing basketball	225
10.	In school	370
11.	Sleeping	450

Well, now we know that the life of our child is dominated by sleeping and school.

Before reading on, restore your original data:

```
. restore
```

5.4 Commands for dealing with missing values

The easiest way to deal with missing values is to use `generate` and `replace`, which indicate a missing value with a dot. This dot can be used like any numeric value. For example, by typing

```
. replace income = . if income == 0
```

you change all occurrences of the value 0 in the `income` variable to the *missing value*.

In addition to the dot, there are 26 other codes for missing values, namely, `.a`, `.b`, ..., `.z`. Observations with these codes are also excluded from statistical analyses. These codes are used to distinguish between different reasons for values being missing. For example, many surveys distinguish between the answer "do not know" and an explicit refusal to answer. In this case, it makes sense to code "do not know" as `.a` and explicit refusal to answer as `.b`. This way, both types of missing data are excluded from statistical analyses, but the information on the different causes is kept in the dataset.

You can assign the special codes for missing values the same way you assign the single dot. For example, the variable income contains some very small values, which seem to be unreliable. You can set all values below $10 to .a with

```
. replace income = .a if income <= 10
```

Often you may want to change multiple instances of the same value to missing in a dataset. For example, in the original GSOEP data, refusals to answer are coded as -1. In data analysis, refusals are usually treated as missing values. Therefore, it makes sense to change -1 in all variables to missing simultaneously. To do so, you can set up a foreach loop (see section 3.2.2) or use the mvdecode command:

```
. mvdecode _all, mv(-1=.a)
```

mvdecode replaces the values of a *varlist* according to the rule specified in the mv() option. The rule can be read as, "Make the value before the equal-sign equal to the missing-value code after the equal-sign". If you do not specify a missing-value code, the simple dot is used. In our example, we used _all instead of a variable list, which executes the command on all variables in the data file. However, our example does not change the data, as we have already set -1 to missing. Before you use _all, you should verify there are no variables in your dataset for which -1 is a valid value. In our case, we know no variables include -1 as a value to be used in statistical computations.

Of course, you can change more values to missing. In the GSOEP, the response "does not apply" is always coded as -2, which means, "This question has not been asked of this respondent". Usually you would want to change this value to missing, as well:

```
. mvdecode _all, mv(-2=.b)
```

Sometimes you might want to restore a missing definition. You can do this using replace or mvencode. mvencode has the same syntax as mvdecode and does the opposite. For example, by typing

```
. replace income = 0 if income == .
```

you change the missing value of income to the numeric value 0. By typing

```
. mvencode _all, mv(.a=-1)
```

you translate the missing value .a of all variables to -1.

These examples also show how problematic assigning missing-value codes can be. Above we have assigned all incomes below $10 to the missing-value code .a. This assignment is irreversible, so there is no way to get the initial values of income back from the recoded version of income. Such problems arise often in assigning missing values to the data, although they also can occur in recoding variables.

To avoid this problem, we recommend that you duplicate the variables and apply the missing-value statements only to the copies. And remember, you are always in luck if you work with do-files in the way we suggested in chapter 2.

5.5 Labels

You have not finished generating a variable until you label it. Labels identify the contents of a variable. There are three main ways to label a variable: the *variable name*, the *variable label*, and the *value labels*.[11]

The variable name is not a label in the strictest sense. It is the name you give the new variable in `generate`. You can use variable names up to 32 characters long, but you typically will not use variable names of that length. Often you will need to type the names repeatedly, which will be tedious if the names are long.

There are two approaches used for choosing variable names. Often you will find datasets with *logical* variable names, where a logical key is used to name the variables. You will quite often find variable names composed of the letter `v` and the number of a question in a survey. Take the variable `v1` for example. This variable has the answers to the first question in our survey. The variable `v2` would contain the answers to the second question, and so on. In the GSOEP, logical variable names are used to express the year of the survey, the survey type, and the question number (see section 10.4.1). However, in a day-to-day data analysis, descriptive variable names can be quite helpful. Descriptive names directly indicate the contents of the variable. In our example, the variable `gender` contains the gender of the respondents.

We recommend that you use logical variable names when entering data from a questionnaire (see chapter 10) but that you use descriptive variable names when preparing variables for a specific analysis.

Short variable names may not describe the contents of a variable clearly. To provide more information, you can use the variable label. For examples of variable labels, look at the right column of the output of `describe`. You also can find variable labels in the output of some statistical procedures.

Variable labels serve no purpose in Stata other than for users to understand the contents of a variable. That said, leaving out variable labels can be very annoying to users, especially in a dataset with logical variable names.

To label a variable, value, or an entire dataset, you use the `label` command.[12] If you want to use `label` to label a variable, you need to specify a keyword. Type `label variable`, then type the name of the variable you want to label, followed by the label itself. You can use up to 80 characters for the label.

To label the variable `worried`, for example, you would use the following command:

```
. label variable worried "Care"
```

You do not need the quotation marks for labels that do not have special characters—dashes, blanks, commas, etc. Some characters, such as letters with accents, are displayed

11. There is one more concept in Stata that is less widely used—that of "notes". We have used notes in `data1.dta` to store the variable name and the record type of the variable in the GSOEP database. You can look at this by typing `notes`. See `help notes` for more information.
12. For information about labeling datasets, see section 10.5.

properly only if you use a font that can display those characters. If you are concerned about portability, you may not want to use such characters.

As stated, variable labels are intended to help the Stata user. The same is true of value labels, which indicate the meanings of the values of a variable. This is important for variables with nominal scaling. The numbers 1 and 2 in the variable for gender (gender) are meaningless until we know that 1 stands for "male" and 2 stands for "female". To store such meanings in the dataset, we define value labels.

We again use the command label, but this time with two steps:

1. Define a label that contains the values and their meanings.
2. Attach the label to the variable.

Consider the variable minor generated on page 76. To label the values of minor, we first define the contents of the label by typing label define, followed by the name of the label we want to define (at most 32 characters) and then the definitions of the values. We do this last step by specifying a value and stating its meaning:

```
. label define minor_lb  0 "adult" 1 "minor"
```

Again quotation marks are needed only if the value labels contain special characters, but we recommend that you always use them. Also avoid using letters with accents and the like. You can use labels up to 32,000 characters long to specify the meaning of a value, but we doubt that labels that long are useful.[13]

Defining a value label has no effect until we attach the contents of the label to a variable. This is done with label values *varname labelname*:

```
. label values minor minor_lb
```

Now the values of the variable are connected to the definitions specified in the label. The output of statistical commands, such as tabulate, then presents the meaning of the numbers instead of the numbers themselves:

```
. tabulate minor

      minor |      Freq.     Percent        Cum.
------------+-----------------------------------
      adult |      3,286       98.38       98.38
      minor |         54        1.62      100.00
------------+-----------------------------------
      Total |      3,340      100.00
```

This two-step process may seem unnecessarily complicated, but it has its advantages. The main advantage is that you can use the same value label to label the values of more than one variable. In our dataset, we have connected all variables indicating features of dwellings with the label janein. This label defines 1 as meaning "yes" and 2 as

13. For compatibility with older versions of Stata and other programs, you might want to restrict yourself to 80 characters. Always put the most significant part of a label into the first few characters.

meaning "no". Whenever variables share the same values, we can use the value label `janein` again.

There is an even easier way to link value labels and variables. Define the value label `yesno`:

```
. label define yesno 0 "no" 1 "yes"
```

As already described, you can link `yesno` with `label value` to any variable. However, for new variables you can use `generate`. To do so, add a colon and then the name of the value label after the name of the new variable. This way you can generate and label a variable in one command:

```
. gen married:yesno = marital==1 if marital < .
. tab married
```

married	Freq.	Percent	Cum.
no	1,480	44.31	44.31
yes	1,860	55.69	100.00
Total	3,340	100.00	

Finally, you can use the command `label list` to get information about the contents of a value label. To look at the contents of the value label `yesno`, you would type

```
. label list yesno
```

If you do not specify a label name, `label list` returns a list of the contents of all the labels we have.

A helpful alternative to `label list` is `numlabel`, which allows you to include the value in a value label. For example, the value label can be "1 yes" instead of "yes". This way, commands such as `tabulate` show both the label and its numeric value. The following command makes this change for any value labels in the dataset:

```
. numlabel _all, add
. tab1 np940*
```

Type `help numlabel` for more information.

5.6 Storage types, or the ghost in the machine

There is a frustrating problem that you will encounter sooner or later. Before we can explain this problem, we must make a short technical introduction. Stata distinguishes alphanumeric variables (*strings*) from numerical variables (*reals*). Strings contain letters and other characters (including numerals that are not used as numbers). Reals are numbers. For both types of variables, Stata distinguishes between different storage types:

```
. help data types
```

The storage types differ in the amount of memory they use. To store an observation in a variable of storage type `byte`, Stata uses exactly 1 byte of memory. For the storage type `double`, Stata uses 8 bytes. To save memory, you should use the most parsimonious storage type. You can provide a keyword with `generate` to specify the storage type of the new variable.[14] If you do not specify a keyword, Stata defaults to using floats for numbers or a string of sufficient length to store the string you specify, so you will usually not need to care about storage types. However, there is one exception. To illustrate, try typing the following:

```
. generate x = .1
. list if x == .1
```

You would expect a list of all observations now, as the variable `x` is 0.1 for all observations. But what you got is nothing.

Here is what has happened: Stata stores numbers in binary form. Unfortunately, there is no exact binary representation of many floating-point numbers. Stored as floating-point variables, such numbers are precise up to about seven digits. The number 0.1 is stored as 0.10000000149011612 as a floating-point variable, for example. Likewise the number 1.2 is stored as 1.200000047683716. The problem therefore is that the variable `x` does not really contain 0.1. The variable contains instead 0.10000000149. On the other hand, when Stata calculates, it is always as precise as it can be. That is, Stata does calculations with a precision of about 16 digits. For this reason, Stata handles the number 0.1 in a calculation as 0.1000000000000000014.... If you compare this 0.1 with the 0.1 stored as a `float`, as in the above `if` qualifier, the two numbers are not equal. To avoid this problem, you can store numeric variables as `double`. You can get around the problem we described by rounding the decimal in the `if` qualifier to `float` precision:

```
. list if x == float(.1)
```

Type `help data types` for more information.

5.7 Exercises

1. Load a subset of the National Health and Nutrition Examination Study (NHANES) into Stata memory by typing

   ```
   . webuse nhanes2, clear
   ```

   (If you get the error message "`no room to add more observations`", please type `clear`, followed by `set memory 30m` before using the above command; see section 10.6 for details.)

14. The keywords `byte`, `int`, `long`, `float`, and `double` cannot be used as variable names for this reason (see section 5.1.1).

2. Create the variable `men` with values 0 for female observations and 1 for male observations. Label your variable with "Men y/n" and the values of your variables with "no" for zero and "yes" for one.
3. Correspondingly, create fully labeled dummy variables (indicator variables coded with 0 and 1) for "being in excellent health" and "being 70 years of age or more".
4. Assuming that the formula to calculate the body mass index (BMI) is

$$\text{BMI} = \frac{\text{Weight in kg}}{\text{Height in m}^2} \tag{5.1}$$

   create a fully labeled BMI variable for the NHANES data.
5. Create a fully labeled version of BMI, where the values have been classified according to the following table:

Category	BMI
Underweight	< 18.5
Normal weight	18.5–24.9
Overweight	25.0–29.9
Obese	30.0–39.9
Severely obese	> 40

6. Create a fully labeled version of BMI, where the values of BMI have been classified into 4 groups with approximately the same number of observations.
7. Create a fully labeled version of BMI, where the values of BMI have been classified into three groups defined as follows: Group 1 are all observations with a BMI lower or equal to one standard deviation below the gender specific average. Group 2 are all observations with values higher than group 1, but lower than one standard deviation above the gender specific average. Finally, group 3 are all observations with values above group 2. Please create this variable with help of `egen`.
8. Create the variable of the last problem without using `egen`.
9. With the help of the command `egen`, create a variable that enumerates all possible covariate combinations formed by the variables `region` and `smsa`.
10. Create the variable of the last problem without using `egen`.
11. Create a variable that indicates for each respondent how many people were interviewed in the same strata where the respondent lives.

6 Creating and changing graphs

In modern data analysis, graphs play an increasingly important role. Unfortunately, some authors regard data analysis using graphs as disjoint from traditional *confirmatory* data analysis. In contrast to confirmatory data analysis, which presents and tests hypotheses, graph-based data analysis is often perceived as a technique solely for generating hypotheses and models. However, as Schnell (1994, 327–342) convincingly spells out, this division between exploratory and confirmatory data analysis is misleading. It seems more sensible for us to regard graphs as tools for data analysis in general, and maybe even primarily for hypothesis-driven data analysis. We therefore use many graphical features in the chapters on distributions (chapter 7) and regression and logistic regression (chapters 8 and 9). There we discuss how to use and interpret the different graphs.

To take full advantage of the applications shown in these later chapters and create your own graphs, you need to understand Stata's basic graphics capabilities, which we explain in this chapter. Naturally, we will not cover every possibility, but after reading this chapter, you should be able to understand the logic behind Stata's graphics features. We strongly recommend that you read [G] **graph intro**, the online help, and last, but not least, the excellent book on Stata graphs by Mitchell (2008).

The examples in this section use the single-person households from the dataset `data1.dta`. Therefore, type[1]

```
. use data1
. keep if hhsize==1
```

6.1 A primer on graph syntax

The syntax of the graph commands is different from that of most other Stata commands, which we will explain briefly here, though we will go into more detail as we go:

- A Stata command for creating graphs comprises two elements: the `graph` command and a graph type. Here, for example, `box` is the graph type:

  ```
  . graph box rent
  ```

1. Make sure that your working directory is `c:\data\kk2`; see page 3.

- For the twoway graph type, a plottype must also be specified. Here is an example with the plottype scatter:

```
. graph twoway scatter rent sqfeet
```

For the twoway graph type, you can leave out graph to save typing. For the plottypes scatter and line, you can even leave out twoway. The following commands are therefore identical to the one given above:

```
. twoway scatter rent sqfeet
. scatter rent sqfeet
```

- The plottypes of the twoway graph type can be overlaid. Here is an example with scatter and lfit:

```
. graph twoway (scatter rent sqfeet) (lfit rent sqfeet)
```

Here both types are set in parentheses. However, you can also separate the plottypes with || as in the following example, where we also leave out graph and twoway

```
. scatter rent sqfeet || lfit rent sqfeet
```

- Occasionally, you will find options such as xlabel(#20, angle(90)) or xscale(range(0 300) reverse alt) in your graph commands. That is, options of the graph command can contain suboptions or a list of options.
- The overall look of a Stata graph is specified by a graph scheme, so changing the graph scheme can change the look of the graph considerably:

```
. set scheme economist
. scatter rent sqfeet
```

To obtain the same graphs as shown in the book, you must switch to the s2mono scheme:

```
. set scheme s2mono
```

6.2 Graph types

For the most part, the graph command is composed of the same building blocks used in other Stata commands. However, you must include a subcommand for the graph type. When creating statistical graphs, you must first decide on the graph type, after which you can use the options to design the graph.

6.2.1 Examples

Among others, the following graph types are available in Stata:

- Bar charts

```
. graph bar sqfeet, over(area, label(angle(45))) title(bar chart)
```

- Pie charts

```
. graph pie, over(area) title(pie chart)
```

- Dot charts

```
. graph dot (mean) sqfeet, over(area) title(dot chart)
```

- Box-and-whisker plots (box plots)

```
. graph box sqfeet, over(area, label(angle(45))) title(box and whisker plot)
```

- Twoway (graphs in a coordinate system). This graph type allows for various plottypes, such as scatterplots, function plots, and histograms.

```
. graph twoway scatter rent sqfeet, title(scatterplot)
. graph twoway function y = sin(x), range(1 20) title(function plot)
. graph twoway histogram rent, title(histogram)
```

- Scatterplot matrices

```
. graph matrix np0105 rent sqfeet, title(scatterplot matrix)
```

You will find examples for each of these graph types in figure 6.1.

(*Continued on next page*)

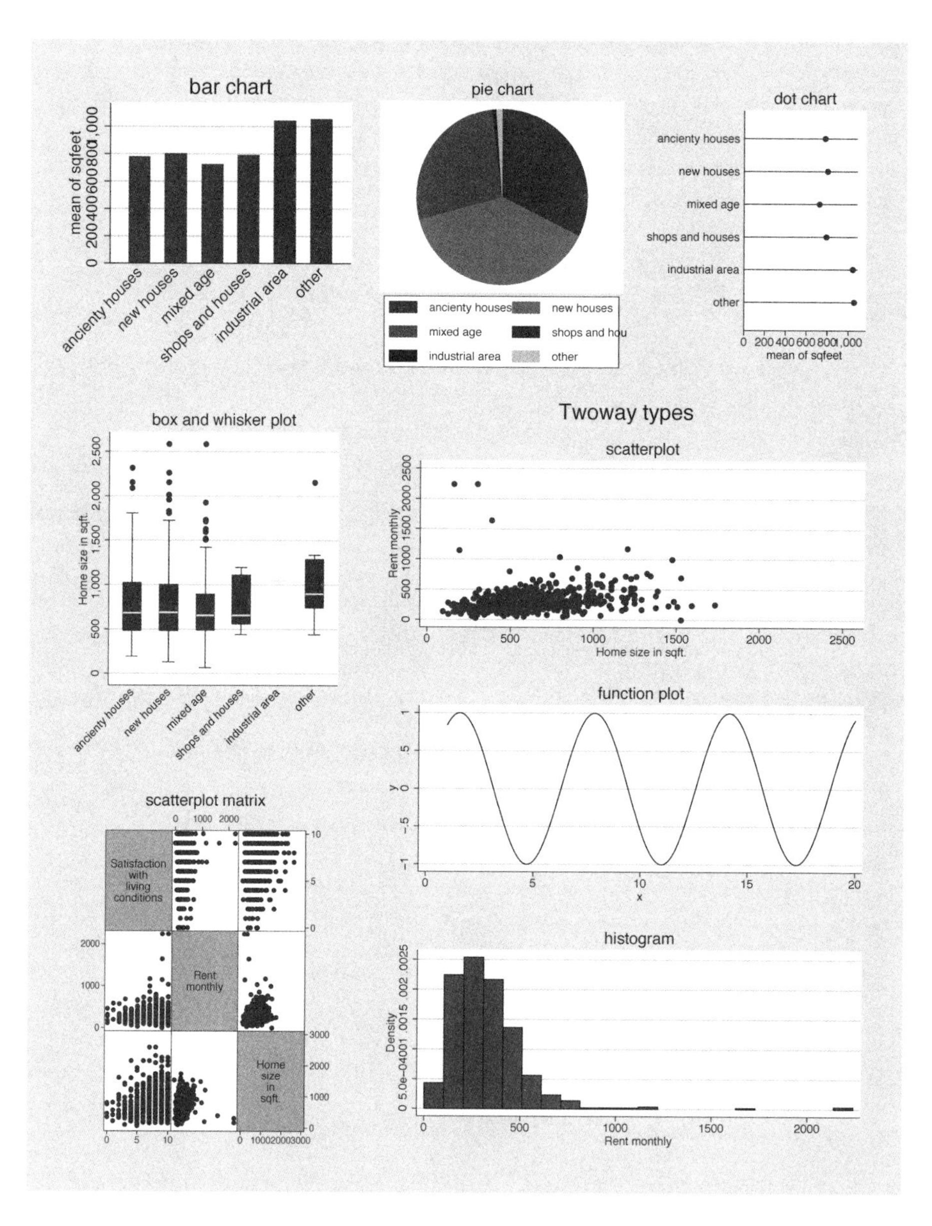

Figure 6.1. Types of graphs

grtypes.do

6.2.2 Specialized graphs

In addition to the basic graphs shown in figure 6.1, there are roughly 50 statistical graph commands, such as distributional diagnostic plots, plots designed for regression diagnostics, time-series graphs, and graphs for survival analysis. All these graphs use the basic graph commands and require specific statistical interpretations or special data preparation. Some are only available immediately after statistical procedures, whereas others are designed to help you prepare a statistical analysis, such as

```
. gladder inc if inc > 0
```

which displays how the variable income would be distributed after nine different power transformations (see section 8.4.3). For a list of all statistical graphs commands, type `help graph other`; you will see several applications of different graph types in the rest of this book.

Some software packages include 3-D graphs, meaning illustrations that appear to be three-dimensional. In 3-D graphs, rectangles of a bar chart may be depicted as blocks, lines in line graphs as tapes or snakes, and the circular segments from pie charts as slices of pie or cake. Graphs such as these might be helpful in business presentations, but most of them are not suitable for presenting statistical data. Therefore, there are no such illustrations in Stata.

6.3 Graph elements

The appearance of graphs is defined by a series of elements, shown in figure 6.2 (see Cleveland 1994, 22–25). These elements can be roughly subdivided as follows. The rest of this chapter discusses these elements.

- Elements that control the display of data, including the shape, color, and size of the "marker symbols", as well as lines, bars, and other ways to display data.
- Elements that control the size and shape of the graph, including the "graph region" and "plot region". The graph region is the size of the entire graph, including titles, legends, and surrounding text. The plot region is the space that can hold data points. Just inside the plot region is the data region. The plot region's size is determined by the axes, whereas the data region's size is determined by the minimums and maximums of the variables being plotted.
- Elements which convey additional information within the graph region, including, for instance, reference lines for crucial values, marker symbol labels, or any other text in the plot region.
- Information outside the plot region, which affects the appearance of the axes that border the graph region on the left (y axis), bottom (x axis), top (upper axis), and right (right axis). The appearance of information outside the plot region is controlled by various elements, e.g., "tick lines", "axis labels", and "axis titles".

Often displayed here are a "legend", which explains the symbols in the graph, and the title and description of the graph.

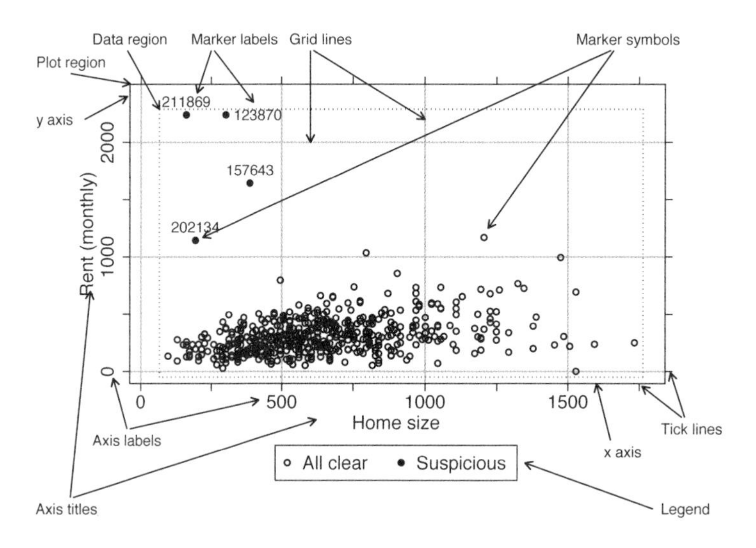

Figure 6.2. Elements of graphs

grcleveland.do

You can change all these graph components in various ways using two different tools:

- The options available for the `graph` command as well as the options available for each of the specialized graph or plottypes. This would be in line with the usual way of operating Stata through its command language.
- The "Graph Editor". The Graph Editor allows you to change the graph using the mouse. You start the Graph Editor by selecting **File** > **Start Graph Editor** within an open Graph window. Likewise you end the Graph Editor by selecting **File** > **Stop Graph Editor**. You cannot enter commands in the Command window as long as the Graph Editor is open.

Which of the two tools is the better one? It is difficult to give an answer to this question that applies to all cases, but we generally prefer the first option, largely because of its possibility for replication. While the Graph Editor does have a feature known as the Graph Recorder to keep track of your edits, we prefer to store graph commands in a do-file. This makes re-creating graphs, perhaps with revised data, trivially easy.

On the other hand, we do acknowledge that there are layout choices that are difficult to achieve using the command line (if not to say impossible). This is especially true for adding text within a graph. The seemingly easy task of adding an arrow with a label to a graph, e.g., as seen in figure 6.2, requires considerable programming that only command line junkies will find reasonable.[2] For tasks of this kind, the Graph Editor is of tremendous help. We would recommend the following: See if you can create and edit your graph to your liking with the command line options. Use the Graph Editor only if you get stuck or the effort far exceeds the output.

Sticking to this principle we will put heavy emphasize on command line options for the rest of the chapter. We will introduce the Graph Editor only in places where it seems particularly useful, such as section 6.3.3, where we explain how to add text within the plot region.

One last note: Stata offers many design possibilities, but you should follow some basic rules and accepted standards in designing statistical graphs. An excellent summary of the design of statistical graphs in general can be found in Cleveland (1994, 23–118).

6.3.1 Appearance of data

When first constructing your graph, you specify how your data appear in the graph by indicating the basic graph and plottype. For instance, `twoway line` plottype generates a line, `twoway scatter` plottype generates round symbols, and `bar` graph type generates vertical bars.

The `twoway` graph type has far more design possibilities, first of all because you specify a variable list. For twoway graphs, the *last* variable in a variable list always forms the x axis. All other variables form the y axis. For example, the following scatterplot uses a variable list with *two* variable names, (`rent` and `sqfeet`). Therefore, `rent`, the variable named first, forms the y axis, and `sqfeet`, the variable named last, forms the x axis.

2. Should you count yourself to this group, take a look at the do file `grcleveland.do`, which is part of our data package.

```
. scatter rent sqfeet
```

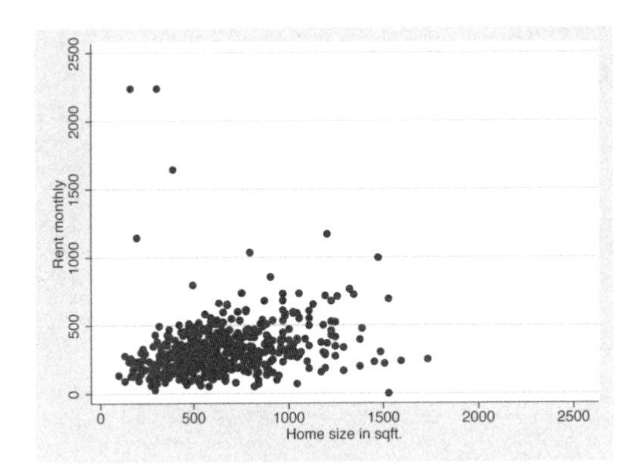

If you specify more than one y variable, the data points for the different y variables are marked with various symbols so that all data ranges are visible. For example, we can plot the rental rates of West and East Germany separately:

```
. generate rent_w = rent if state <= 9
. generate rent_e = rent if state >= 10 & state < .
. scatter rent_w rent_e sqfeet
```

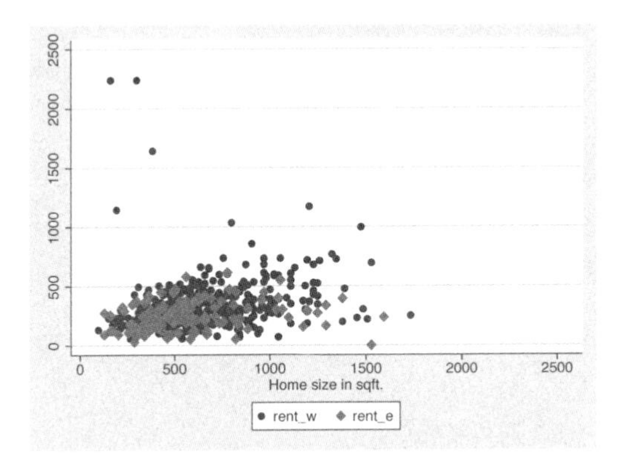

You can change the shape, size, and color of marker symbols, including the borders of the markers. All options affecting the markers begin with the letter `m` and contain a code word for the part of the marker to be changed. For example, the `msymbol()` option changes the marker symbol's shape, the `msize()` option changes its size, and the `mlabel()` option changes its label. The next section describes the most important marker options in more detail; for more information, see [G] ***marker_options*** or type `help marker_options`.

Choice of marker

The msymbol option specifies the symbol used to mark observations. You specify a shape inside the parentheses, which in many cases can be abbreviated to a single letter. Below are the basic shapes for the plot symbols and their one-letter abbreviations.

```
circle    O
triangle  T
diamond   D
plus      +
x         X
point     p
none      i
```

With the exception of point and none, these basic shapes can be further modified: for example, the letters sm before a basic shape transform a large symbol into a small one. The code word smcircle therefore indicates a small circle. Alternatively, you can specify a small circle using the lowercase letter o. Similarly, t and d, respectively, stand for smtriangle and smdiamond. Also you can add _hollow to the name to use hollow marker symbols. If you used code letters instead of the symbol names, an h is added to the other code letters; thus, circle_hollow or Oh indicates a hollow circle. The following command results in a graph that uses a hollow diamond as a marker symbol instead of a circle:

```
. scatter rent sqfeet, msymbol(diamond_hollow)
```

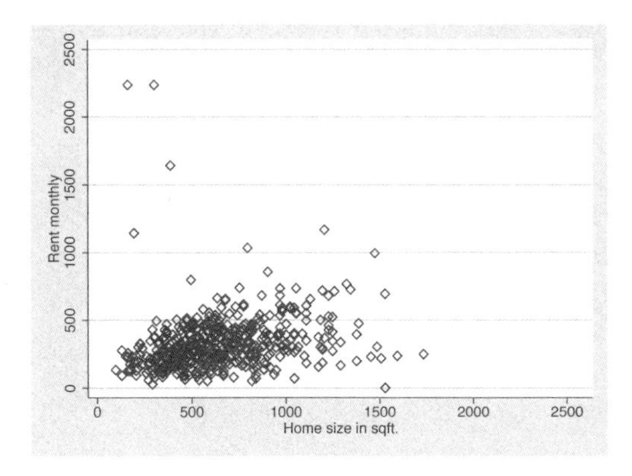

Depending on the Stata version you have, more marker symbols may be available. For a list of available marker symbols, type

```
. graph query symbolstyle
```

If you have designated many y variables in your graph command, the first y variable stands for the first series, the second y variable stands for the second series, and so on. In this case, Stata automatically uses different marker symbols for each series. To change the marker symbols for individual series, specify one letter for each series in the parentheses of `msymbol()`, separated by a space. The first code letter stands for the first y variable, the second letter for the second y variable, and so on. Typing

```
. scatter rent_w rent_e sqfeet, msymbol(+ dh)
```

assigns a large plus sign to the values for `rent_w` and a small hollow diamond for the values for `rent_o`:

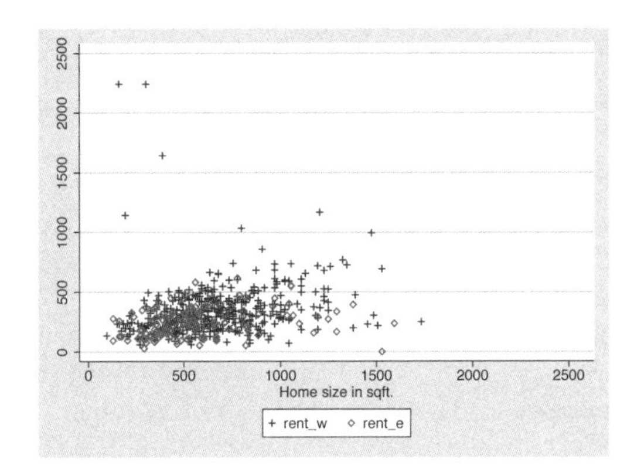

Marker colors

You can change the color of the markers using the `mcolor()` option. The command for changing the color of the marker symbols is nearly the same as that for changing marker symbols. You simply enter the desired color for each data area in the parentheses. In the following example, we use a large lavender-colored triangle for the rent in the West and large lime-colored plus sign for the rent in the East.

```
. scatter rent_w rent_e sqfeet, msymbol(T +) mcolor(lavender lime)
```

You can obtain a list of predefined colors and gray tones with `help colorstyle`, or you can define your own colors. You can change a color's intensity by multiplying the color by a factor (e.g., `mcolor(green*.8)`); you can also mix colors by specifying either RGB or CMYK values. To specify the RGB values, you type three numbers, standing for red, green, and blue, respectively, between 0 and 255.

```
. scatter rent_w rent_e sqfeet, msymbol(T +) mcolor(lavender ``255 0 0'')
```

You can also specify four numbers between 0 and 255, representing the CMYK values: cyan, magenta, yellow, and black.

You can set a plot symbol's outline and fill colors separately. You can also set the color of the marker symbol fill using `mfcolor()` and outline using `mlcolor()`. You can also change the outline's line thickness and line style; see the descriptions of the `mlstyle()`, `mlpattern()`, and `mlwidth()` options under `help marker_options` or in [G] ***marker_options***.

Marker size

As well as using the `sm` preset marker sizes, you can more finely adjust the size of the marker symbol using the `msize()` option. You can enter either absolute or relative sizes. An absolute size is one of 12 predefined sizes that range from `vtiny` (very tiny) to `ehuge` (extremely huge); see `help markersizestyle`. Relative sizes allow you to multiply existing symbol sizes by any number, e.g., `msize(*1.5)` for a 1.5-fold increase in symbol size. You can also make the diameter of the symbol relative to the size of the graph. In the example below, we specify `msize(*.5 2))`. The triangle for rent is thus half as large is it was displayed previously, and the circle for rent in the East has a diameter that is 2% of the height of the graph.

```
. scatter rent_w rent_e sqfeet, msymbol(th oh) msize(*.5 2)
```

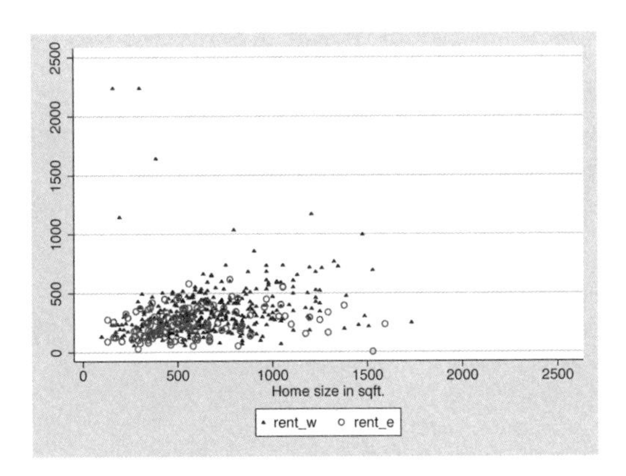

Lines

With line graphs, instead of having one marker symbol for each data point of the graph, the data points are connected with a line. Line graphs are often used in time series, so our example will use the `ka_temp.dta` file, which contains the average yearly temperatures in the German city of Karlsruhe for the period 1779–2004.

```
. preserve
. use ka_temp, clear
```

For this line graph, we could use the `connect()` option with the `scatter` plottype. To do this, you must enter a line type in the parentheses of `connect()`. In the following example, we use the `direct` line type, which connects the points of a graph with a straight line. We will also specify the invisible marker symbol (`i`) and use the `sort` option—more on that later (page 115).

```
. scatter mean year, msymbol(i) connect(direct) sort
```

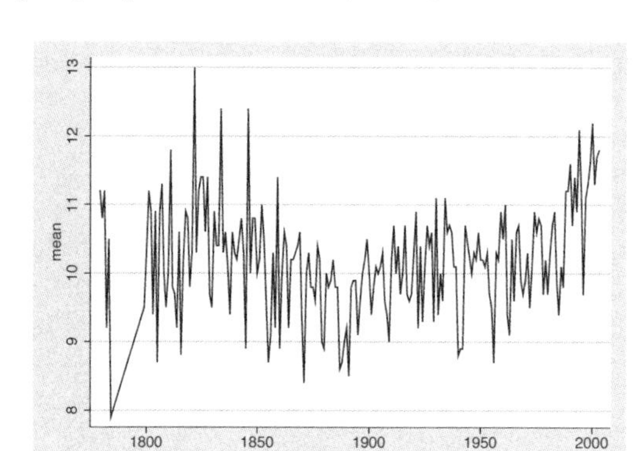

You can generate the same graph in a much easier way. The `line` plottype is a synonym for the scatterplot with the invisible marker symbol and the `connect(direct)` option; thus, you can use the `line` command:

```
. line mean year, sort
```

Accordingly, the plottype `connected` is a synonym for the scatterplot with the `connect(direct)` option and visible plot symbols. The following commands therefore generate the same graph:

```
. scatter mean year, connect(direct) sort
. twoway connected mean year, sort
```

Regardless of how you generate your line graph, you can always specify many y variables. Using the `connect()` option, you can designate the line type for each y variable. You can also specify a line type that is analogous to the `msymbol()` option, which again can also be abbreviated. Here we provide an overview of the available line types; for a complete list, type `graph query connectstyle`.

`direct` or `l` connects successive data points with a straight line.

`ascending` or `L` connects successive data points with straight lines if the x variable shows increasing values.

`stairstep` or `J` connects successive data points with a steplike line, which is first horizontal and then vertical.

`stepstair` connects successive data points with a steplike line, which is first vertical and then horizontal.

`none` or `i` meaning that no line should be drawn (invisible).

The following graph shows examples of these two line types, as well as the invisible line. We will generate a graph where the average temperatures for July are connected with a steplike line and the average temperatures in January use a straight line. The average yearly temperatures will remain unconnected:

```
. scatter jan mean jul year, connect(l i J) msymbol(i oh i) sort
```

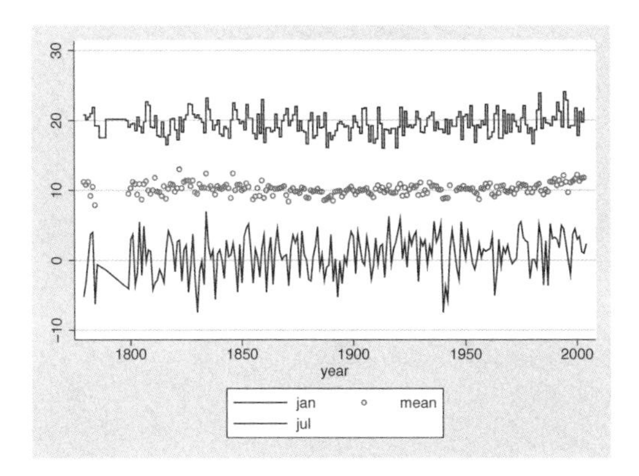

As well as changing the line type, you can also modify the lines' color, thickness, and pattern; for details, type `help connect_options`. Mostly, these options work the same way as the equivalent options for marker symbols. We strongly recommend that you read [G] **graph twoway line** to better understand these options.

Why sort?

We have used the `sort` option in all the line graphs. It is not necessary, but we have good reason for doing so. All line types connect successive data points *in the dataset*, not successive along the x axis. This feature occasionally leads to complications. To demonstrate this, try sorting the observations in the dataset by the average yearly temperature:

```
. sort mean
```

You should then generate the following graph:

(*Continued on next page*)

```
. line jan year
```

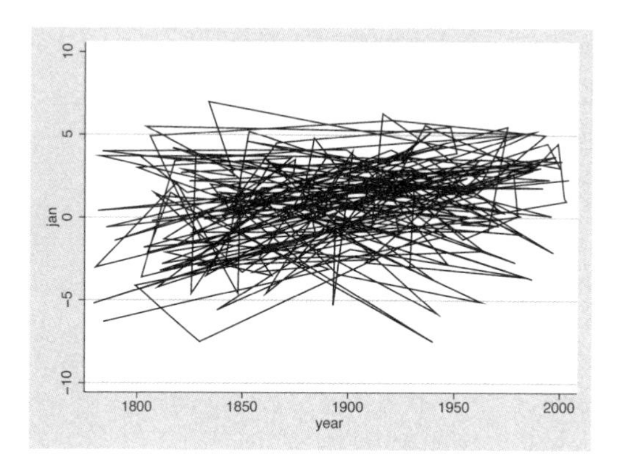

This is not the type of graph that you would associate with a time series. Yet in principle, this is the same process we used in generating the graphs earlier. After all, the data points are being connected with straight lines! However, the points are connected in the order that they appear in the dataset; i.e., the line goes from the first data point in the dataset to the second data point in the dataset. However, given the way that the data are sorted, the first point in the dataset is the one with the lowest average yearly temperature and not the one with the smallest value on the x variable.

In time-series plots, and indeed in most other graphs where the points should be connected with lines, the points should be connected in the order of the values of the x axis. You can do this by doing a preliminary sorting of the data or by using the `sort` option. We use the `sort` option, but this is a matter of taste. However, regardless of your method, when you generate a line graph, make sure you have adequately sorted the data. Also make sure that you can reproduce the graph regardless of any inadvertent resorting of the original data.

Before you continue reading, you should reload your original dataset using the command

```
. restore
```

6.3.2 Graph and plot regions

The conclusions drawn from the content of a graph are often affected by the design of the data region. For example, the smaller the data region is in proportion to the graph region, the larger any correlations in a scatterplot will appear. The smaller the relationship between the height and width of a graph region, the weaker any changes over time will appear to be, and so on. Therefore, you should design the data region carefully. Pay specific attention to these three elements: the size of the graph region—the relationship between the height and width of the graph, the margin around the plot region, and the scale of the two axes.

Graph size

You control the overall size of the graph using the options `xsize()` and `ysize()`, with the height and width of the graph (including all labels and the title) in inches specified in the parentheses. You normally will not change the graph size, as it is better to do this when printing it or importing it into a presentation or desktop publishing program. However, the options are of interest because they allow for changes in the ratio between height and width (the so-called aspect ratio) and therefore in the overall design of the graph and plot regions.

The default size of a graph in scheme `s2color` is 4×5.5, i.e., 4 inches high and 5.5 inches wide (1 inch $\sim$ 2.5 cm). If you change only the width or the height, the design of the graph region and the plot region will also change. Here are two sufficiently extreme examples.

```
. scatter rent sqfeet, xsize(1)
. scatter rent sqfeet, ysize(1)
```

Plot region

In the scheme currently available for the graphs, the data region is slightly smaller than the plot region; i.e., there is a small margin between the smallest and largest data points and their allocated axes. This has the advantage that no data points are drawn exactly on the axis. You change the default setting using the `plotregion()`, which you can use to change various plot region characteristics (such as color and shading; see [G] ***region_options***). For example, to change the margin around the plot region, you would use the `margin()` suboption. You enter the margin size in the parentheses using a codeword, such as `tiny`, `vsmall`, or `medium`. Here is an example that does not use a margin:

```
. scatter rent sqfeet, plotregion(margin(zero))
```

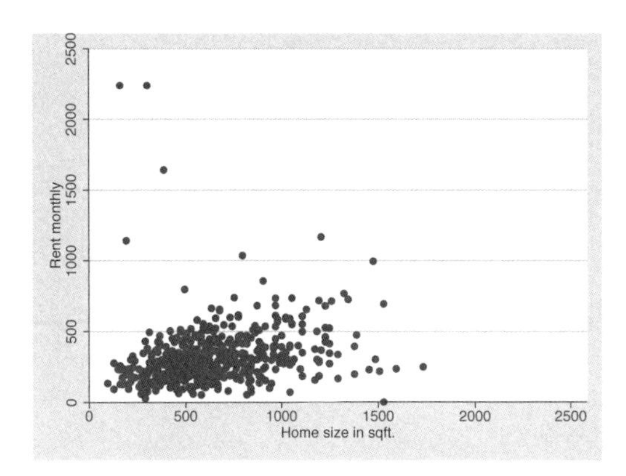

The larger the margin, the smaller is the plot region (relative to the graph region) and vice versa. For information about specifying the plot region margin, type `help marginstyle`.

Scaling the axes

Axis scaling affects the size and shape of the data region, so it may heavily influence the conclusions drawn from a graph. Within the options `yscale()` and `xscale()`, you can determine how axes are scaled (arithmetic, log, reversed), the range of the axes, and the look of the axis lines. To change the upper and lower boundaries of an axis, you use the `range()` suboption, specifying in the parentheses a number for the lower boundary followed by a second number for the upper boundary of the range of that axis.

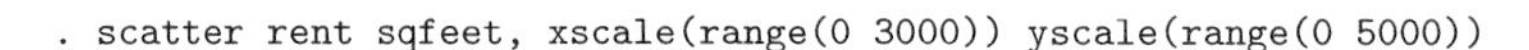

```
. scatter rent sqfeet, xscale(range(0 3000)) yscale(range(0 5000))
```

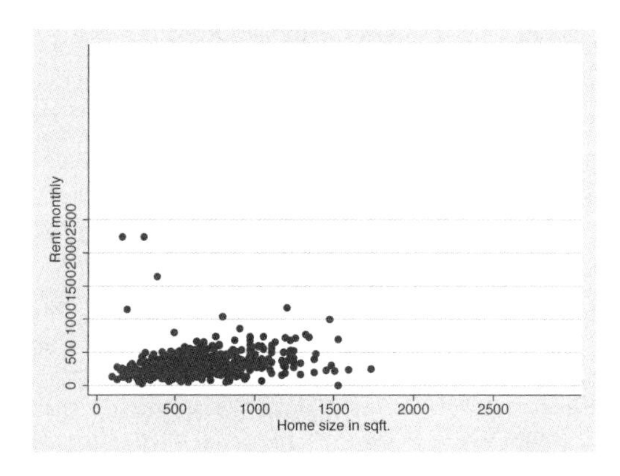

The lower the value of the lower boundary and the higher the value of the upper boundary, the smaller the data region will be in relation to the plot region. Cleveland (1994, 92–95) suggests that the data region take up as much of the plot region as possible, although many authors recommend always starting the axis at zero (Huff 1954). Cleveland reasons that you can assess the extent of any differences between the values of the y variables more easily. When analyzing data, it is often helpful to have a graph that provides as much space as possible for the data region and therefore allows for an accurate reading of the individual values.

Stata's default settings takes Cleveland's suggestions into account. The lower boundary of each axis is the minimum of the respective variable that forms the axis. Correspondingly, the upper boundary of each axis is defined by the highest value of the variable it is based on. When changing the range of the axis, you can make the data region smaller than the default setting, but you cannot make it bigger. This means that you can have more white space between the data region and the axis, but you cannot cut out data points. If you specify a smaller axis upper boundary than the maximum of the variable this axis is based on, Stata will ignore the request and the graph will be generated with the default setting with no error messages. This feature ensures that all data points can always be drawn.

However, sometimes you may want to enlarge the data region beyond the default size. For example, when a certain area of a scatterplot is densely occupied, you may need to enlarge that area so that you can explore the relationships between observations in that part of the graph. Here you can reduce the data to be displayed by adding an `if` qualifier to the graph command.

```
. scatter rent sqfeet if sqfeet <= 1000 & rent <= 1000
```

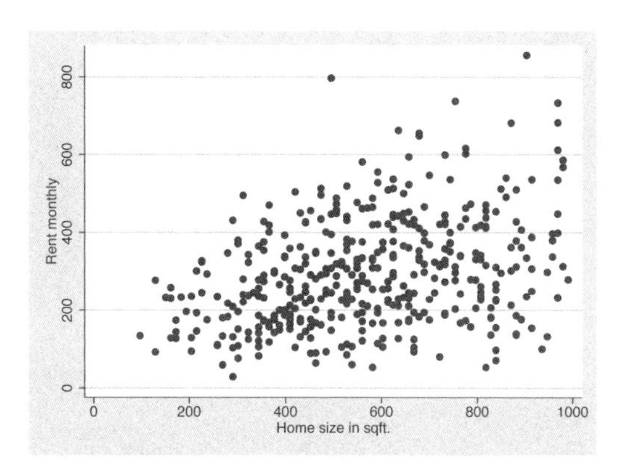

You can also change the appearance of the plot region by transforming the variables to be plotted. After all, a transformation simply converts the unit of measure into a different unit. For example, temperatures in degrees Celsius are transformed into degrees Fahrenheit by $F = C \times 1.8 + 32$. Therefore, you can always change the axis unit of your graphs by using a variable regenerated by the respective transformation (see chapter 5). You can also perform two standard unit transformations, taking the log and mirroring, by using suboptions in `xscale()` and `yscale()`:

`xscale(log)` and `yscale(log)` for logarithmic scales.

`xscale(reverse)` and `yscale(reverse)` to draw axes from the data maximum to the data minimum.

Also among the suboptions of `yscale()` and `xscale()` are some that do not affect the appearance of the displayed data. These include omitting the axes altogether and the placement of the axes; a complete list of these options can be found in `help axis_scale_options`. We will explain this in more detail in section 6.4.3.

6.3.3 Information inside the plot region

A glance at the graphs in any publication shows that the plot region is often used for information that goes well beyond the actual data. Quite often, this is merely so-called "chart junk" (Tufte 2001, 107), i.e., visual elements that often hinder the actual aim of the graph, namely, the depiction of data. Usually it is sensible to limit the use of the graph region to displaying data. Nevertheless, you may find it useful to draw lines at important points and use individual labels, as long as you do this sparingly.

Reference lines

You can add reference lines to the plot using the `xline()` and `yline()` options, typing in the parentheses a list of numbers (section 3.1.7) where the lines will be drawn.

The following command draws vertical and horizontal lines at the position 1000. Note the difference between the reference lines and the "grid lines" drawn along the y axis in the graph scheme `s2mono`. Grid lines are drawn at points where the axes are labeled. These grid lines are linked to the options for labeling axes and are discussed in further detail in section 6.3.4.

```
. scatter rent sqfeet, xline(1000) yline(1000)
```

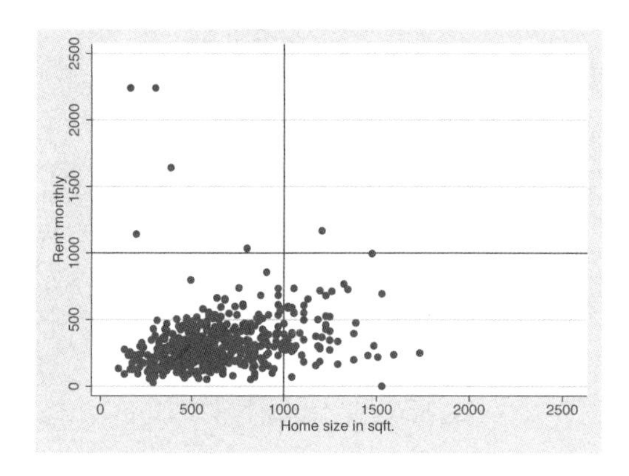

You can use reference lines to visualize internally saved results, such as the mean or median (see chapter 4). Here is an example of a horizontal line on the mean value of the monthly rents:

```
. sum rent if rent < .
. local y = r(mean)
. scatter rent sqfeet, yline(`y')
```

Labeling inside the plot region

Use labels *inside* the plot region sparingly because they can easily be mistaken for data points, making it harder to interpret the data; they may also hide the actual data points. Information about the meaning of plotted data should usually be provided outside the plot region, such as in a legend. Sometimes it makes sense to put it inside the plot region, such as when you want to label marker symbols with the content of a variable or place text at a given x–y coordinate. Let us begin by labeling the marker symbols.

To label the markers, we use the `mlabel()` option with a variable inside the parentheses, which will display the content of the variable next to the marker symbol for each observation. In the following example, the markers are labeled with the individual person's ID.

```
. scatter rent sqfeet, mlabel(persnr)
```

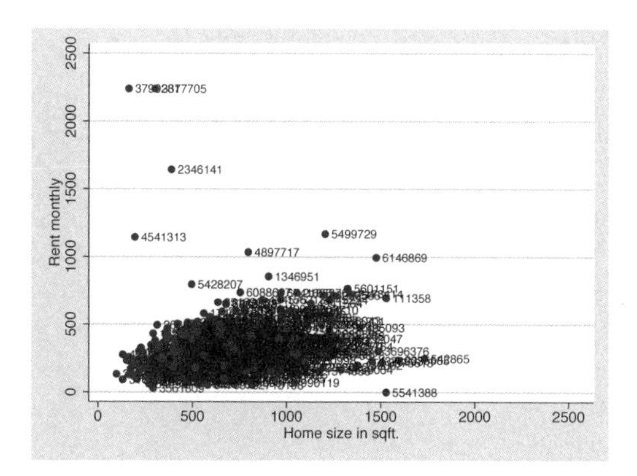

Even with moderate sample sizes, labeling markers usually means that labels are placed on top of each other and on top of the data points, making the graph illegible. In data analysis, it is often sufficient to be able to read the labels for observations that lie far apart. For presentations, however, you should use marker labels only with small sample sizes, such as with aggregate data. The following example illustrates this. In this scatterplot, the average values for "life satisfaction" (`lsat`) are plotted against average values for "annual income" (`inc`) for 19 survey years from the German Socio-Economic Panel. Here we use the survey year as the marker label:

```
. preserve
. use data2agg, clear
. scatter lsat inc, mlabel(wave) mlabposition(12) mlabsize(small)
```

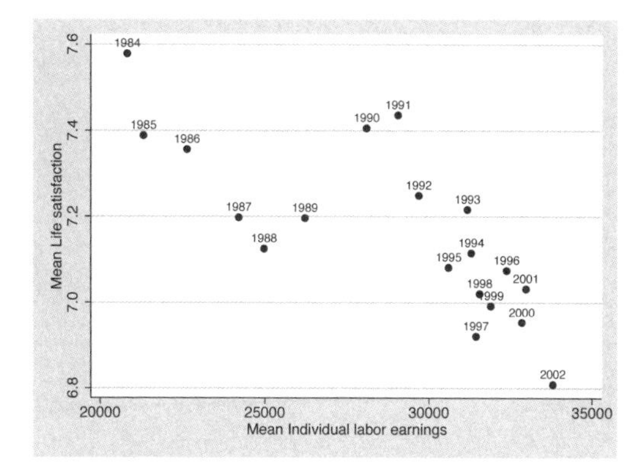

Note especially the two options `mlabposition()` and `mlabsize()`. The `mlabsize()` option allows us to adjust the size of the marker label. You can enter the same arguments in the parentheses that you would for the `msize()` option described above (see section 6.3.1). The `mlabposition()` option allows us to change the position of the marker label. The number entered in the parentheses is called the "clock position". If you think of the space surrounding a marker as the face of a clock, the number entered reflects the given position on the face of the clock. The number 12 is equivalent to the position directly above the marker symbol; the number 6 is directly below the marker symbol, and so on.

Unfortunately, some of the values are so close together that the annual figures are slightly in each other's way (here 1999 collides with 1998). In such graphs, changing the position of the marker label for various symbols is often unavoidable. Here we use the `generate` command and two `replace` statements to build the variable `pos` that indicates the position of the individual labels. Again we use clock position. The newly built variable is then specified in the `mlabvposition()` option.

```
. gen pos = 12
. replace pos = 3 if inlist(wave,1984,1985,1999)
. replace pos = 9 if inlist(wave,1986,1988,1989,1990,1995,1998,2001)
. scatter lsat inc, mlabel(wave) mlabvposition(pos)
```

Using a user-written `egen` function, `_gmlabvpos`, you can automatically generate a variable like `pos`. For more information on user-written programs, see chapter 12.

You can also place labels inside the plot region using the `text()` option. The `text()` option allows you to enter text at any x–y coordinate in the plot region. This is of great use if, for instance, you wish to label a line graph directly inside the plot region.

```
. sort wave
. local coor = lsat[1]
. line lsat wave, text(`coor' 1984 "Happiness", placement(e))
. restore
```

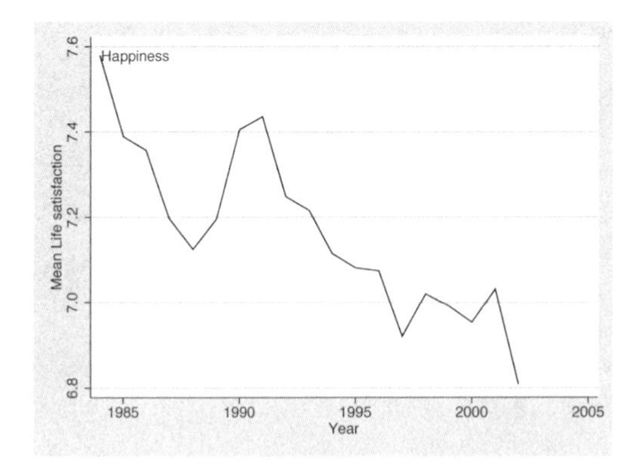

This adds the label at $x = 1984$ and $y =$ the value of `lsat` for the first observation. The `placement()` option tells Stata how to orient the text relative to the point you specify; see [G] ***added_text_options*** or `help added_text_options`.

Using the Graph Editor

It is much easier to place text within the plot region with the help of the Graph Editor instead of command options. To start the Graph Editor, select **File > Start Graph Editor** from the menu bar within the Graph window. On the right side, you will see the Object Browser, which allows you to select single elements in the graph. On the left side of the graph, you see a vertical menu bar containing tools that you can use to change the graph. The first tool with the little arrow pointer is automatically activated when you launch the Graph Editor. It is used to mark and move objects.

To get a sense of the different tools, click on each of them one by one. You will see that the gray area above the graph changes each time you select a new tool. Each tool is connected to the Contextual Toolbar where you can further specify the functionality of the tools.

To add text, use the Add Text tool, which is marked with a capital **T**. After selecting the tool (with a mouse click), you can click on any position within the graph. This brings up a new window in which you can enter the text that should be added to the graph. Clicking on **Apply** allows you to examine the change that will be applied to the graph without closing the dialog.

Drawing lines and arrows with the Graph Editor is just as easy. You can use the third tool in the vertical menu bar, indicated by a diagonal line. After selecting this tool, you can use the context menu to decide on the characteristic of the line you are about to draw. You can for example decide on the width of the line, or choose whether to have arrow heads on one of both ends of the line. Once you specify the characteristics, you can click on the graph at the position you want to start the line and drag it to its endpoint. When you let go of the mouse, the line will be part of the graph.

Figure 6.3 shows the graph window of the Graph Editor after we used the tools described.

(*Continued on next page*)

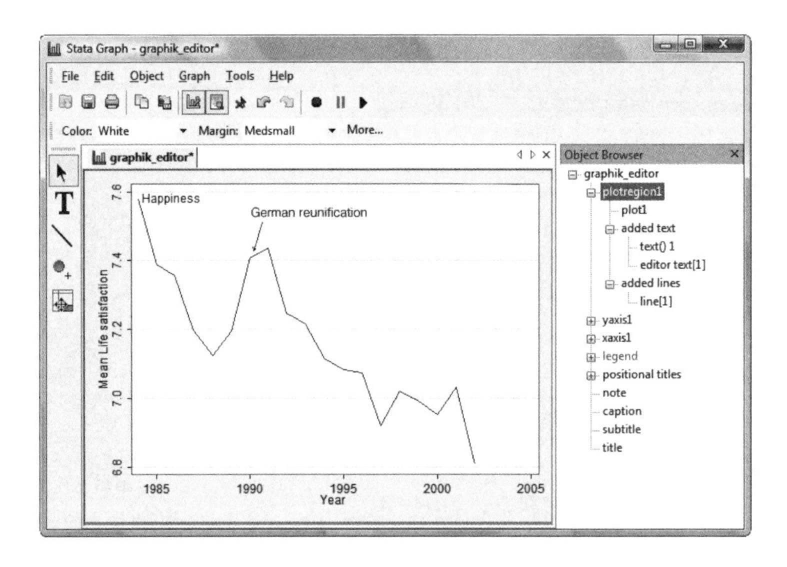

Figure 6.3. The Graph Editor in Stata for Windows

Before you continue reading, please close the Graph Editor without saving the graph. The changes you made are now lost. That is fine for our purpose. If you do want to keep the changes, you should save the graph using **File < Save As...**. When you save the graph, keep in mind that you only save the final product, not the path that lead you there.[3]

6.3.4 Information outside the plot region

As well as using labels inside the plot region, you also have many possibilities for graph design outside the plot region, and changes you make there may not affect the data presentation. Nevertheless, some changes outside the plot region can affect the plot region and therefore the data region. This is primarily because the information outside the plot region takes available space from the plot region. In extreme cases, this can affect the shape and size of the plot region, making it harder to interpret its content. Also some of the following options may affect the axis scaling.

Labeling the axes

Axis labels are the values that are placed next to the tick lines along the axes (see figure 6.2 on page 108). Although the default setting labels the axes reasonably well, you can change the settings using the options `xlabel()` and `ylabel()` for the x and y axes, respectively. Within the parentheses, you can indicate one of three ways to specify the values to be labeled:

3. The item **Recorder** in the **Tool** menu allows you to store the changes that you have made to your graph. Storing changes allows you to redo the same changes to another version of the same graph.

- Specify a list of numbers (see section 3.1.7):

```
. scatter rent sqfeet, ylabel(0(400)2800)
```

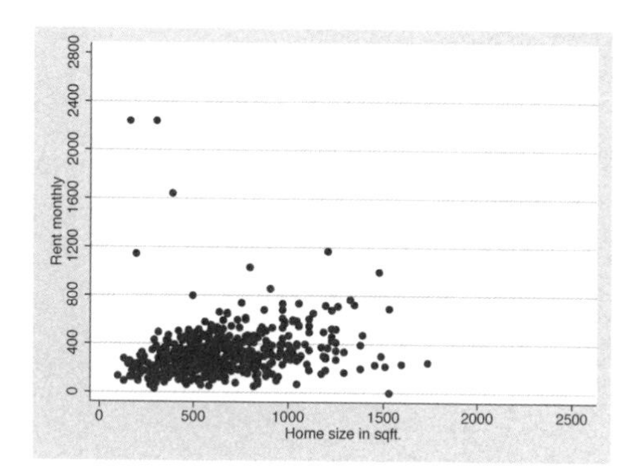

A list of numbers whose value area goes beyond that of the minimum or maximum values of the data will change the scaling of the axes.

- Specify an approximate number of values to be labeled. This number is placed after the pound sign:

```
. scatter rent sqfeet, xlabel(#15)
```

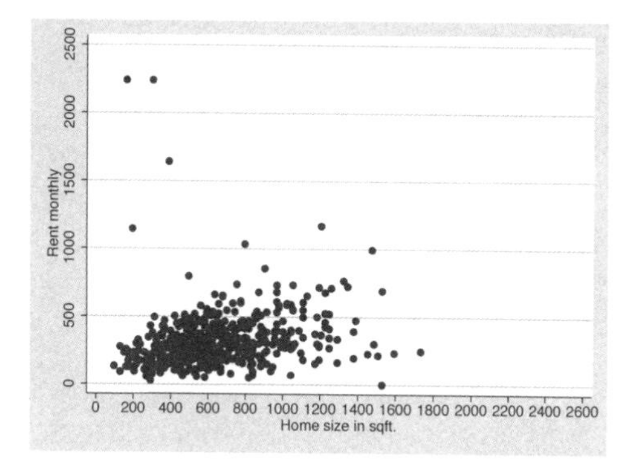

By using this method you are specifying only an *approximate* number of values to be labeled. Stata still tries to find reasonable values. So, do not be surprised that in our example we end up having 14 value labels instead of 15.

- Specify no labels with the keyword `none` or labeling the minimum and maximum with `minmax`.

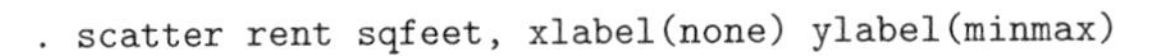

```
. scatter rent sqfeet, xlabel(none) ylabel(minmax)
```

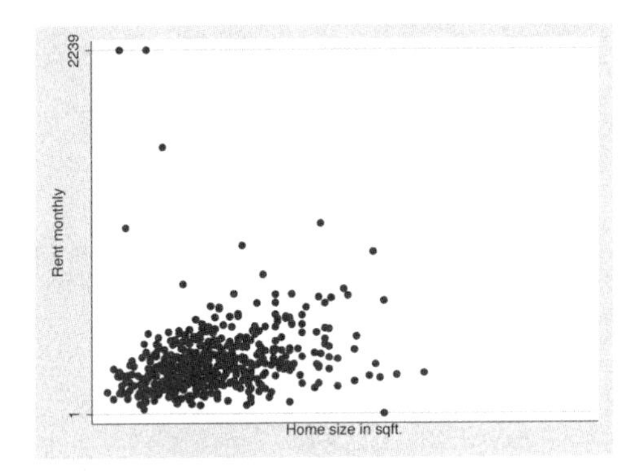

The axes are normally directly labeled with the numbers in `xlabel()` or `ylabel()`. However, you can also enter text after one of more values of the number list. The number will then be displayed as text in the graph. The text must be set in quotation marks. Here is an example:

```
. sum sqfeet if rent < .
. local x = r(mean)
. scatter rent sqfeet, xline(`x') xlabel(0 500 `x' "Mean" 1000(500)2500)
```

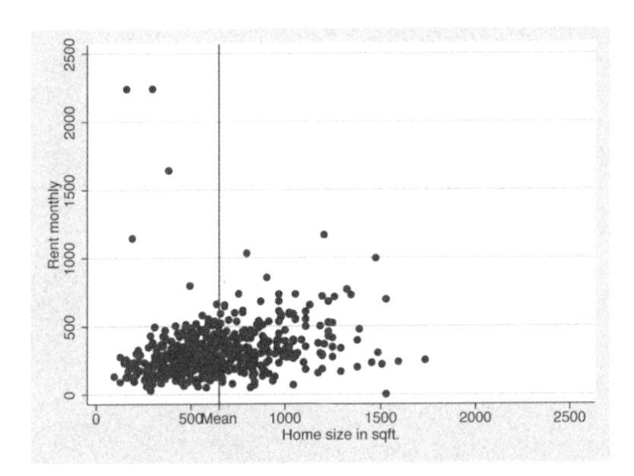

The `xlabel()` and `ylabel()` options also have a series of suboptions that may be entered into the parentheses and separated by a comma. For example, depending on the chosen graph scheme, grid lines are often drawn onto the labels. You can suppress grid lines by using the `nogrid` suboption.

The suboption `valuelabel` is part of the `xlabel()` and `ylabel()` options and enables you to replace the number with the value label assigned to the variable forming the axis. In the following example, we will simultaneously use the `angle(45)` suboption for the x axis. This places the axis label at a 45° angle to the axis. For the y axis, we specify `nogrid`. `nogrid` is a suboption and is therefore entered after a comma within `ylabel()`.

```
. scatter sqfeet htype, xlabel(1(1)7, valuelabel angle(45)) ylabel(,nogrid)
```

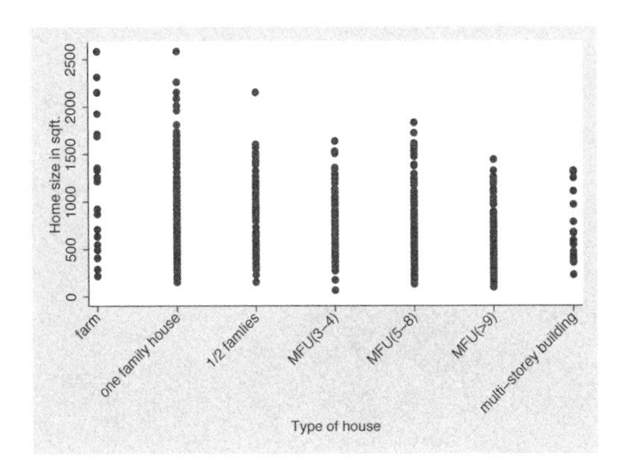

For a complete list of available suboptions, type `help axis_label_options` or see [G] ***axis_label_options***.

Tick lines

Tick lines are the little lines between the axis and the axis labels and are usually drawn by Stata near the labels. You can modify tick lines using the following options:

- `xtick()` for tick lines along the x axis and `ytick()` for the y axis.
- `xmtick()` for small tick lines along the x axis and `ymtick()` for the y axis.

To specify the number of tick lines, you can use the three possibilities we introduced within the previous section: a list of numbers, an approximate number of tick lines, or keywords. For small tick lines, you need two pound signs to specify the number of small tick lines, and the number you specify refers to the number of small tick lines between each of the larger tick lines.

(*Continued on next page*)

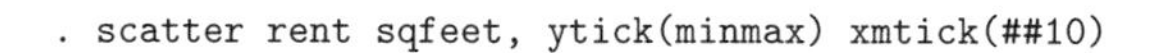

```
. scatter rent sqfeet, ytick(minmax) xmtick(##10)
```

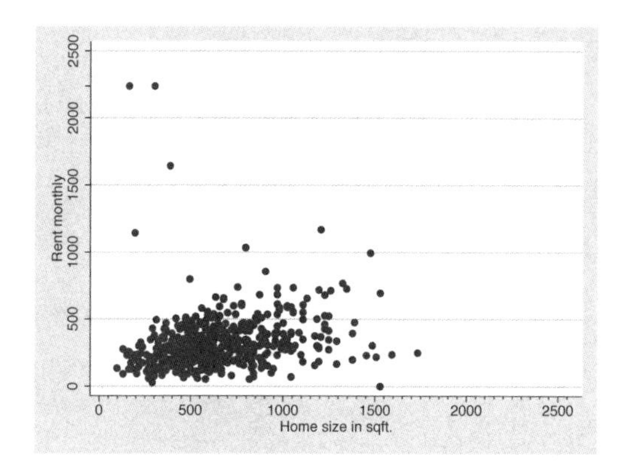

Axis titles

You can provide every axis in a graph with a title, one or more lines long, by using the `xtitle()` and `ytitle()` options. You type the title, set in quotation marks, within the parentheses.

```
. scatter rent sqfeet, ytitle("Rent (Monthly) in USD")
> xtitle("Home Size in Square Feet")
```

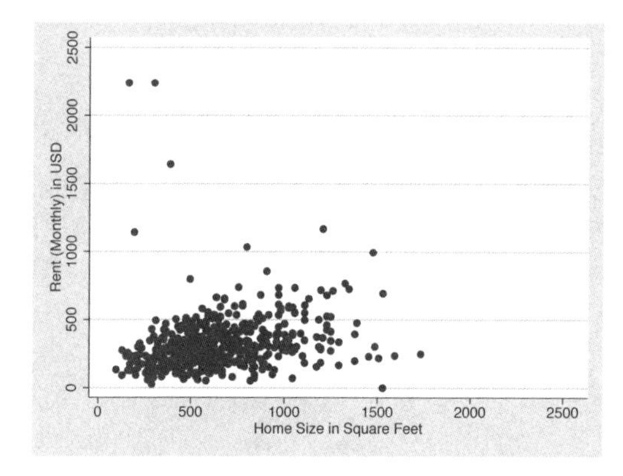

When specifying an axis title with more than one line, set each line in quotation marks separately. The first piece of quoted text will appear on the first line, the second on the second line, and so on.

```
. scatter rent sqfeet, ytitle("Rent (Monthly)" "in USD")
> xtitle("Home size" "in sqft.")
```

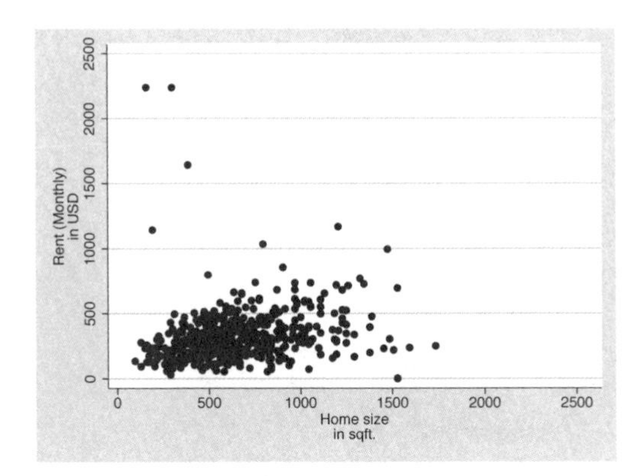

You can easily modify the shape, appearance, and placement of the title using textbox options in the parentheses of the `ytitle()` and `xtitle()` options. For detailed information, type `help textbox_options` or see [G] ***textbox_options***. The example below is primarily a teaser:

```
. scatter rent sqfeet, xtitle("Home size" "in sqft.", placement(east) box
> justification(right))
```

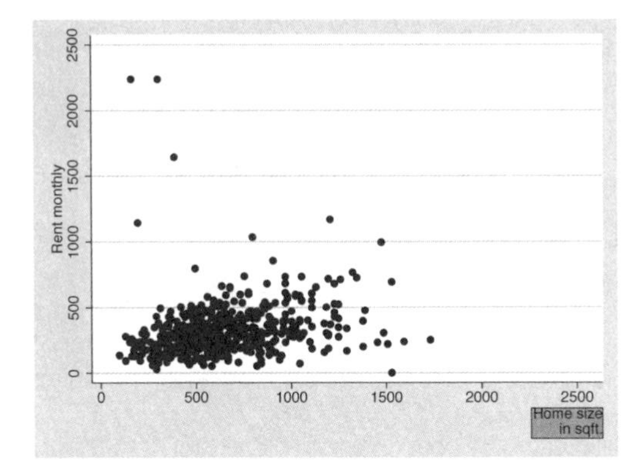

The legend

If a graph contains more than one y variable with visible markers, Stata automatically creates a legend that displays the variable names or labels used in a given data area. You can edit the legend using the `legend()` option, in which you specify the content and position of the legend. Here we will limit ourselves to specifying the position. However, there are many other things you can add to a legend and many other options for controlling how legends look; see `help legend_option` for the details.

First, take a look at the graph resulting from the following command before we consider the individual options more closely:

```
. scatter rent_w rent_e sqfeet, legend(cols(1) ring(0) position(1))
```

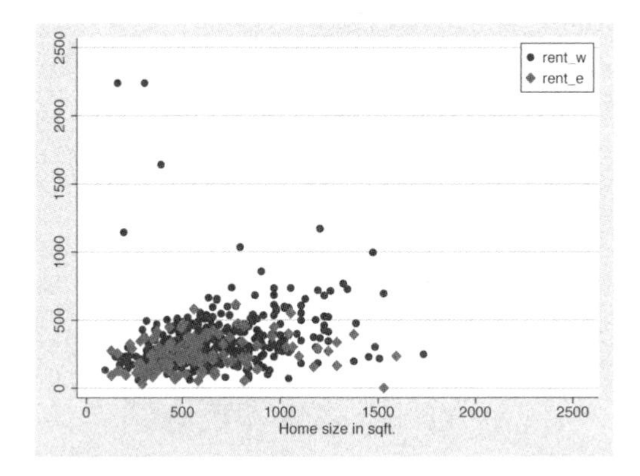

In this example, we made three changes to the legend. First, we entered all entries in the legend in one column by using the `cols(1)` option. To use this option, you specify the number of columns that the legend should have in the parentheses. You can specify the number of rows by using `rows()` option.

Then we changed the position of the legend using the `ring()` option. The legend is now located inside rather than outside the plot region. The ring position 0 assigns a position inside the plot region; a ring position more than 0 means a position outside the plot region.

Finally, we have set the position of the legend inside the plot region. The number specified in the position option refers to the position of numbers on a clock face, as we discussed earlier (see section 6.3.3).

Graph titles

The following options allow you to control the title and other label elements:

- `title()` for the graph title;
- `subtitle()` for a subtitle for the graph;
- `note()` for some explanatory text in a small font, which is frequently used for references or bibliographies;
- `caption()` for text beneath the graph, which can make the graph easier to interpret.

In each case, you enter text in the parentheses. If the text consists of more than one line, each line must be entered in quotation marks, as described in section 6.3.4. Here is an example:

```
. scatter rent sqfeet, title("Rent by Home Size") subtitle("Scatterplot")
> note("Data: GSOEP" "Randomized Public Use File")
> caption("This graph is used to demonstrate graph titles."
> "Please note how text is divided into several lines")
```

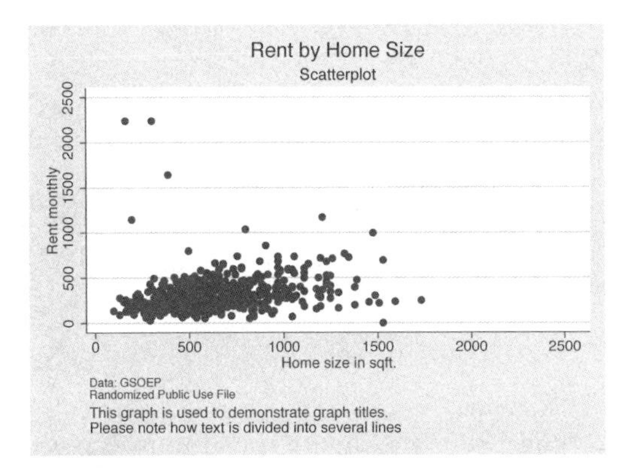

You can use several suboptions to change the appearance of the title; for more information, see [G] ***title_options*** and [G] ***textbox_options***.

6.4 Multiple graphs

In Stata, you can create multiple graphs in several different ways. By "multiple graphs", we mean graphs that consist of different graph parts, in particular

- `twoway` graphs that are plotted on top of each other,
- graphs that are broken out with the `by()` option and are then displayed together, and
- varying graphs that are combined using the `graph combine` command

We will now quickly introduce these three types of graphs.

6.4.1 Overlaying many twoway graphs

You can overlay as many types of `twoway` graphs as you want in the same coordinate system. In the following example, three graphs are placed on top of each other: a scatterplot; a "linear fit" (or "regression line"; see chapter 8) for the same data, but restricted to the old federal states in West Germany (`rent_w`); and finally, a "linear fit" that is restricted to the new federal states in East Germany (`rent_e`).

```
. twoway || scatter rent sqfeet || lfit rent_w sqfeet || lfit rent_e sqfeet
```

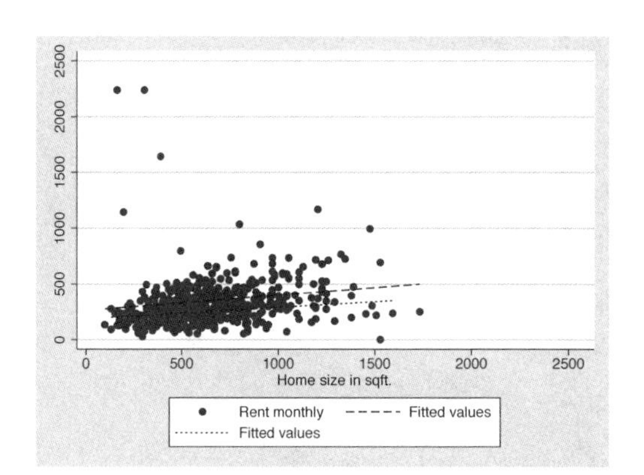

To overlay twoway graphs, you consolidate the graphs in a single `twoway` command, separated by parentheses or two vertical lines. In this book and our own work, we use two vertical lines because there are already so many parentheses in the graph syntax. The two vertical lines are particularly readable in do-files containing the individual graphs one after another with line breaks commented out (see section 2.2.2).

If you are combining several twoway graphs, you can specify options that correspond to the respective graph types, as well as twoway options that apply to all the graphs to be combined. Generally, the syntax for overlaid twoway graphs is as follows:

<u>`tw`</u>`oway`

`||` <u>`sc`</u>`atter` *varlist*`,` *scatter_options*

`||` <u>`lfit`</u> *varlist*`,` *lfit_options*

`||` *plottype varlist*`,` *plottype_options*

`||` `,` *twoway_options*

The first and the last two vertical bars are superfluous. However, we tend to use them for long graph commands to enhance readability. This syntax structure can be illustrated with an example:

```
. twoway || scatter rent sqfeet, msymbol(oh)
>        || lfit rent_w sqfeet, clpattern(dot)
>        || lfit rent_e sqfeet, clpattern(dash)
>        || , title("Scatterplot with Regression-Lines")
>           legend(order(2 3) label(2 "West") label(3 "East"))
```

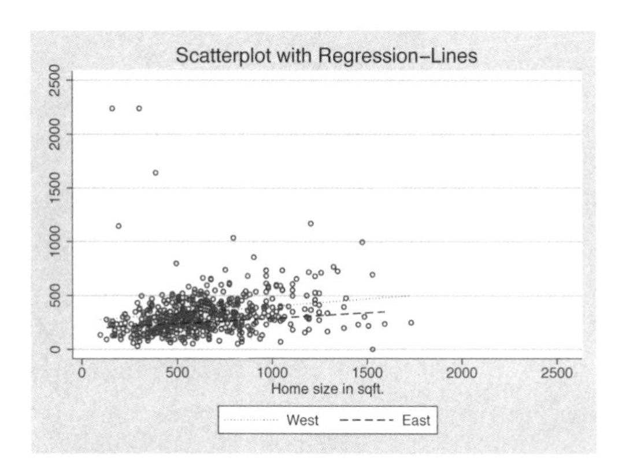

6.4.2 Option by()

The by() displays separate graphs for each group defined by the variable in the parentheses. If more than one variable is entered in the parentheses, graphs are provided for every combination of the chosen variables. If you also specify the total suboption, another graph is displayed without separating it by group. Other suboptions control the positioning (e.g., rows(), cols()), displaying or fading out of individual axes (e.g., [no]ixaxes), or the appearance of the margins between the individual graphs. For the list of suboptions, see help by_option or [G] ***by_option***. One example should be enough at this point:

```
. scatter rent sqfeet, by(state, total)
```

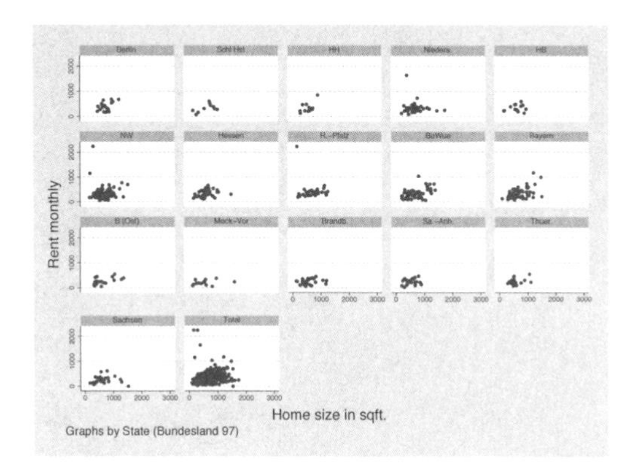

6.4.3 Combining graphs

Stata allows you to combine as many graphs into a joint graph as you want. For this, you first save the individual graphs and then combine them using **graph combine**. We will demonstrate this using a display of **rent** by **sqfeet**, separated by respondents from East and West Germany:

```
. scatter rent_w sqfeet, name(west, replace)
. scatter rent_e sqfeet, name(east, replace)
. graph combine west east
```

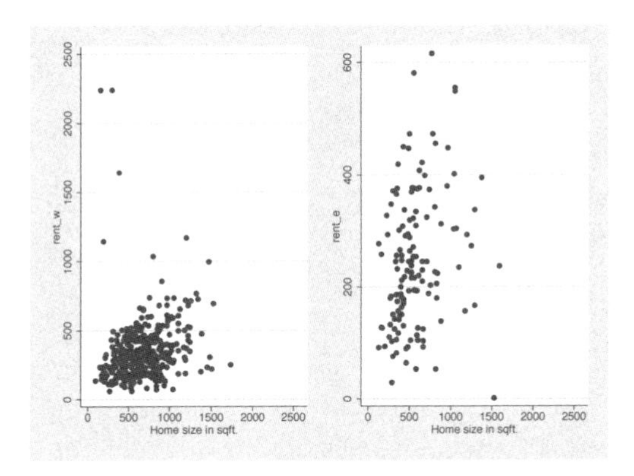

To save both graphs, we use the `name()` option, which specifies the name under which the graph will be saved in the computer's memory. The `replace` suboption tells Stata to delete any graphs already saved under this name. We then combine the two graphs using `graph combine`.

The `graph combine` command has a series of options for controlling how the combined graph is to be displayed. To begin with, it is important to set the number of rows and columns in the combined graph. The individual graphs are placed in the combined graph in rows and columns in a matrix-like fashion. The positioning of the individual graphs depends on how many rows and columns the matrix has. In the matrix above, one row and two columns were used. Here you will see what happens if we instead use two rows and one column:

```
. graph combine west east, rows(2)
```

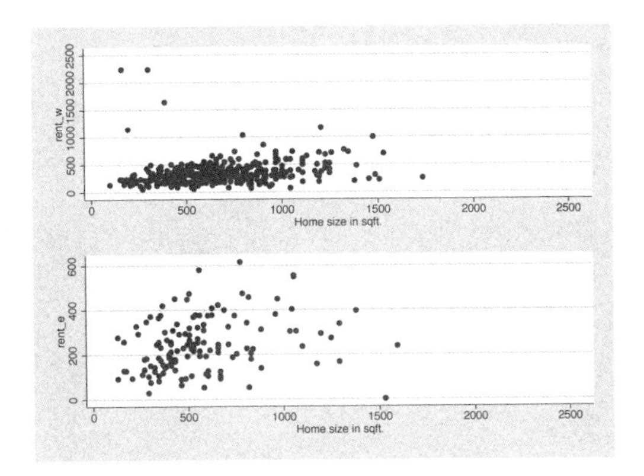

The number of individual graphs you can put in a multiple graph is limited only by printer and screen resolutions. If the main issue is the readability of the labels, you can increase their size with the `iscale()` option. The default font size decreases with every additional graph. With `iscale(1)`, you can restore the text to its original size; `iscale(*.8)` restores the text to 80% of its original size.

If you want the individual graph parts of the combined graph to have different sizes, you will have to save the graphs with different sizes before combining them. However, you cannot use the `xsize()` and `ysize()` options discussed in section 6.3.2, as these sizes are not taken into account by `graph combine`. Instead, you will have to use the forced size options `fysize()` and `fxsize()`, which tell Stata to use only a certain percentage of the available space. For example, the `fxsize(25)` option creates a graph that uses only 25% of the width of the available space; correspondingly, a graph created using the `fysize(25)` option uses only 25% of the available height.

Here is a slightly more advanced example of `graph combine` using the forced size options.

(Continued on next page)

```
. twoway scatter rent sqfeet, name(xy, replace) xlabel(, grid)
> ylabel(, grid gmax)
. twoway histogram sqfeet, fraction xscale(alt) xlabel(, grid gmax) fysize(25)
> name(hx, replace)
. twoway histogram rent, fraction horizontal yscale(alt)
> ylabel(0(500)2500, grid gmax) fxsize(25) name(hy, replace)
. graph combine hx xy hy, imargin(0 0 0 0) hole(2)
```

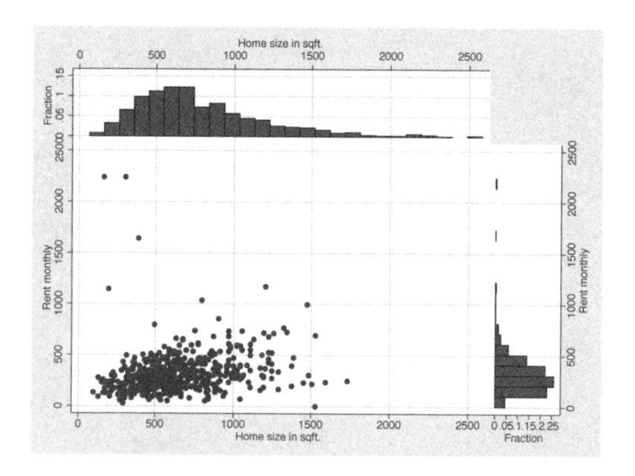

For more details on creating such graphs, see [G] **graph combine**. The Graph Editor has further capabilities regarding the positioning of graph elements. The repositioning is best done with the fifth tool (from the top of the list) the "grid edit tool". A description of its functionality can be found in [G] **graph editor**.

6.5 Saving and printing graphs

To print a Stata graph, you type

```
. graph print
```

The graph displayed in Stata's Graph window is then printed directly. If you create many graphs in a do-file, you can also print them with that do-file by typing the `graph print` command after every graph command. The `graph print` command also has many options that can be used to change the printout of a graph (see `help pr_options`).

You can also print graphs that are not (or are no longer) in the Graph window, but you must first save the graph to memory or to a file on the hard drive. You already learned how to save graphs to memory in the previous section. There you saved numerous graphs with the `name()` option, which we then combined with the `graph combine` command. To print out these graphs, you first display them with `graph display` and then print them using `graph print`. Let us try this with the graph saved as `east` above (see page 134):

```
. graph display east
. graph print
```

Saving graphs to memory is often useful, but they are lost when you close Stata. To print graphs that you created a few days ago, you must have saved them as a file using the `saving()` option. It works the same way as the `name()` option, except it saves the file to the hard drive. You type the `filename` under which the file is to be saved in the parentheses. By default, Stata uses the `.gph` file extension. To overwrite an existing file of the same name, include the `replace` option in the parentheses. The option `replace` is commonly used when creating graphs with do-files.

Be careful when you use the Graph Editor. When you exit the Graph Editor you will be asked if you want to save the changes you made to the graph. In most cases, you would want to answer yes. If you do so, you should make sure that you use a *new* filename. If you save the graph with its old name you might be in trouble next time you run the do-file that created the graph originally. Running the do-file will re-create the original graph and therefore overwrite the changes you made with the Graph Editor. You would need to start over.

All saved files can be, at a later point in time, printed or edited with the Graph Editor. To do so, you simply call the saved graph on the screen using the command `graph use` and start printing or editing. The command

```
. graph combine hx xy hy, hole(2) imargin(0 0 0 0) saving(combined, replace)
```

saves the graph under the name `combined.gph` in the current working directory. If a file with the same name already exists in the working directory, it is overwritten since you specified the `replace`. You can now close Stata, shut down the computer, or display another graph ...

```
. graph display east
```

... and regardless of what you do, you can print the graph saved in a file by typing

```
. graph use combined
. graph print
```

Finally, you will usually be exporting a Stata graph to a word processing program or presentation program rather than printing it. For this you can use the much-loved *copy-and-paste* procedure, where you first copy the graph displayed in a Graph window and then paste it into the respective document. If you are dealing with several graphs, it is better to save the graph in a suitable file format on the hard drive and then import to the desired program when needed.[4]

To save a graph in a different file format, use the `graph export` command. You type the filename with the appropriate file extension after the command. Table 6.1 lists the formats that are available to you.

4. When doing this and working with do-files, it is a good idea to document your work (see chapter 2).

Table 6.1. Available file formats for graphs

Extension	File format	Restriction
`.ps`	PostScript	
`.eps`	Encapsulated PostScript	
`.wmf`	Windows Metafile	Windows only
`.emf`	Windows Enhanced Metafile	Windows only
`.pict`	Macintosh Picture Format	Macintosh only
`.pdf`	Portable Document Format	Macintosh only
`.png`	Portable Network Graphics	
`.tif`	Tagged-Image File Format	

Microsoft applications normally handle files saved in WMF or EMF formats well. The same applies to most other software that runs under Windows operating systems. PostScript and Encapsulated PostScript work well on Unix systems, and you should use them if you write reports using LaTeX. Macintosh users will usually prefer PDF or PICT files. In any case, if you want to save the graph in the Graph window as a WMF file called `mygraph1.wmf`, use

```
. graph export mygraph1.wmf
```

Use the other file formats as you wish.

6.6 Exercises

1. Get data from the National Health and Nutrition Examination Study (NHANES) using the following command:

   ```
   . webuse nhanes2.dta, clear
   ```

   (If you get the error message "`no room to add more observations`", please type `clear`, followed by `set memory 30m` before using the above command; see section 10.6 for details.)

2. Using the NHANES data, produce a scatterplot of weight in kg (`weight`) by height (in cm). Use hollow circles as the marker symbol.

3. Change the title of the vertical axis to "Weight (in kg)", and add a note for the data source of your graph.

4. Add reference lines to indicate the arithmetic means of weight and heights.

5. Add an axis label to explain the meaning of the reference lines.

6. Use blue marker symbols for male observations and pink marker symbols for female observations, and construct a self explanatory legend. Remove the reference lines.

7. Plot the data for men and women separately and produce a common figure of both plots placed on top of each other. Take care that the note on the data source does not appear twice in the figure.
8. Construct a graph similar to the previous one, but this time with reference lines for the gender specific averages of weight and height.
9. Create a variable holding the body mass index [BMI; see (5.1) on page 102], and classify the observations according to the table on page 102. Change the previous graph such that the colors of the marker symbols represent the categorized BMI.
10. Add the unique person identifier (`sampl`) to the symbols for the male and female observation with the highest BMI.
11. Export your graphs so that they can be imported into your favorite word processing program.

7 Describing and comparing distributions

So far, we have dealt with the basic functionality of Stata. We have used data mainly to show you how Stata works. From now on, we will do it the other way around: we will show you how to use Stata as a tool to analyze data *and* understand the analysis.

We begin with the process of describing distributions. An example for such a description is the presentation of political election returns on television. Consider the announcement that a candidate for the presidency is elected by, say, 80% of American voters. You will definitely regard this description as interesting. But why? It is interesting because you know that this candidate has won the election and that he will be the next president of the United States. Moreover, you know that 80% of the votes is quite a success because you know that previous elections have never seen such a result.

Now consider the following description of a distribution: In Baden Württemberg, a state in the southwest of Germany, 28% of the inhabitants live in single-family houses. What do you think about this information? You might wonder what the proportion would be in the area where you live—which means you need some knowledge about the distribution to figure out if 28% is high or low.

Generally speaking, a description of a single distribution is satisfying as long as we know something a priori about the distribution. Here we can compare the actual distribution with our knowledge. If we do not have such knowledge, we need to collect information about the distribution from somewhere else and compare it with the distribution. A full description of a distribution therefore often involves comparing it with a comparable distribution.

There are many ways to describe distributions. One criterion—not exactly a statistical criterion, but in practical applications quite an important one—for choosing the description is the number of categories. Distributions with just a few categories can often be fully described with tables, but there are also some graphical tools for them. Describing distributions with many categories is more difficult. You will often evaluate statistical measures, but in most cases, graphical tools are better.

In what follows, we will first distinguish between distributions with few categories and those with many categories. Then we will treat those two cases separately.

To follow this chapter, load our example dataset:[1]

```
. use data1
```

7.1 Categories: Few or many?

Let us begin with some vocabulary. In what follows, we use the letter n for the number of observations, the uppercase letter Y for a variable, and y_i, $i = 1, \ldots, n$ to denote the values of the variable. The value of the variable Y for the first observation is y_1, for the second observation it is y_2, etc. Take our dataset `data1.dta` as an example, which has $n = 3340$. One of the variables in it (`np9402`) contains the party affiliation of each respondent. If we choose this variable as Y, we can look at the values $y_1, \ldots, y_{3340}$ with

```
. list np9402
```

Obviously, this list of answers generally is not the information we were looking for; we are simply overwhelmed by the amount of information scrolling by. You should break the list as described on page 8.

To get a better understanding, you can begin by finding out how many different numbers occur in Y. We will call these numbers *categories* and denote them by a_j. Thus j is an index that runs from the lowest to the highest category ($j = 1, \ldots, k$). The number of different categories of a variable usually is much smaller than the number of observations. In Stata, we can find out the number of categories with `inspect`. For example, we get the number of categories for party affiliation as follows:

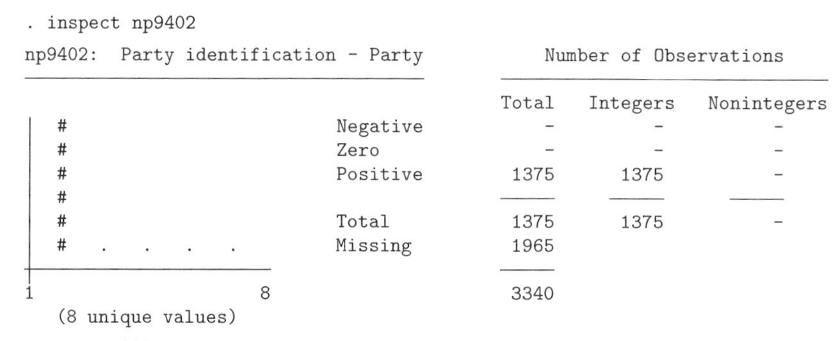

```
. inspect np9402
np9402:  Party identification - Party          Number of Observations
                                                Total   Integers   Nonintegers
|  #                            Negative           -          -             -
|  #                            Zero               -          -             -
|  #                            Positive        1375       1375             -
|  #
|  #                            Total           1375       1375             -
|  #    .    .    .    .        Missing         1965
+----------------------------
1                         8                     3340
   (8 unique values)
     np9402 is labeled and all values are documented in the label.
```

Party identification in our dataset has $k = 8$ categories (`8 unique values`). All categories are positive integers (`Integers`) between 1 and 8. There are only 1,375 respondents with one of those eight categories. The rest have another category: the missing value.[2]

We will distinguish between variables with few categories and variables with many categories by setting the threshold at approximately 6–10 categories. But do not take

1. Your working directory must be `c:\data\kk2`; see page 3.
2. Read more on missings in section 5.4, as well as on pages 12 and 310.

this threshold too literally. With graphical tools, you can take "few" much more broadly than you can with tables. Often you will get a good description of a variable with few categories by using a technique designed for variables with many categories, and vice versa. The most important criterion for describing a variable is that the main properties of the distribution stand out. You should try out different approaches.

7.2 Variables with few categories

7.2.1 Tables

Frequency tables

The most important way to describe a distribution with few categories is with a one-way frequency table, which lists the absolute and relative frequencies of all categories, a_j, of a variable. The absolute frequency, n_j, is the number of observations in the category a_j. The relative frequencies, f_j, are the ratios of the absolute frequencies to the entire number of observations:

$$f_j = \frac{n_j}{n}$$

In what follows, we will often use the word "fraction" for relative frequency.

In Stata, one-way frequency tables are produced with the command `tabulate`, which can be abbreviated to `tab` or even `ta`. You specify the variable for which the table is displayed by typing the variable name after the command. The following command shows the one-way frequency table for partisanship:

```
. tabulate np9402

      Party |
 identificat |
 ion - Party |      Freq.     Percent        Cum.
-------------+-----------------------------------
         SPD |        586       42.62       42.62
         CDU |        425       30.91       73.53
         CSU |         96        6.98       80.51
         FDP |         35        2.55       83.05
     B90/Gr. |        127        9.24       92.29
         PDS |         68        4.95       97.24
         Rep |         28        2.04       99.27
      Sonst. |         10        0.73      100.00
-------------+-----------------------------------
       Total |      1,375      100.00
```

The first column of this frequency table shows the different categories, a_j, of party identification (Germany has a *multiparty system*). The second column shows the absolute frequencies, n_j, and the third column shows the fraction as a percentage ($f_j \times 100$). The final column is for the cumulated relative frequencies, which we will not discuss here.

To illustrate how to interpret all those figures, suppose that the answers to the question about partisanship are votes in an election for a "parliament of respondents". The Social Democratic Party (SPD) has won $n_1 = 586$ votes, or $f_1 = 586/1375 = 42.62\%$. Four parties have been chosen by more than 5% of the respondents. No single party has won more than 50% of the votes. But together with the Green Party (B90/Gr.), the Social Democrats have won $586 + 127 = 713$ votes (about 52%) and might therefore take office in this hypothetical parliament.

The frequency table has been calculated for 1,375 observations, but the entire dataset has 3,340 observations. The difference stems from observations with missing values. In our dataset, people who have not answered a question are set to the missing value for that variable. By default, the `tabulate` command excludes missing values, so the above table is produced only for respondents who have answered the question on party identification. To include the missing values, use the option `missing`:

```
. tabulate np9402, missing
```

In the following tables and graphs, we will usually not explicitly include missing values. Some of the commands we will describe allow the inclusion of missing values with the option `missing`, just like the example above. If not, you can always replace missing values with numeric values (see section 5.4.)

More than one frequency table

Using `tabulate` with two variables will give you a two-way frequency table. To generate more than one one-way frequency table with a single command, you need `tab1`, which generates a frequency table for each variable of a *varlist*. Here are some examples:

```
. tab1 np9402
. tab1 np9402 gender
. tab1 np94*, mis
. tab1 gender - emp
```

Comparing distributions

Examining a single distribution is rarely useful. Often you need to compare a given distribution with the same distribution from some other time, population, or group.

If you have sampled your distribution in different groups, you can do this by simply producing a one-way table for each of the groups by using the `by` prefix. For example, the following command produces a one-way table for men and women separately:

```
. by gender, sort: tabulate np9402
```

We have compared one distribution (the distribution of party identification) based on different levels of another distribution (gender). A more technical way to say this is that we have shown the distribution of party identification *conditioned* on gender.

A more illustrative way to show a distribution conditioned on another distribution is a two-way table, also known as a cross table or contingency table. A two-way table displays the distribution of one variable, say Y, for the categories of another variable, say X, side by side.

To produce such a two-way table with Stata, you simply include a second variable name in the *varlist* after `tabulate`. Here is an example—the two-way frequency table of party identification by gender:

```
. tabulate np9402 gender

     Party |
 identifica|
    tion - |        Gender
     Party |       men      women |     Total
-----------+----------------------+----------
       SPD |       317        269 |       586
       CDU |       208        217 |       425
       CSU |        55         41 |        96
       FDP |        14         21 |        35
   B90/Gr. |        60         67 |       127
       PDS |        31         37 |        68
       Rep |        17         11 |        28
    Sonst. |         4          6 |        10
-----------+----------------------+----------
     Total |       706        669 |     1,375
```

The first variable forms the rows of the table, and the second variable forms the columns. The body of the table shows the distribution of party identification for each level of the gender variable.[3] For example, in the "men" column, there are 317 SPD supporters and 208 CDU supporters. In the "women" column, there are 269 SPD supporters and 217 CDU supporters. You have already seen these numbers in response to the previously entered command `by gender, sort: tabulate np9402`.

In addition to the number of observations with specific variable combinations, you also find the row and column sums of those numbers. The rightmost column shows the overall distribution of party identification, which you have already seen as a result of the command `tabulate np9402` (page 143). The bottom row displays the same for `gender`.

Although it is easy to understand the meaning of the numbers in the two-way frequency table above, you should not use such tables to compare distributions between different groups. Instead you should use the fractions within each group, sometimes called the conditional relative frequencies, for the comparison. Under the condition that $X = a_j$, the conditional relative frequency of the variable Y is calculated by dividing each absolute frequency by the number of observations within the group $X = a_j$.

3. Or, the other way around, the distribution of gender for each level of party identification.

Thus, for our example, when `gender==men`, the conditional relative frequency distribution of party identification is given by dividing the number of male supporters of each party by the total number of men.

Stata calculates the conditional fractions in two-way tables with the options `row` and `column`. The fractions of party identification conditioned on gender are calculated with the option `column`, giving us the fractions for the row variable conditioned on the *column* variable. Here is an example where we also use the option `nofreq` to suppress the output of the absolute frequencies:

```
. tabulate np9402 gender, column nofreq
```

Party identifica tion - Party	Gender men	women	Total
SPD	44.90	40.21	42.62
CDU	29.46	32.44	30.91
CSU	7.79	6.13	6.98
FDP	1.98	3.14	2.55
B90/Gr.	8.50	10.01	9.24
PDS	4.39	5.53	4.95
Rep	2.41	1.64	2.04
Sonst.	0.57	0.90	0.73
Total	100.00	100.00	100.00

The numbers shown in this table are the same as those calculated with the command `by gender, sort: tabulate np9402` on page 144. As you can see, 45% of the German men support the SPD, whereas only 40% of the German women do. That is, German men lean somewhat more to the SPD than German women do. You will also learn that women lean somewhat more toward the Greens, the CDU, and the FDP than men do.

As well as the fractions conditioned on the column variable, you can also calculate the fractions of the column variable conditioned on the *row* variable using the `row` option. You can use that option in addition to, or instead of, the `column` option. You could use any of the following commands:

```
. tabulate np9402 gender, row
. tabulate np9402 gender, row col
. tabulate np9402 gender, row nofreq
. tabulate np9402 gender, row col nofreq
```

Summary statistics

The `tabulate` command has options for calculating overall statistics for the differences in the distributions between groups. We will not explain these calculations here, but we will simply list the options and the names of the statistics as an overview. For a complete description, see the cited references. The formulas can be found in [R] **tabulate twoway**:

- `chi2` Pearson's chi-squared (Pearson 1900)
- `gamma` Goodman and Kruskal's gamma (Agresti 1984, 159–161)
- `exact` Fisher's exact test (Fisher 1935)
- `lrchi2` likelihood-ratio chi-squared (Fienberg 1980, 40)
- `taub` Kendall's tau-b (Agresti 1984, 161–163)
- `V` Cramér's V (Agresti 1984, 23–24)

More than one contingency table

`tabulate` allows up to two variable names. If you list three variable names, you will get an error message. There are two reasons why you might want to try `tabulate` with more than two variable names: to produce a three-way table or to produce more than one two-way table with a single command. You can produce three-way frequency tables using the `by` prefix (section 3.2.1) or the `table` command (section 7.3.2).

To produce more than one two-way table with one command, use `tab2`, which produces two-way tables for all possible combinations of the variable list. Therefore, the command

```
. tab2 np9401 np9402 np9403 gender
```

is equivalent to typing

```
. tabulate np9401 np9402
. tabulate np9401 np9403
. tabulate np9401 gender
. tabulate np9402 np9403
. tabulate np9402 gender
. tabulate np9403 gender
```

However, in many cases you will want to display only some of the tables. For example, you might want to tabulate each of the three party-identification variables conditioned on gender. In this case, using a `foreach` loop (section 3.2.2) would be more appropriate.

7.2.2 Graphs

During data analysis, graphs are seldom used for displaying variables with just a few categories because tables are usually sufficient. However, in presentations you will often see graphs, even for variables with few categories. Most often these are special types of histograms or atypical uses of bar charts or pie charts. Stata can be used for all these presentation techniques. Stata also produces dot charts (Cleveland 1994, 150–153), which are seldom used but powerful.

Before we explain the different chart types, we emphasize that we will use only very simple forms of the charts. To learn how to dress them up more, refer to chapter 6.

Histograms

In their basic form, histograms are graphical displays of continuous variables with many outcomes (see section 7.3.3). In this form, histograms plot the frequencies of groups or intervals of the continuous variable. Grouping the continuous variable generally requires that you choose an origin and the width of the intervals.

When dealing with variables with just a few categories, you do not need to make this choice. Instead, you can plot the frequency of each category with the `histogram` command and the `discrete` option.[4] First, we draw a histogram of the variable for party identification (`np9402`):

```
. histogram np9402, discrete
```

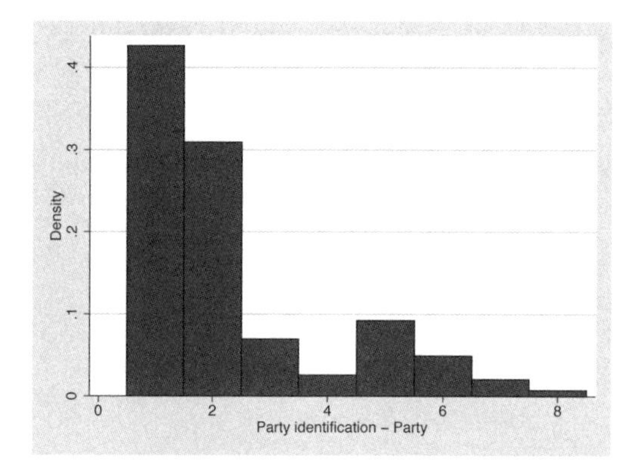

What do we see here? Essentially, we see some rectangles, or *bars*, each of which represents *one* category of the variable for party identification. Namely, the first bar represents the first category of party identification (the SPD), the second bar represents the second category (the Christian Democratic Union or CDU), and so on. As party identification has eight categories, there are eight bars in the graph.

The height of the bars varies. According to the label on the y axis, the height of the bars represents the density. In the special case of the histograms discussed here, this density equals the fraction. You can use the option `fraction` to see fractions instead of densities

```
. histogram np9402, discrete fraction
```

but this changes only the label on the y axis, not the numbers or the relative height of the bars. In the more general case, the density is not equal to the fraction (see section 7.3.3).

4. The name `discrete` arises from the fact that variables with few categories are usually discrete. Especially in small samples, continuous variables can also have few categories. Here it makes sense to specify `discrete` even for continuous variables.

The height of the bar represents the fraction of the category. The higher the bar, the higher the fraction of supporters for the party represented by the bar. You can see that the SPD supporters are most frequent in the data, followed by the CDU supporters, and so on. In fact, you see the same results as those in the table in section 7.2.1, but this time graphically.

The graph is a bit hard to read, as we do not know from the display which bar represents which party. Therefore, you should use the option `xlabel`, which we described in section 6.3.4. Moreover, there are two options specific to histograms, which we introduce here: `gap()` and `addlabels`. With nominal scaled variables, you might prefer a display with gaps between the bars. This can be achieved with the option `gap()`, where a number inside the parentheses specifies the percentage to reduce the width of the bar. For example, `gap(10)` reduces the width of the bar by 10%, creating a gap between the bars. Strictly speaking, with `gap()`, your graph can no longer be called a histogram. Finally, some people like to have numbers with the exact fractions above the bars. The `addlabels` option is their friend. Let us use these options now:

```
. histogram np9402, discrete fraction gap(10) addlabels
> xlabel(1(1)8, valuelabel)
```

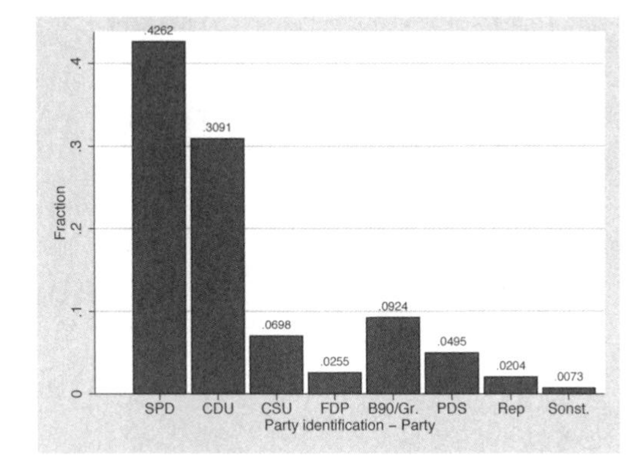

Conditional distributions can be shown using the `by()` option, which we have already described in section 6.4.2. The argument is the name of the variable on which you want to condition your distribution. For example,

```
. histogram np9402, discrete fraction by(gender) gap(10)
> xlabel(1(1)8, valuelabel)
```

displays the distribution of party identification for men and women separately:

(Continued on next page)

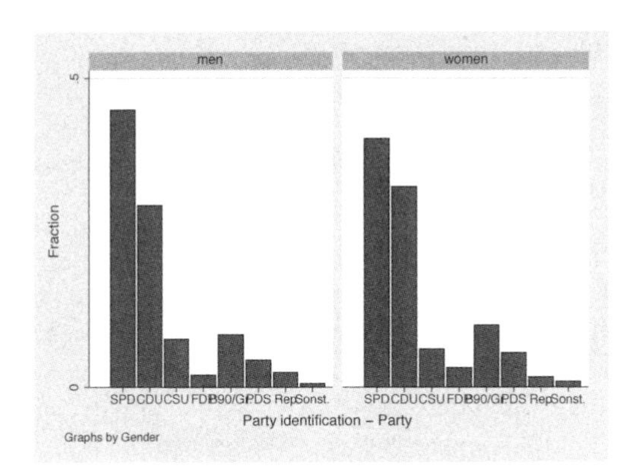

Again we see that among the GSOEP respondents, men lean more toward the SPD than women do. Also we see that the overall answer pattern is quite similar for men and women.

Bar charts

Bar charts and histograms are frequently confused, as both are often used to show fractions of discrete variables, although they are not intended for that task. Histograms are intended to show fractions of continuous variables, and bar charts are intended to display summary statistics of one or more variables.

Using techniques in an unintended manner can lead to surprises, and that often happens with bar charts. Look what happens if you naively try to use a bar chart to display the fractions of party affiliation:

```
. graph bar np9402
```

As we mentioned, bar charts are intended to display summary statistics of one or more variables. This is done by plotting a bar with a height proportional to the size of the summary statistic for each variable to be plotted. Stata allows you to plot a variety of summary statistics with a bar chart, e.g., the mean, the number of nonmissing observations, the sum. The default is the mean, so the bar in our bar chart represents the mean of party affiliation, which in this case is useless or at any rate a waste of space.

To get what we want, we need to generate what we will call a set of *dummy* variables. Generally, dummy variables are variables with values of 0 and 1. For example, we could create a dummy variable that is 1 for all respondents who favor the SPD and 0 for all others (see page 76 for details on generating dummy variables). The mean of this dummy variable would be equal to the fraction of SPD partisans. Therefore, if we used this dummy variable in our bar chart, we would get a bar with height equal to the fraction of SPD partisans, which would not be useless but would still be a waste of space. We also need bars for all the other parties, so we need to define a dummy variable for each category of party affiliation and use all these dummies in our bar chart.

The easiest way to generate such a set of dummy variables is to use the `generate()` option of the `tabulate` command. Typing

```
. tabulate np9402, generate(pid)
```

produces the dummy variable `pid1` for the first category of party identification, `pid2` for the second, `pid3` for the third, and so on. This set of dummy variables can be used for the bar chart:

```
. graph bar pid*
```

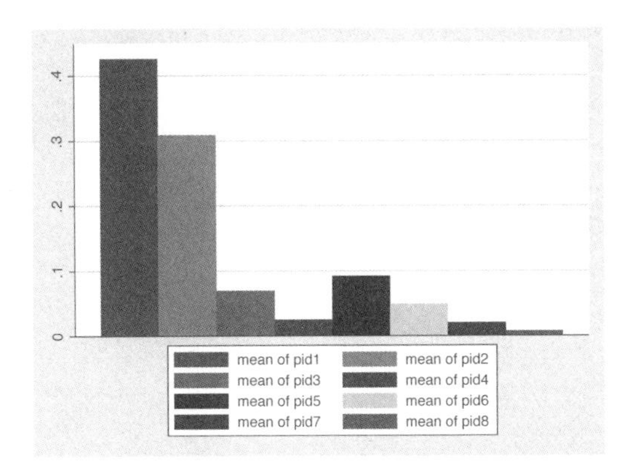

This bar chart displays the fractions of each category of party identification—much like the histogram in the previous chapter. Now you know why bar charts and histograms are so frequently confused. But be aware that we have used both techniques in a somewhat nonstandard way, by graphing a variable that is not continuous.

To show conditional distributions, you can choose the option `over()` or `by()`. Put the name of the variable on which you want to condition your distribution inside the parentheses. The option `over()` displays the fraction of party identification conditioned on its argument in a single graph. The `by()` option gives you one graph for each distribution, arranged side by side in one display.

Pie charts

In the mass media and certain business presentations, you often see pie charts used to present distributions of variables with few categories. In the literature on graphical perception, pie charts are often criticized because the reader needs to decode sizes and angles to interpret a pie chart, which is not easy for humans to do (Cleveland 1994, 262–264). We share this critical view of pie charts, so we will keep our description of them short.

In Stata, pie charts are implemented much like bar charts. Stata pie charts show slices of a pie with a size proportional to a summary statistic. Consequently, we again need to produce dummy variables to show the fractions of the categories. To compare distributions, you must use the option `by()`:

```
. graph pie pid*, by(gender)
```

Dot charts

Dot charts were introduced by Cleveland (1984) as a graphical method to display data with labels. Dot charts have shown their strength in experiments on graphical perception—not only for variables with few categories but also for variables with many categories. However, dot charts are seldom used either in scientific literature or in the mass media. One reason for this is that dot charts are rarely implemented in statistical packages or spreadsheet programs.

In principle, dot charts are similar to bar charts, except dot charts use a dot to display the value of a statistic instead of a bar. Thus dot charts are implemented like bar charts in Stata. You can replace the graph subcommand `bar` with `dot` to produce a dot chart. As in bar charts, however, you need to use dummy variables to show relative frequencies. Here is a first attempt:

```
. graph dot pid*
```

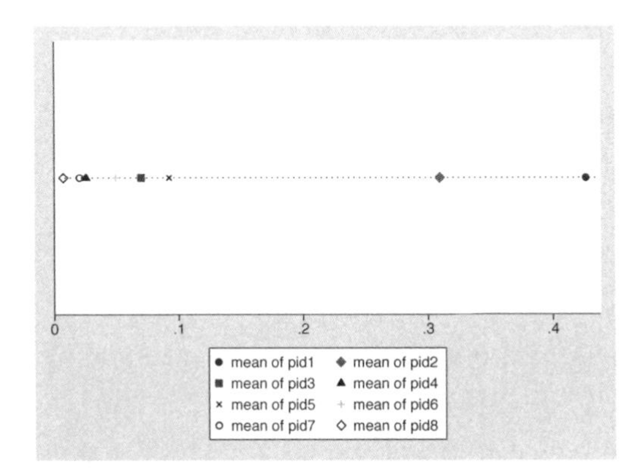

This is not a good-looking display. But before refining the graph, you should understand what you see. The graph shows a marker symbol, on one dotted line, for each summary statistic. The rightmost marker symbol (the filled dot) shows the mean of the variable `pid1`, which happens to be the fraction of respondents leaning toward the SPD. The other marker symbols can be interpreted similarly.

You might get a nicer graph if you did not put all the symbols on one line, but on different lines. We will do so presently, but first note from the previous graph that you *can* put more than one symbol on one line and that this is possible only because we have used symbols instead of bars. Later on, when describing the distributions of continuous variables, we will use this feature—let us call it *superposition*—to compare distributions between groups. But for now, we need to find a way to put the markers on different lines. This is most easily done using the option `ascategory`, which specifies that the variables listed in the *varlist* be treated as a grouping variable. See how it works:

```
. graph dot pid*, ascategory
```

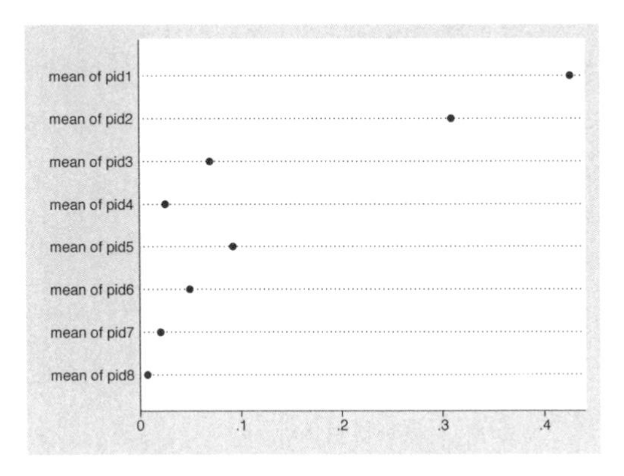

Using this as a starting point, we can immediately use the option `over()` or `by()` to compare the distribution between groups:

```
. graph dot pid*, ascategory over(gender)
. graph dot pid*, ascategory by(gender)
```

7.3 Variables with many categories

You can describe distributions of variables with many categories using tables, summary statistics, or graphs. To display such variables with tables, you need to group the variables first. Therefore, we begin by showing you the nuts and bolts of grouping variables in section 7.3.1.

In section 7.3.2, we explain the most common summary statistics and show you how to use them to describe and compare distributions, both with tables and graphically.

Finally, we show you some graphical methods intended especially for the description of distributions with many categories in section 7.3.3.

7.3.1 Frequencies of grouped data

Variables with many categories usually cannot be described in tables. If you type

```
. tabulate income
```

you will immediately understand why: such tables can become huge, and the frequencies for each category are likely to be too small to provide any useful information. One solution is to *group* such variables, that is, recode the variable by assigning the information on some similar categories to one category. Taking `income` as an example, this could mean using the frequencies of the intervals $[0, 1000)$, $[1000, 2000)$, etc., instead of the frequencies for every category of income. This would lead to a frequency distribution of grouped data.

You can arrive at a description of grouped data in two steps:

1. Generate a new variable that is a grouped version of the original variable.
2. Display the new variable with techniques for the description of variables with few categories.

You can generate the grouped version of the variable using the methods described in chapter 5, as well as some specific tools for this task. But first we would like to make some general remarks on grouping variables.

Some remarks on grouping data

Grouping variables leads to a loss of information, which may emphasize or hide certain properties of distributions. Grouping variables therefore is an important step of data analysis. You should carefully decide how to group your variables and inspect the consequences of your decision. There are suggested rules for grouping variables, for example, "Always use intervals of equal width" or "Do not group nominal scaled variables". However, in practice such rules are not helpful. The research goal always determines whether the loss of information due to a specific grouping is problematic. If the goal is to detect data errors, group the moderate values into one category and leave the extreme values as they are. If the goal were to compare the distribution of income of unemployed people between Germany and France, you might want to group all incomes above a specific level into one category and form finer intervals for lower incomes. The same is true for nominal scaled variables. In a very large population sample of German inhabitants, the variable for nationality can easily include 20–30 categories. For most data analysis, you will want to group that variable, for example, by differentiating only between Germans and foreigners. However, whether that or any grouping procedure will be useful depends on the research topic.

Special techniques for grouping data

In grouping variables we need to differentiate between nominal scaled variables and other variables. For nominal scaled variables—such as `nation`—the order of the categories has no special meaning. To group such variables, you need to decide if any category can be grouped with another. Therefore, in performing the grouping, you are confined to the techniques described in chapter 5. For variables with categories that have a quantitative meaning, where observations with low values are also low in a real sense, you can always say that *neighboring* values are similar to some extent. Therefore, you can often refer to entire ranges of categories, and Stata has a set of special techniques for that task.

Grouping by quantiles

Sometimes you will want to group a distribution into intervals that contain nearly the same numbers of observations. We call this grouping by quantiles (see also section 7.3.2). In Stata, you can group by quantiles with the command `xtile`. To generate a new variable with, say, four groups with almost the same number of respondents, you simply type

```
. xtile inc_4 = income, nquantiles(4)
```

If you had used the option `nquantiles(10)`, the variable `inc_4` would have contained 10 intervals with nearly the same number of respondents, and so on.

Since the new variable has only four categories, you can simply use the techniques described in the last section, for example,

```
. tabulate inc_4
```

Grouping into intervals with same width

Grouping variables by quantiles leads to variables having almost the same number of observations in each category, whereas the widths of the intervals of each category differ. That is, the difference between the upper bound of a category, c_j, and the upper bound of the category below, c_{j-1}, is not constant for all categories.

Instead of grouping by quantiles, you can also try to get intervals with the same width for each class. For example, you could group income so that all respondents with income between $0 and $1,000 form the first interval, respondents with income between $1,001 and $2,000 form the second interval, and so on, until the maximum income is reached.

More generally, you group the values of a variable into k categories with $d_j = c_j - c_{j-1}$ equal for all categories using either the `recode()` function or the `autocode()` function.[5]

Let us begin with an example of the `recode()` function. With that function the grouping of income into intervals with equal width can be done like this:

```
. generate inc_g1 = recode(income,1000,2000,3000,4000,5000,6000,12500)
```

This generates the new variable `inc_g1`, which becomes 1,000 for all observations with an income of $1,000 or below (that is, between $0 and $1,000), 2,000 for all observations between $1,001 and $2,000, and so on. All nonmissing observations with an income above $6,000 become $12,500 on `inc_g1`.

The one-way frequency table of the grouped income variable `inc_g1` can be displayed as usual with

```
. tabulate inc_g1
```

The general syntax of the `recode()` function is `recode(`*exp*`,`*x1*`,`*x2*`,`...`,`*xk*`)`, where *exp* is an expression (see section 3.1.6) and *x1*, *x2*, ..., *xk* are arbitrary numbers in ascending order. The function gives the value *x1* if $exp \leq x1$, *x2* if $exp \leq x2$, and *xk* if $exp > x(k-1)$. To get an intuitive understanding of the `recode()` function, you might want to try out some simple examples using `display`:

```
. display recode(1000,1,2,3)
. display recode(1000,800,1200,3000)
. display recode(ln(734),5,6,7,8)
```

As always, you can use a variable name for *exp*, as we did when generating the variable `inc_g1`.

5. Do not confuse the `recode()` function with the command `recode` (section 5.2.1). Stata functions are general elements of Stata commands. You can use the `recode()` function wherever an expression is allowed (see chapter 3).

When you understand `recode()`, `autocode()` is easy. The `autocode()` function is a sort of shorthand for the `recode()` function if you want to have intervals with equal widths. The syntax of the `autocode()` function is

`autocode(`*exp*`,`*k*`,`*min*`,`*max*`)`

The `autocode()` function internally splits the interval from *min* to *max* into *k* intervals with equal widths and returns the upper limit of the interval that contains the value of the expression *exp*. To group the variable `income` into 13 intervals, you can use the `autocode()` function as follows:

```
. generate inc_g2 = autocode(income,13,0,12500)
```

Grouping into intervals with arbitrary widths

You can group a variable into intervals with arbitrary widths by simply imposing the upper limits of the arbitrary intervals on the list of numbers in the `recode()` function:

```
. generate inc_g3 = recode(income,400,500,750,1000,2000,5000,12500)
. tabulate inc_g3
```

7.3.2 Describing data using statistics

Summary statistics are often used to describe data with many categories. With summary statistics, you can describe distributions parsimoniously. Generally, you will distinguish between summary statistics for the position and for the dispersion of distributions. Summary statistics for the position describe which values the observations *typically* have, and summary statistics for the dispersion describe how different the values for the observations are. Generally, you should use at least one summary statistic of each type to describe a distribution.

We will begin by briefly describing the most important summary statistics for distributions. Then we will present two commands for calculating these statistics: `summarize` and `tabstat`.

Important summary statistics

The arithmetic mean

The most common summary statistic for determining the central position of a distribution is the arithmetic mean, often simply called the *average*. The arithmetic mean is a summary statistic for determining the position of variables with interval-level scales or higher, that is, for variables like income, household size, or age. You can also use the arithmetic mean for dichotomous variables that are coded 0 and 1 (dummy variables) since the arithmetic mean of such variables gives the fraction of observations that are coded 1. Sometimes you might also use the arithmetic mean for ordinal variables, such as life satisfaction or the intensity of party identification. But strictly speaking, this approach is wrong, and you should not take the resulting numbers too seriously. You definitely should not use the arithmetic mean for categorical variables that take on more than two values, such as marital status or type of the neighborhood, as the result makes no sense.

The standard deviation

The most common summary statistic for determining the dispersion of a distribution is the standard deviation. The standard deviation can be thought of as the average distance of the observations from the arithmetic mean. This interpretation is not entirely correct but may give you an idea of the notion. To calculate the standard deviation, you first need to calculate the arithmetic mean. This means that you can calculate the standard deviation only for variables for which you can calculate the arithmetic mean.

In some ways, the arithmetic mean and the standard deviation can be seen as sibling measures for describing a distribution. Usually when you describe the position of the distribution with the arithmetic mean, you will use the standard deviation to describe the dispersion of the distribution. But the description of a distribution in terms of its mean and standard deviation can be problematic. To understand why, take a look at the data in figure 7.1. Each line in this graph gives you the values of one of four artificial distributions. As you see, the distributions of all four variables are different. The first variable has what statisticians call a normal distribution, the second variable has a uniform distribution, and the third and forth variables are skewed to the right and to the left, respectively. Nevertheless, all four variables share the same summary statistics: they have a mean of 5.5 and a standard deviation of 3.05. Given only the description with summary statistics, you might have concluded that all four distributions are identical.

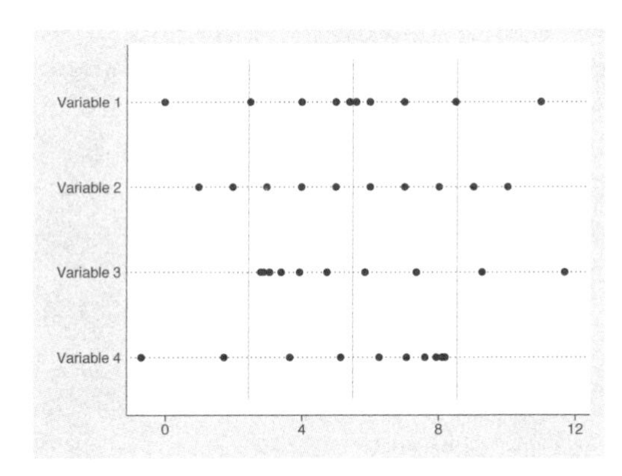

Figure 7.1. Distributions with equal averages and standard deviations

grmeans.do

To guard against such misinterpretations, you should consider other summary statistics (or describe your data graphically).

Quantiles

Quantiles are the most important companions to the arithmetic mean and the standard deviation. The p-quantile ($x_{[p]}$) splits a distribution into two parts such that the first part contains $p \times 100$ percent of the data and the second part contains $(1-p) \times 100$ percent. In particular, the .5-quantile—the "median"—separates the data so that each part contains 50 percent of the observations.

To calculate the quantiles, you use the position, i, of each observation in the sorted list of a distribution. The p-quantile is the value of the first observation with position $i > np$. If there is an observation with $i = np$, you use the midpoint between the value of that observation and the value of the following observation. To find the .5-quantile of a distribution with 121 valid observations, you need to search for the 61st ($121 \times 0.5 = 60.5$) observation in the sorted list and use the value of the distribution for this observation. For 120 observations, you would choose a value between the values of the 60th and 61st observations.

To find the quantiles, you use the position of the observations in the sorted data. The values of the categories are of interest only insofar as they determine the order of the sorted list. It does not matter if the highest value is much higher than all others or just a bit higher. In this sense, quantiles are *robust* against outliers.

In practice, the most important quantiles are the *quartiles*: the quantiles with $p = .25$ (first quartile), $p = .5$ (*median* or second quartile), and $p = .75$ (third quartile). From the quartiles, you can also learn something about the skewness and the dispersion of a distribution.

- If the distances of the first and third quartiles from the median are equal or almost equal, the distribution is symmetric. If the first quartile is closer to the median than the third quartile is, we call the distribution "skewed to the right". If it is the other way around, we call the distribution "skewed to the left".
- The difference between the third quartile and the first quartile is called the *interquartile range*. This value tells us the range of values that are the middle 50% of the observations.

By the way, the quartiles for the distributions from figure 7.1 are shown in table 7.1. The results clearly show the different shapes of the distributions.

(*Continued on next page*)

Table 7.1. Quartiles for the distributions

Variable	1st Quartile	Median	3rd Quartile
1	4.00	5.50	7.00
2	2.98	5.50	8.02
3	3.07	4.33	7.35
4	3.65	6.67	7.93

The summarize command

You already know the most important command for calculating summary statistics: `summarize`. In its basic form, `summarize` calculates the mean and the standard deviation:

```
. summarize income
```

From this output, you can see that the mean monthly income of the German population in 1997 was $1,349.21 and that (roughly speaking) the difference between the true income and the mean income was about $1,245.70 *on average*.

In addition to the mean and the standard deviation, `summarize` also reports the minimum (lowest value) and the maximum (highest value) of the distribution. In our example, the lowest income of all respondents is $0, and the highest income is $12,438.

`summarize` also calculates a set of quantiles if you specify the `detail` option:

```
. summarize income, detail
```

The tabstat command

The `tabstat` command also displays and calculates summary statistics. It is a generalization of `summarize`, as it allows you to specify a list of statistics to be displayed by using the option `statistics()`. Inside the parentheses you specify the names of the statistics to be displayed. The default is `statistics(mean)`, which displays the arithmetic mean, but you can also use other statistics and even multiple statistics. For example, typing

```
. tabstat income, statistics(count mean sd min max)
```

displays the number of nonmissing observations (`count`), the mean (`mean`), the standard deviation (`sd`), the minimum (`min`), and the maximum (`max`)—replicating the output of `summarize`.

The following command shows the minimum, the three quartiles (`p25 p50 p75`), and the maximum—what is sometimes referred to as the "five-number summary" of a distribution:

```
. tabstat income, statistics(min p25 p50 p75 max)
```

For a list of names for statistics you can calculate with `tabstat`, see `help tabstat`.

Comparing distributions using statistics

Stata provides several tools for comparing different distributions through summary statistics:

- a combination of `summarize` with the prefix `by`,
- the `summarize()` option of `tabulate`,
- the `by()` option of `tabstat`,
- the `table` command, and
- graphs of summary statistics.

We will not go into the combination of `summarize` and the `by` prefix since we have already described this tool in section 3.2.1. Most likely, you have already used `summarize` with `by` several times by now. If not, you should read section 3.2.1 now. This combination is easy to use and quite powerful: you can use it to obtain most of the results of the other tools we have described. The main advantage of the other tools is the attractive arrangement of the results.

For those already familiar with inference statistics, we should point out that Stata has a variety of built-in commands. Among the frequently used built-in commands are `ttest` to test the equality of means, `prtest` to test the equality of proportions and `ranksum` to test the hypothesis that two independent samples are from populations with the same distribution. In addition, Stata offers a series of exact tests like the Kolmogorov–Smirnov test for equality of distribution functions (`help ksmirnov`). Accessible descriptions of these tests can be found in statistic textbooks such as Moore and McCabe (2005) or Conover (1999).

The summarize() option of tabulate

The `summarize()` option of `tabulate` is used to display the arithmetic means and standard deviations of a distribution, conditioned on the values of one or two other variables. For example, to show the average and standard deviation of income conditioned on gender, you can use

```
. tabulate gender, summarize(income)
```

The `summarize()` option includes the means and standard deviations of the variable to be summarized. This tells you that the mean income for men is about $1,671 and the mean income for women is about $1,053.

You can also use the summarize() option in two-way tables. The advantages of comparing distributions this way may be even more convincing. Suppose that you want to know the income inequality between men and women in each German state separately—you want to know the mean income conditioned on gender *and* state. In this case, you might produce a two-way table containing the mean of income in each of its cells:

```
. tabulate state gender, summarize(income) nostandard nofreq
```

We used the options nostandard and nofreq to suppress the output of standard deviations and frequencies.

The statistics and by() option of tabstat

The summarize() option of tabulate allows you to compare only means and standard deviations of a distribution between different groups. tabstat is statistically more flexible. You can use the by() option of tabstat to specify that the statistics be displayed separately for each unique value of a variable given within the parentheses. Therefore, to produce the five-number summary of income conditioned on gender, you could type

```
. tabstat income, statistics(count q max) by(gender)
```

You can use only one variable name in the parentheses of the by() option. The table command is one way to overcome this restriction.

The table command

The table command is a generalization of the techniques we have described so far. With table, you can display many types of statistics, including means, standard deviations, and arbitrary quantiles, in tables having up to seven dimensions. In this respect, the advantages of table are obvious, but it is not as fast and its syntax is slightly more complicated.

The syntax of table has two parts: one to format the table and one to specify the contents of the cells. As with tabulate, the table can be laid out simply in rows

```
. table np9502
```

or in rows and columns

```
. table np9502 np9501
```

However, in table you can also specify a third variable, which defines a *supercolumn*. The supercolumn works as if you had specified tabulate with a by prefix, except that the different tables are displayed side by side. This way, you get a three-way table:

```
. table np9502 np9501 gender
```

Finally, you can specify up to four *superrows*, which work like supercolumns, except that the different parts of the table are displayed one below each other—but still in one table. Superrows are specified as a variable list with the option `by()`:

```
. table np9502 np9501 gender, by(state)
```

The table forms the groups on which we want to condition our variable of interest. But to describe the variable, we still need two pieces of information:

- The variable to be described and
- The summary statistics by which we provide the description.

Both pieces of information are specified with the option `contents()`. Within the parentheses of `contents()`, you first specify the name of the summary statistic and then the name of the variable. The `contents(mean income)` option, for example, would fill the cells of the table with the arithmetic mean of income for each of the groups of the table.

As an example, we will reproduce some of the tables we produced earlier. With `table`, the comparison of income conditioned on gender from page 161 would look like this:

```
. table gender, contents(count income mean income sd income)
```

Accordingly, you would get the mean income conditioned on gender and state (page 162) by typing

```
. table state gender, contents(mean income)
```

What if you distrust the mean and want to use quartiles instead? Voilà:

```
. table gender, contents(p25 income p50 income p75 income)
```

Suppose that you want to compare the monthly rent among apartments having different features. How much do the monthly rents of apartments having a kitchen, a bath or shower, and a garden differ from those without those amenities?

```
. table kitchen shower garden, contents(mean rent)
```

In this table, you can see that the former rents for \$379 on average, whereas the latter rents for \$1,192. To clarify this somewhat strange result, look at the number of observations on which the averages within each cell are based:

```
. table kitchen shower garden, contents(mean rent n rent)
```

The more cells a table has, the more important it is to check that the number of observations is large enough to get reliable estimates of the summary statistic. With a small number of observations, results are sensitive to outliers. As we can see now, the rent of \$1,192 in the table above is based on just two observations. From

```
. list rent renttype ybirth htype if kitchen==2 & garden==2 & shower==2
```

we see that one of those observations is a respondent who lives as a tenant in an apartment building and is recorded as having to pay $2,239. This might very well be a data error.

Our examples do not fully explore the range of possibilities from using tables. There are further available summary statistics, and there are many tools for improving the appearance of the table; see [R] **table** and `help table` for further insights.

Graphical displays of summary statistics

Stata has three special tools for displaying summary statistics graphically: bar charts, dot charts, and pie charts. Technically, all three work similarly in Stata, but dot charts have been shown to be the most powerful way to graphically display statistical information (Cleveland 1994, 221–269). So we concentrate on the presentation of dot charts. For basic bar or pie charts, it is usually enough to replace the `graph` subcommand `dot` with `bar` or `pie`, respectively.

In our first example, we compare the arithmetic mean of income among the German states:

```
. graph dot (mean) income, over(state)
```

The command begins with `graph dot`, which is the command for making a dot chart. Then we indicate our choice of summary statistic inside the parentheses. As the arithmetic mean is the default, we could have left `mean` out. For a complete list of available summary statistics, refer to `help graph_dot`. After indicating the summary statistic, we entered the name of the variable we want to describe. Finally, inside the parentheses of the option `over()` we put the name of the variable on which we want to condition the summary statistic. Here is the result:

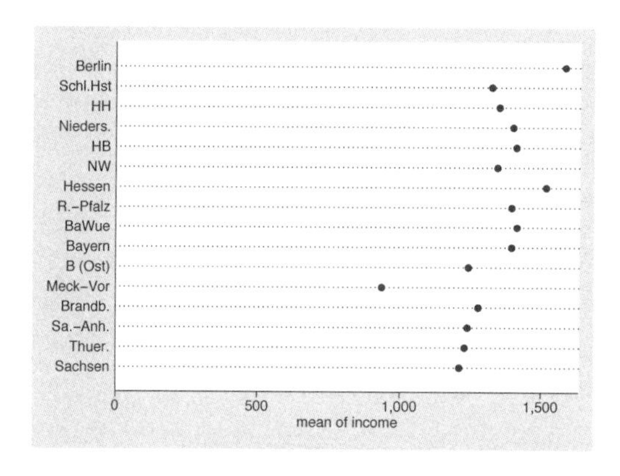

The graph shows a dot for the mean income of each German state. It can be seen rather quickly that Berlin and Hessen have the highest mean income of all German states and that Mecklenburg-Vorpommern has the lowest. After some inspection, German readers will see that the average income in the five East German states (the former Communist ones) is lower than in West Germany.

However, the graph is hardly optimal—even if it is not meant for publication. In our graph, we have followed the rule "Alabama first", which is a well-known rule for displaying data *badly* (Wainer 1984). "Alabama first" means that you display your data in alphabetical or some other order that is unrelated to the content. If you do not want to display data badly, you should give your data a meaningful order. You can do this most easily by specifying the `sort()` suboption within the `over()` option. Inside the parentheses of `sort()` you specify a sort order, either by the contents of a variable or by a function of these contents. We choose to have our data sorted by the average income: `sort((mean) income)`. `sort` is a suboption under `over()`, not an option under `graph dot`, so you place it within the `over()` option.

```
. graph dot (mean) income, over(state, sort((mean) income))
```

To condition the summary statistic on more than one variable, we can use the `by()` option. Another method specific to dot charts is *superposition*, which we use in our next example to show the mean income in the different states conditioned on gender.

With dot charts, you can plot more than one summary statistic on one line. If you enter a second variable, a second marker symbol appears on each line, representing the second variable. To show conditional means for men and women separately, you therefore need two income variables: one for men and one for women. You can generate these variables easily with

```
. generate inc1 = income if gender == 1
. generate inc2 = income if gender == 2
```

or in one command with

```
. separate income, by(gender)
```

Now you can use the new *separated* variables for the dot chart instead of the original one. You need to choose a sort order for the states. We decided to sort by the mean income of males, but you could also use the mean income of females, the overall mean income, or any other useful order. Your research goals, however, determine what is useful.

(Continued on next page)

```
. graph dot (mean) income1 income2, over(state, sort((mean) income1))
```

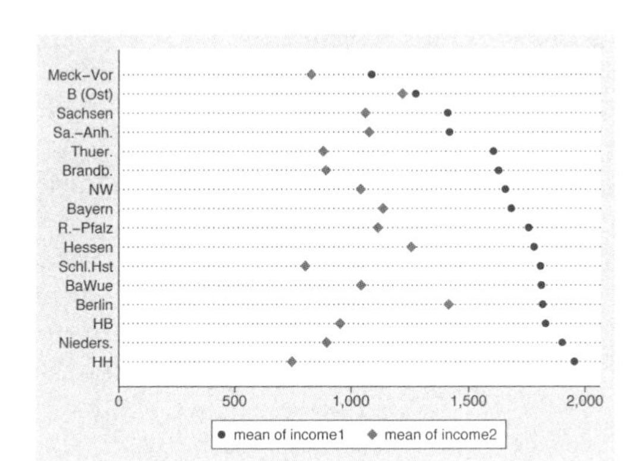

7.3.3 Graphs

Summary statistics describe variables by highlighting a certain property of the distribution and excluding most information about it according to certain *assumptions*. For example, the description of a distribution with the mean and the standard deviation excludes, among many other things, all information about the skewness of the distribution. If all the distributions are symmetric (or at least equally skewed), the means and the standard deviations provide useful descriptions for comparing them. But if the assumptions underlying a summary statistic are not true, the summary statistic does not provide an informative description.

This section provides an overview of techniques for describing data that require fewer assumptions than those we have previously mentioned. Fewer assumptions mean that more information must be communicated, and for this we use graphs. But these graphs differ from those discussed earlier, which we used to display summary statistics that were calculated under certain assumptions. Here we use graphs to describe the distribution with as few assumptions as possible.

Box plots

In section 7.3.2, we introduced the five-number summary of distributions, which uses the minimum, maximum, and quartiles to describe the data. Box plots can be seen as graphical representations of the five-point description with some enhancements.

Here is an example:

```
. graph box income
```

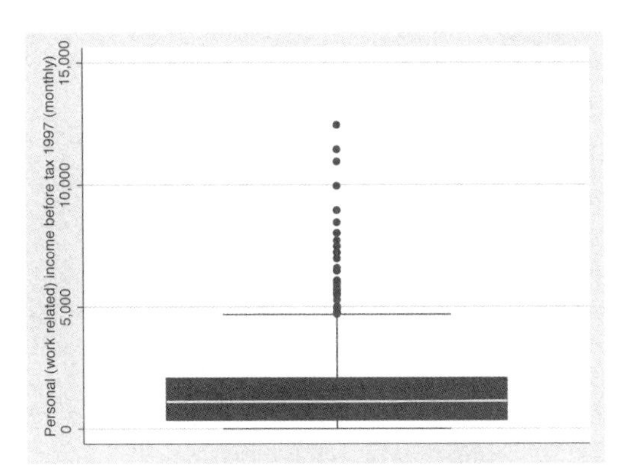

Box plots in general are composed of a *box*, two *whiskers*, two *fences*, and some marker symbols. The box is the rectangle in the middle of the graph. The lower border of the box is the first quartile; the upper border is the third quartile. The line in the middle of the box is the median. The height of the box therefore shows the interquartile range.

The whiskers are the two vertical lines below and above the box, which end with small horizontal lines called the fences. The upper fence is the highest value of the distribution that is smaller than or equal to the third quartile plus 1.5 times the interquartile range. The lower fence is the lowest value of the distribution that is greater than or equal to the first quartile minus 1.5 times the interquartile range.

Observations below the lower fence or above the upper fence are regarded as outliers and are plotted with single plot symbols.

Box plots provide much information about a distribution. You can directly infer the position of the distribution from the position of the box along the y axis. Fifty percent of the observations have values between the lower and upper bounds of the box. You can infer the dispersion of the distribution by the size of the box, by the length of the whiskers, and by the positions of the outliers. Moreover, you can see the symmetry or skewness of the distribution: in symmetric distributions, the distances between the median and the lower and upper bounds of the box are equal. Also the whiskers should be of equal length, and the outliers above and below should be equally spaced. If the outliers, the whiskers, or the box are squeezed at the bottom, the distribution is skewed to the right. If they are squeezed at the top, it is skewed to the left. Here the distribution of income is skewed to the right.

Box plots can be used effectively to compare distributions across different groups using the option `over()` or `by()`. You simply put the names of the variables on which you want to condition your distribution inside the option's parentheses. Here is the distribution of income conditioned on state. Since the value labels for `state` are rather long, we use `alternate` within `over()` to avoid overlapping.

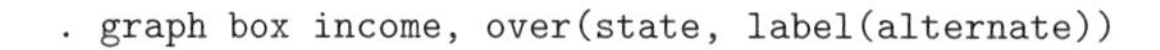

```
. graph box income, over(state, label(alternate))
```

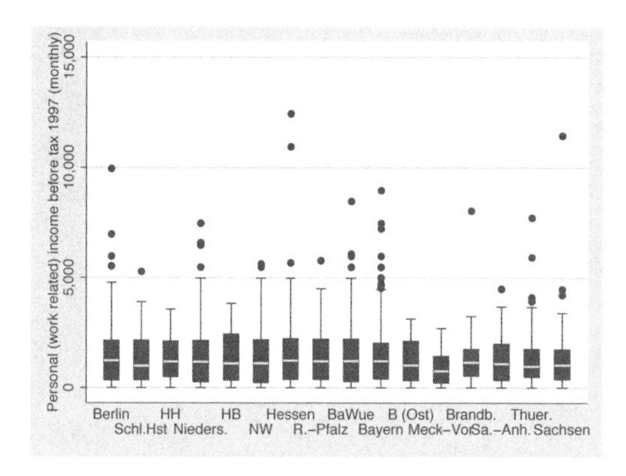

Histograms

The most common graphical display for distributions with many categories is the histogram, which is a graphical display of grouped frequency tables. In histograms, a rectangle of height

$$\widehat{f}_j = \frac{n_j}{d_j} \tag{7.1}$$

and of width d_j is drawn for each interval, j, of the grouped distribution, where n_j is the absolute frequency and $d_j = c_j - c_{j-1}$ is the width of the interval. The quantity $\widehat{f}$ is called the *density*. The density $\widehat{f}_j$ is not the relative frequency f_j. Relative frequencies are always between 0 and 1, whereas densities can be any positive real number. In histograms, the areas of the rectangles are proportional to the relative frequencies of the intervals, whereas their height is proportional to the data density within those intervals.

The implementation of histograms in Stata is confined to a special case: Stata draws histograms where the widths of all rectangles are equal. In other words, Stata draws histograms of distributions that are grouped with `autocode()` (see page 157). For this reason, the heights of the rectangles can also be interpreted as fractions within the intervals. The taller the rectangle, the larger the fraction within the corresponding interval.[6]

6. The ado package `eqprhistogram`, described by Cox (2004), implements histograms that may have rectangles of different widths. For more on installing external ado packages, see section 12.3.

You obtain a histogram by using the **histogram** command. The command is easy to use: you just specify the command and the name of the variable you want to display. The graph again shows the skewed income distribution.

```
. histogram income
```

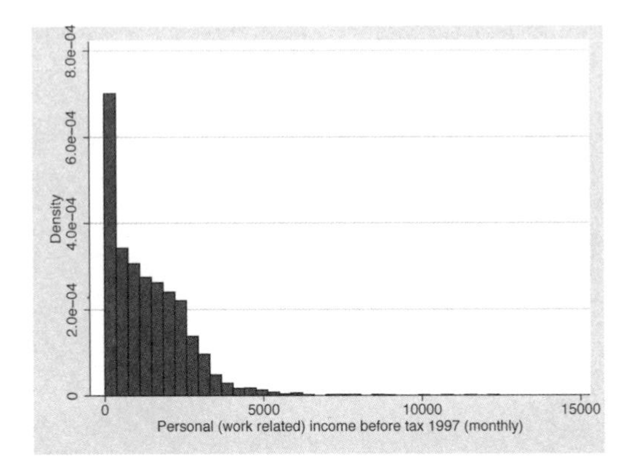

Often the histograms you get from this simple command will suffice. But be careful, as the appearance of histograms depends on your choice of the origin and the width of the displayed intervals. If you do not specify them, Stata chooses for you. Stata always chooses the minimum value of the distribution as the origin and infers the width of the intervals from a formula for the *optimal* number of bars to be displayed. For our distribution, Stata decided to use 30 bars. Then Stata internally grouped the income variable into 30 intervals with equal widths and drew bars with heights proportional to the fraction within these intervals. You can do this on your own by using the `autocode()` function and **histogram** with the **discrete** option:

```
. generate inc30 = autocode(income,30,0,25000)
. histogram inc30, discrete
```

The optimal number of groups for histograms is debatable (Emerson and Hoaglin 1983). The goal is to generate a histogram that shows all the important features of the distribution but ignores the unimportant ones. As rules of thumb for the optimal number of groups, $10\log_{10} n$, $2\sqrt{n}$, and $1 + \log_2 n$ have been proposed. For 3,034 valid observations these rules propose histograms with 30, 90, and 13 groups, respectively. Stata uses

$$\min(\sqrt{n}, 10\ \log_{10} n)$$

which generates 30 bars.

As we have said, the appearance of histograms depends on the width of the intervals, which is a function of the number of bars. With the `bin()` option, you can choose the number of bars yourself by indicating the number of bars you want inside the parentheses.

It is always a good idea to try out some other numbers. Here are the commands to generate histograms with 13 and 90 bars:

```
. histogram income, bin(13)
. histogram income, bin(90)
```

Here 12% of the respondents have no income at all, and some respondents have very high incomes. To focus on the interior part of the distribution, we can use `if` qualifiers in the usual way (see section 1.3.10):

```
. histogram income if income > 0 & income < 5000
```

You can compare different distributions with histograms by using the `by()` option. The following example shows the income distribution conditioned on gender:

```
. histogram income, by(gender)
```

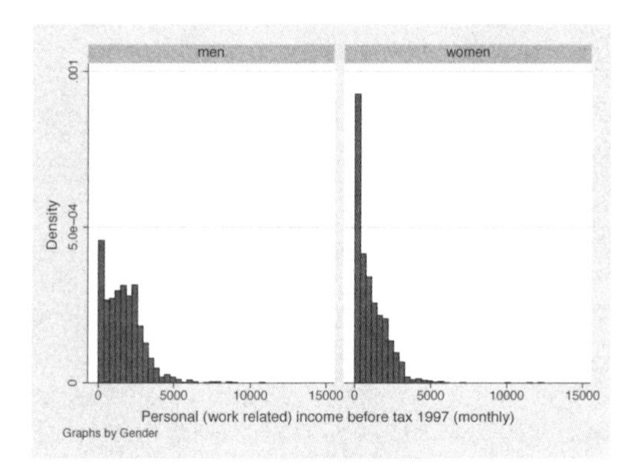

We do not want to interpret these graphs here. Instead we remind you that the appearance of histograms depends on the origin and width of the bars. Moreover, the necessary grouping of data often leads to surprising results—take a look at the histograms of income by state, for example. A graphical technique to circumvent some of these problems uses kernel density estimators, which we describe below.

Kernel density estimation

When displaying a distribution with a histogram, we are interested in the density at arbitrary points of the distribution. That is, we are interested in the density of persons with a value of about x. Histograms approximate the density by choosing an interval that contains x and calculating the fraction of persons within this interval. You can see a disadvantage of this procedure in figure 7.2.[7]

7. We took the figure with permission from Fahrmeir et al. (1997, 99).

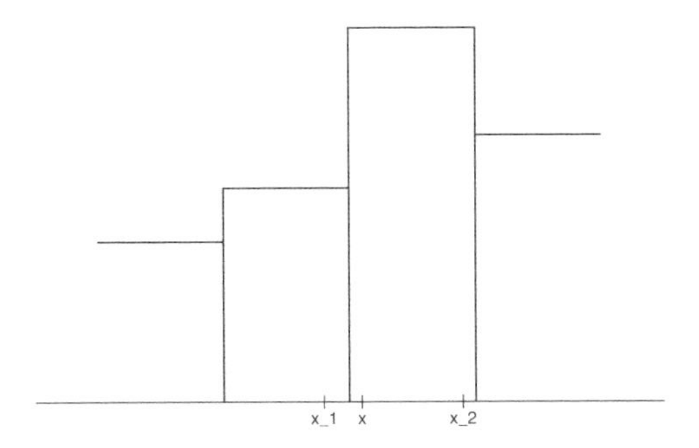

Figure 7.2. Part of a histogram

grfahrmeir.do

The figure displays a part of a histogram. There are three values of interest in the figure, x_1, x_2, and x. Suppose that you are interested in the density of the distribution at point x. The height of the bar at point x is based on the fraction of observations within the interval around x. Clearly, observations with a value of x_1 are not counted in calculating the height of the bar at point x. But the observations at point x_2, which are farther away from x than x_1 is, are counted. To calculate the fraction of observations with a value of about x, it would be better to use an interval that has x as its midpoint. That is, we should use an interval that ranges from $x - h$ to $x + h$. The density at x would be the fraction within that interval divided by its width ($2h$).

We can do the same for any arbitrary value of a distribution. If we graph each of these densities along the y axis, we get what is called a *sliding histogram*. Sliding histograms are special cases of kernel density estimators, which generally estimate the density at x by

$$\widehat{f}_{(x)} = \frac{1}{nh} \sum_{i=1}^{n} K \left[\frac{x - x_i}{h} \right] \tag{7.2}$$

In the case of a sliding histogram, we define

$$K[z] = \begin{cases} \frac{1}{2} & \text{if } |z| < 1 \\ 0 & \text{otherwise} \end{cases}$$

This looks rather complicated, but it is not. To fully understand this formula, consider circumstances where $z = |(x - x_i)/h|$ is smaller than 1. Assume that you want to calculate the density at, say, 6. Further assume that a value is approximately 6 when it is not more than 2 above or below 6; that is, the quantity h is 2. Finally, assume that your distribution has values from 0 to 10.

Substitute $x_i = 9$ into $(x - x_i)/h$:

$$z = \frac{6 - 9}{2} = -1.5$$

For $x_i = 9$, the absolute value of z is greater than 1. Therefore, K becomes 0. If you try other values for x_i, you will see that $|(x - x_i)/h|$ becomes lower than 1 whenever x_i is inside the interval from $6 - h$ to $6 + h$. As a result, K takes the value 1/2 for all values within that interval and zero elsewhere. The sum in (7.2) is therefore

$$\frac{\text{Number of observations within } [x - h, x + h)}{2}$$

If you denote the numerator—the absolute frequency within the interval j—by n_j, you can write (7.2) as

$$\begin{aligned} \widehat{f}_{(x)} &= \frac{1}{nh} \cdot \frac{n_j}{2} \\ &= \frac{f_j}{2h} = \frac{f_j}{d_j} \end{aligned}$$

Equation (7.2) therefore is just a slightly more complicated way to express (7.1).

Equation (7.2) is nice because of $K[z]$, which is the kernel from which the name "kernel density estimate" stems. Kernel density estimates are a group of estimates for the density at point x, which are all calculated with (7.2). They differ only in the definition of $K[z]$. The rectangular kernel used for sliding histograms is only one case. Its main feature is that all observations within the observed interval are equally important for estimating the density. More common are kernels that treat observations closer to x as more important. However, with many observations the different kernel density estimates do not differ very much from each other.

In Stata, kernel density estimates are implemented in the command `kdensity`. For the easiest case, simply enter the name of the variable for which you want to estimate the densities. Here we generate a graph displaying densities that were calculated with an *Epanechnikov kernel*:

```
. kdensity income
```

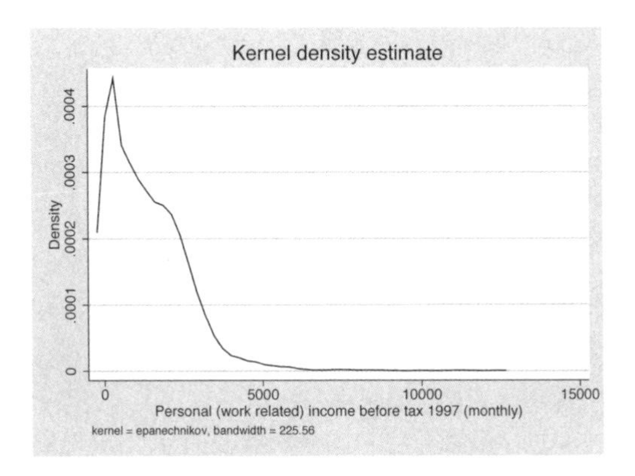

There are eight different kernels available. They are listed in the online help and are more fully described in [R] **kdensity**. The `kernel(rectangle)` option, for example, calculates the densities with the rectangular kernel:

```
. kdensity income, kernel(rectangle)
```

As stated, the choice of kernel typically does not affect the estimation of the density curve much. Both commands display the skewed income distribution.

Often the width of the interval ($2h$) used for the density estimation is more critical than the choice of kernel. If you do not provide a value, Stata calculates some *optimal* value, which some consider to be slightly too high for skewed or multimodal distributions. The `bwidth()` option can be used to change the interval half-width (h). If you scale down the optimal value to, say, $h = 100$, you will get a less smooth display:[8]

```
. kdensity income, bwidth(100)
```

To compare distributions with kernel density estimators, you can draw the curves for different distributions on one graph. You can do this in two steps:

1. Estimate the density for each distribution separately.
2. Display the results of the first step on one graph.

Let us demonstrate these two steps by comparing the logarithmic income of men and women. First, you need a variable for the logarithmic income:

```
. generate linc = log(income)
```

Observations with an income of \$0 are set to missing for this new variable.

8. See Marron (1988) for a discussion of different algorithms for selecting the interval width.

The command kdensity does not calculate the density at each possible value of the distribution. To save computer time, Stata instead calculates the density at only 50 points.[9] When comparing the densities of two distributions, you must therefore take care that you have calculated the densities at the same points. You can use the at() option as a solution to that problem.

The at() option refers to a variable containing the points at which to calculate the densities. Therefore, you need to generate a variable with, say, 50 points evenly spaced along the log-income distribution. Here is one way to do it:

```
. generate xhelp = log(_n * 250) in 1/50
```

This command generates a variable that contains 50 values, from $\log(1 \times 250)$ to $\log(50 \times 250)$ with spaces of $\log(250)$ between.[10]

Now you can do the first step. Let us begin by estimating the densities of log income for men:

```
. kdensity linc if gender == 1, generate(xvals1 fmen) at(xhelp) nograph
```

Again you use the kdensity command, but this time with the generate() option. Using generate(xvals1 fmen) creates two new variables. xvals1 contains the points at which the densities were calculated, and fmen contains the calculated densities. The at(xhelp) option tells kdensity to calculate the densities at the points specified in the xhelp variable, and nograph suppresses the graphical display, as you do not need the graph at the moment. Here we can ignore the xvals1 variable because we used the at() option to specify the points at which to calculate the densities; in other instances where we allow kdensity to pick the points, we would need to use the xvals1 variable to make a graph later.

Now follow the same procedure for the women:

```
. kdensity linc if gender == 2, generate(xvals2 fwomen) at(xhelp) nograph
```

The log-income densities for men and women are now stored as variables in the dataset. In the second step, you simply plot these variables against xhelp in one two-way graph. See chapter 6 for a detailed explanation of the options used here:

```
. graph twoway connected fmen fwomen xhelp, connect(l l) clpattern(l 1)
> msymbol(O O) mfcolor(white black) mlcolor(black black) title(Income by Gender)
> ytitle(Density) xtitle(Log (Income)) legend(label(1 "Men") label(2 "Women"))
```

9. You can change the number of points with the option n().

10. Read section 5.1.3 if you do not fully understand this command.

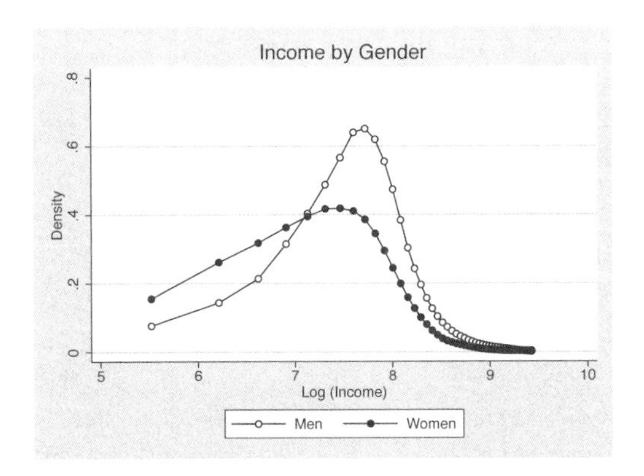

Overlaying density curves allows you to compare different distributions more easily than is possible with histograms: the density curve for men is steeper and higher than the one for women. Moreover, the level of log income with the highest density is farther to the right for men than for women. Thus the incomes of women are somewhat more uniformly distributed at the lower end of the income scale, whereas the incomes of men are more concentrated around the average.[11]

Unlike histograms, kernel density estimates are not affected by the selection of the origin. And they do look much nicer than histograms. However, kernel density estimates share one disadvantage with histograms: they *are* affected by the selection of the interval width. A graphical technique that attempts to circumvent those problems is the quantile plot, which is discussed below.

Quantile plot

Quantile plots display marker symbols for each observation in the dataset, using the value of the observation as the y coordinate and the cumulative fraction of observations in the sorted list of values at that point as the x coordinate. The symbols form a curve from which we can infer the shape of the distribution, relative densities, digit preferences, and outliers.

An example will help make things clearer. Take a look at the following data on the birth years of 10 people:

```
. preserve
. use qplot, clear
. list
```

We have sorted the data by birth year, making it easier to explain. The fraction of any observation is 1/10.

11. Be careful in interpreting the graph. It is useful for comparing the *forms* of the curves, but do not attach significance to the *difference* between them, as that can lead to an erroneous interpretation (Cleveland 1994, 227–230).

To begin, look at the lowest value of birth year. This observation has the value $x_1 = 1901$, which is going to be the y coordinate of the first plot symbol. The x coordinate is the cumulative fraction up to this first observation, which is 0.1. The first marker symbol therefore is plotted at the position $(0.1, 1901)$.

Now look at the second observation. This observation has the value $x_2 = 1902$, which again will be the y coordinate. The x coordinate is the cumulative fraction up to the second observation, which is $0.1 + 0.1 = 0.2$. Hence, the coordinate for the second marker symbol is $(0.2, 1902)$.

We could proceed like this for any other value of the birth year, but we should let Stata do the work for us. For that, we need a formula to calculate the x coordinates. We use $(i - 0.5)/n$, with i being the position of observation in the sorted list of values and n the number of observations. Subtracting 0.5 from in the formula is not necessary—and we did not do it above—but is a convention among statisticians (Cleveland 1994, 137).

Here is the Stata command that calculates the formula and stores the results in a variable. If you have read chapter 5, you should be able to use it with no problem:

```
. generate q = (_n - .5)/_N
. list
```

To generate a quantile plot, you graph both quantities in a two-way graph:

```
. scatter gebjahr q, mlabel(gebjahr) mlabposition(12)
```

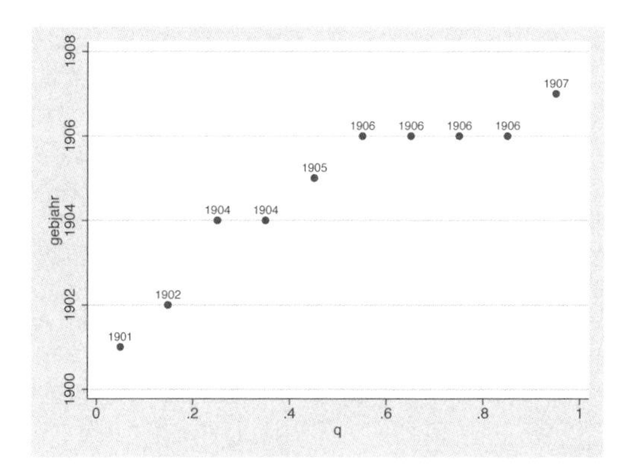

There are four observations for 1906. The marker symbols for these observations form a horizontal line. That may not appear very impressive, but you should remember that equal values always form horizontal lines in quantile plots. Marker symbols that display low slope correspond to data regions with high density. That is, the steeper the curve formed by the marker symbols, the lower is the density of values in the corresponding data.

The power of quantile plots can be demonstrated more compellingly with a larger number of observations. Take the quantile plot of income in our main data as an example. To produce this plot, you can use the command `quantile`, which calculates the x coordinates and displays the graph in one step. Moreover, you can use most of the general options for two-way graphs, including `by()`, to compare distributions between groups.

```
. restore
. quantile income, xline(.25 .5 .75)
```

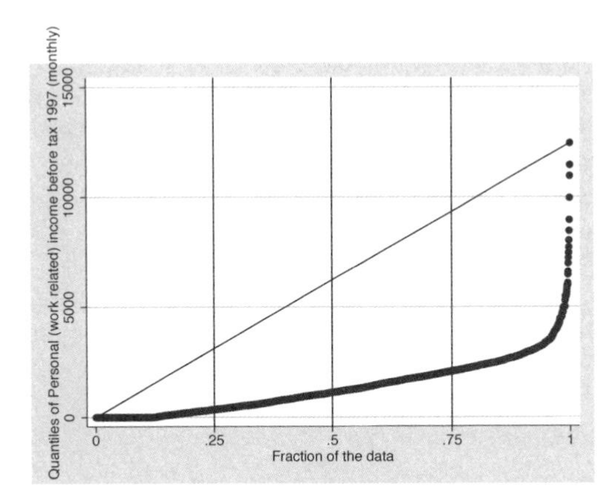

This quantile plot has three different regions. In the first is a horizontal line for zero incomes—we call such entities *digit preferences*. The second, with relatively high density, extends from just above zero up to about $4,000. Above it is a region with low and declining density. As the density is high for low values and low for high values, the distribution is skewed to the right. You can see this also from the fact that all symbols are below—to the right of—the main diagonal. The distribution would be skewed to the left if they had been above the main diagonal.

A main advantage of quantile plots is that they provide a marker symbol for every observation. Single outliers therefore can be quickly identified—like those four respondents who earned more than $10,000 a month.

Finally, you can read off arbitrary quantiles from the quantile plot. The x coordinate is the p-value of the quantile, so the 0.5 quantile—the median—of income is the y-value that corresponds to 0.5 on the x axis. It is about $1,000 here.

Finally, here are two further examples for quantile plots—first, a quantile plot for a uniformly distributed random variable

```
. set seed 731
. generate r1 = runiform()
. quantile r1
```

and second, a quantile plot for a normally distributed random variable:

```
. generate r2 = rnormal()
. quantile r2
```

Comparing distributions with Q–Q plots

A *Q–Q plot*, or *quantile–quantile plot*, is a graphical method for comparing *two* distributions. Q–Q plots are very simple when both distributions have the same number of observations. Then you sort the values of both distributions and then plot the lowest value of the first distribution against the lowest value of the second, and so on. Usually, there are unequal numbers of observations, which is slightly more complicated, but the general idea stays the same. In the first step, you calculate the quantity $(i - .5)/n$ for each observation of the distribution with fewer observations and then calculate the same quantiles for the other distribution. Finally, you plot the lowest value of the smaller dataset against the quantile of the larger dataset that corresponds to the lowest value of the smaller dataset, and so forth. Thus in the case of unequal numbers of observations, there are as many marker symbols as there are values in the smaller of the two datasets.

To compare the income of men and women with a Q–Q plot, you need to generate separate income variables for men and women. You can do this using the `separate` command shown in section 7.3.2. You need this command here only if you did not enter it above.

```
. separate income, by(gender)
```

To generate the Q–Q plot comparing these two distributions, simply enter

```
. qqplot income1 income2
```

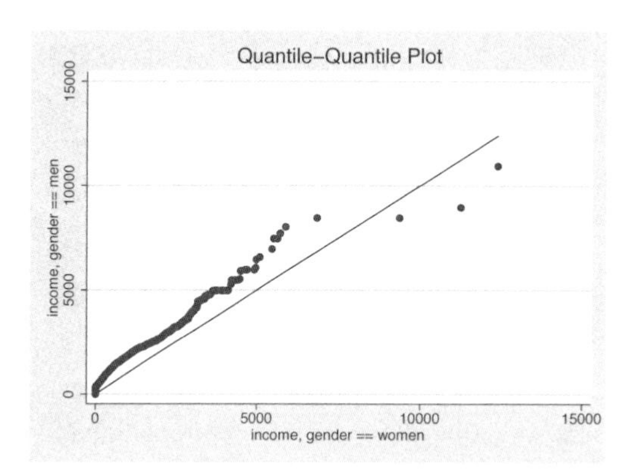

We can see that most data points are to the left of the main diagonal. This means that the values of the variable on the y axis usually are higher than those of the variable on the x axis. Here this means that men earn more than women. We also see that income inequality is somewhat stronger for lower and higher incomes, and somewhat weaker for middle incomes. Moreover, we see that the three women with the highest incomes earn more than the three men with the highest incomes. Finally, we can identify two digit preferences. Women more often earn nothing—or, more precisely, stated that they had zero income on our questionnaire. Men, on the other hand, seem to have an income of $5,000 very often—or at least more frequently than women.

7.4 Exercises

1. Get data from the National Health and Nutrition Examination Study (NHANES) using the following command:

```
. webuse nhanes2, clear
```

   (If you get the error message "`no room to add more observations`", please type `clear`, followed by `set memory 30m` before using the above command; see section 10.6 for details.)

2. Produce a frequency table of health status (`hlthstat`).
3. Produce one-way frequency tables of the following variables with just one command: `psu`, `region`, `smsa`, `sex`, and `race`.
4. Investigate whether men or women in this subsample are of better health using a cross-tabulation of gender and health status.
5. Investigate how health differs between races using cross-tabulation and the chi-square statistic.
6. Produce a table holding the mean weights of observations by race and gender. Do not show any other value than the mean weights.
7. Produce a dot plot for the figures of the previous problem.
8. Produce a histogram of weight with a normal overlay
9. Produce a kernel density plot of weight with a normal curve overlay
10. Compare the distribution of the weights between Whites and Blacks using a Q–Q plot. Do this for men and women separately, and show both plots in one figure.
11. Use a box-and-whisker plot to compare the distribution of the systolic (`bpsystol`) and diastolic (`bpdiast`) blood pressure across race and gender.

8 Introduction to linear regression

In the last chapter, we described the distributions of a few different variables for various subgroups. For example, we compared the distributions of income and political party affiliation for men and women using several techniques. One of those techniques was cross-tabulation, which we used to examine the relative frequencies of votes cast for different groups formed by the values of a second variable—`gender`, in this case (page 146). Applying a different technique to the `income` variable, we compared the distribution of income for men and women using statistics such as means, quantiles, and standard deviations (page 161). In other words, we looked at how income *depends* on gender. Therefore, `income` was our dependent variable, and our *independent* variable was `gender`.

The techniques described in chapter 7 provide a reasonably good representation of your data if you want to compare the distribution of one variable for a few different subgroups formed by a second variable. However, if you are interested in the relationship between two variables with many categories, a scatterplot may be more useful. A scatterplot is a graphical representation of the joint distribution of two variables. When you draw the scatterplot, each observation is plotted in two-dimensional space (along two axes). The coordinates of each point are the values of the variables for that particular observation. The values of the independent variable are graphed on the x axis, whereas the values of the dependent variable are graphed on the y axis.

Three examples[1] of scatterplots can be seen in figure 8.1.

1. The data for these examples are taken from the WHO and UNICEF web sites. Detailed information is provided as notes in the data file (see footnote 11 on page 98). These data are included as Stata datasets in the data package provided with this book.

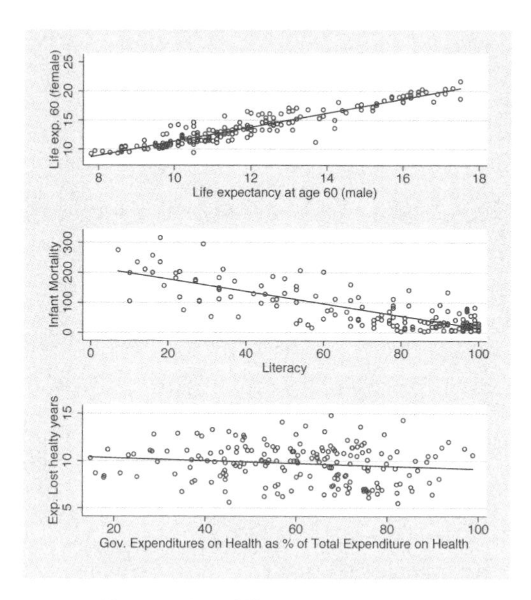

Figure 8.1. Three scatterplots

grscatter.do

The first scatterplot shows data from 188 nations on the life expectancy at age 60 for females plotted against the life expectancy at age 60 for males. The dots are distributed from the lower-left corner to the upper-right corner. This distribution suggests that high life expectancies for males go along with high life expectancies for females. Cases such as these are called *positive relationships*.

The second scatterplot depicts the relationship between infant mortality and female literacy. There we find the data points for 162 nations spreading out from the upper-left corner to the lower-right corner. This means that the higher the female literacy rate in a country is, the lower the observed infant mortality rate. This is called a *negative relationship*.

The third scatterplot shows the relationship between the expected lost healthy years at birth for males and the governmental health expenditures as a percentage of total health expenditures. Here the observations from 190 different countries are distributed fairly evenly over the entire diagram. The relationship between health expenditures and lost healthy years is therefore not obvious. We can, at best, find a weak relationship.

All three graphs contain a solid line that summarizes the relationship between the two variables and is called a *regression line*. In the first scatterplot example, the dots are close to the regression line; there we have a *strong* correlation. In contrast, widely scattered clouds of dots, as in the third example, indicate a *weak* correlation. One way to measure the strength of the correlation is Pearson's correlation coefficient r. A Pearson's correlation coefficient of 0 means that no relationship can be observed between the two variables. Both -1 and $+1$ represent the strongest possible observed relationships, but -1 indicates a negative relationship and $+1$ indicates a positive relationship.

Regardless of the strengths of correlation, there is not necessarily a causal relationship between the variables. The life expectancy of women is not caused by the life expectancy of men. You can instead think of a common cause for both of them. You could hypothesize on the causal link between literacy and infant mortality, but neither scatterplots nor regression lines can test such an assumption (King, Keohane, and Verba 1994; Berk 2004).

Creating scatterplots for different values of r is useful for getting an idea of the relationship. You can practice doing so by using a small demonstration we wrote for Stata.[2] Type

```
. do cplot 0.5
```

and you will see a scatterplot of two variables whose correlation coefficient is 0.5. You can vary the strength of the relationship by changing the number you enter for r after the `do cplot` command.

A simple linear regression analysis aims to characterize the relationship between one dependent variable and one independent variable with a line. A straightforward generalization of this is multiple linear regression analysis, which characterizes the relationship between one dependent and *more than one* independent variables. The term *multivariate regression* is reserved for a technique for more than one *dependent* variables.

We begin by outlining the basic principle behind simple linear regression in section 8.1, though we hold off discussing the relationship between the estimated parameters and the true population parameters in detail until section 8.5. We then extend the model to deal with multiple independent variables in section 8.2. Linear regression analysis requires us to make several assumptions, and section 8.3 introduces several techniques to check those assumptions. Refinements of the basic model are the subject of section 8.4. We then discuss alternative methods of computing standard errors and other extensions of the linear regression model in section 8.6.

Although we will explain some of the statistical background, our main purpose is to show you how to perform regression analysis with Stata. You will need to do more reading to gain a full understanding of regression analysis. Books that work well with our approach are Hamilton (1992) and Fox (1997). You should also read Berk (2004) for a critical discussion of common mistakes.

2. Make sure that your current working directory is `c:\data\kk2`; see page 3.

8.1 Simple linear regression

8.1.1 The basic principle

Here we will introduce terms such as *OLS*, *RSS*, *predicted values*, and *regression*. If you are already familiar with these terms, you may skip this section.

The basic principle of all regression models is straightforward. To describe the relationship between your variables, you are looking for an equation that allows you to predict the values of a *dependent* variable as well as possible with the help of one or more *independent* variables. Consider the following example.

You have a hunch that the size of someone's dwelling is determined by his or her net income. You believe that the higher someone's income is, the larger the home will be. At the same time, you know that apartments are not of zero size if someone's income is zero. Hence, apartments should have a certain minimum size. You could formalize your suspicion about the relationship between income and home size with the aid of a simple equation:

$$\text{home size}_{\text{Lopez}} = \beta_0 + \beta_1 \cdot \text{income}_{\text{Lopez}} + \epsilon_{\text{Lopez}}$$

or you use symbols instead of text:

$$y_{\text{Lopez}} = \beta_0 + \beta_1 x_{\text{Lopez}} + \epsilon_{\text{Lopez}} \tag{8.1}$$

where y and x could be symbols for any variable. In what follows we will consistently use y for home size and x for income, however.

Our equation calculates the home size of the Lopez family, where the Lopez family is an arbitrary household in our dataset. The first term, β_0, is the apartment size when the household income is zero. To this term we add the term $\beta_1 x_{\text{Lopez}}$. The crucial part here is the parameter β_1 which tells us how many square feet each additional dollar of household income can buy. If you multiply β_1 with the actual income of the Lopez family, you will get an estimate on how much larger the Lopez family home is compared with the minimum home size.

Now you might argue that income is not the only variable that affects home size. For example, family size or the ages of family members might play a role, as well. You may, in fact, come up with several factors that might affect home size. If you do not know all the factors, the home size, y, your calculation using the above equation will always deviate from the observed values. This deviation is called the *error term*. In (8.1) the error term is indicated by ϵ_{Lopez} for the Lopez family.

Equation (8.1) above is just another way to write down the hunch that an individual's home size depends on a minimum apartment size, some quantity that rises with income and some other factors. In practice we do not know the value of those parameters. We can only speculate how many square feet that minimum home size β_0 is and how large that quantity β_1 is that must be added to the home size with each additional dollar of income. Estimating those regression parameters (as β_0 and β_1 are often called) is the aim of regression analysis.

Let us assume you would, based on your past housing experience, make a rough estimate for the minimum home size and the effect of income.[3] Based on those estimates, you could go ahead and predict the square footage of other peoples places knowing nothing about those people but their income. To do so, you would replace x_1 with their respective income in the formula

$$\widehat{y}_i = b_0 + b_1 \times x_i \tag{8.2}$$

where b_0 and b_1 symbolize your estimates for the parameters β_0 and β_1. (In statistics Greek symbols commonly refer to unknown parameters and roman letters indicate their estimates that one has derived from empirical sources.) The subscript i indicates that we are now dealing with values from several households not just the Lopez family. The estimated regression coefficients b_0 and b_1 do not have a subscript. They are constant across individuals. Also compared with (8.1) the error term ϵ is missing, which means we ignore all factors besides income that could influence square footage. Instead the predicted home size $\widehat{y}_i$ in (8.2) has a *hat* over y_i, symbolizing that it is an estimate.

Since you did not include any of the other factors beyond income that could influence the households' housing choices, the home sizes you predicted using the above equation will deviate from the actual values of these peoples' homes. If we know an individual's actual home size, y_i, and the predicted size of her home, $\widehat{y}_i$, we can compute the difference:

$$e_i = y_i - \widehat{y}_i \tag{8.3}$$

This deviation e_i between the actual value and the predicated value is called the *residual*. Looking at the equations we can say that our prediction will get better the smaller e_i is. Likewise we could say that—given that our model is correct—the closer our estimated coefficients are to the true values of the parameters, the smaller e_i tends to get. Hence, it is reasonable to replace simple guessing based on day-to-day experience with statistical techniques. You can use several techniques to get estimates for the regression parameters. Here we will limit ourselves to the one that is the simplest and in wide use: ordinary least squares (OLS).

OLS is based on a simple rule: make the difference between the predicted and observed values as small as possible. To put it differently, we want to choose the parameters that minimize the sum of the squared residuals. Thinking back on what we said earlier, that the error term in (8.1) includes all other effects that influence home size, it might seem funny to you that we try to find values for b_0 and b_1 that minimizes the residuals. After all, there could be other important variables like household size, that we do not include in our model. This is true! Trying to minimize the residuals implicitly makes the assumption that the error term is zero in expectation. We will come back to this assumption in section 8.3.1.

3. Maybe you would say that houses are at least 600 square feet and people would spend roughly one third of their net income on housing. If one square foot costs \$5, people could afford an additional square foot with every \$15 of additional income, or, in other words, each additional dollar of income would increase the flat size by $1/15$ ft^2.

To understand what minimizing the sum of the squared residuals means, look at the scatterplot in figure 8.2. Try to find a line that depicts the relationship between the two variables. You will find that not all points lie on one line. You might try to draw a line among the points so that the vertical distances between the points and the line are as small as possible. To find these distances, you might use a ruler.

The goal was to minimize the differences across all points, so looking at one of the distances will not provide you with enough information to choose the best line. What else can you do? You could try adding the distances up for all the points. If you did this, you would notice that negative and positive distances could cancel each other out. To find a way around this problem, you might use the squared distances instead.

If you drew several lines and measured the distances between the points and every new line, the line with the smallest sum of squared distances would be the one that reflects the relationship the best. This search for the line with the best fit is the idea behind the OLS estimation technique: it minimizes the sum of squared residuals (e_i^2). The points on the line represent the predicted values of ($\widehat{y}_i$) for all values of X. If your model fits the data well, all points will be close to the straight line and the sum of the squared residuals will be small. If your model does not fit the data well, the points will be spread out and the sum of the squared residuals will be relatively large.

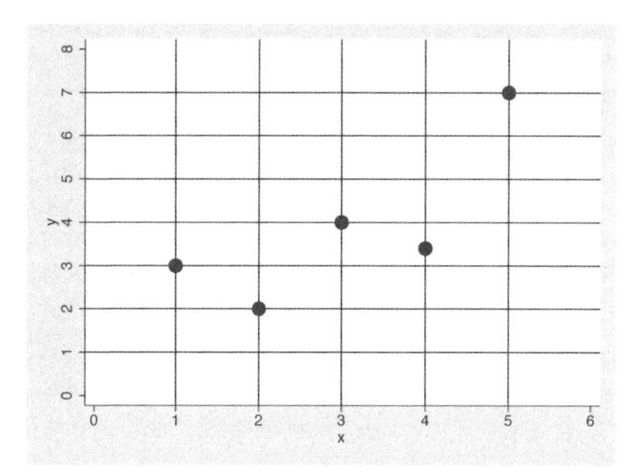

Figure 8.2. Exercise for the OLS principle

grreg1.do

We have prepared a small demonstration of the OLS solution to the regression problem in figure 8.2. Typing

```
. do grreg2.do
```

causes figure 8.2 to be displayed with the regression line.

We can also present the OLS principle a bit more formally. We are looking for those parameters (b_0, b_1) in (8.3) for which the sum of the squared residuals (the residual sum of squares, [RSS]) is at a minimum. Those parameters are the y axis intercepts and the slopes of the lines we drew. A search for the best fit using a trial-and-error technique like the one described above would be time consuming. Using mathematical techniques to minimize the RSS is an easier way to find our parameters that more reliably leads to the correct solution. Mathematically, the RSS can be written as the difference between the observed and predicted values:

$$\text{RSS} = \sum_{i=1}^{n} e_i^2 = \sum_{i=1}^{n} (y_i - \widehat{y}_i)^2 \tag{8.4}$$

Substituting for $\widehat{y}_i$, we can write the above equation as

$$\text{RSS} = \sum e_i^2 = \sum (y_i - b_0 - b_1 x_i)^2 \tag{8.5}$$

Now that we have defined the RSS mathematically, we can use the OLS technique to minimize it.[4] This means that we must find values for b_0 and b_1 for which (8.5) is as small as possible. To do this, we can take the first partial derivatives of (8.5) with respect to b_0 and b_1, set them equal to zero, and solve for b_0 and b_1. At this point, it is not particularly important that you be able to take the derivative yourself. The entire technique is nothing more than a search for the minimum of a function with two unknowns.

If, on the other hand, you wish to review the high school and college math necessary for taking partial derivatives, you can find a helpful review in Hagle (1996, 38–58).[5]

---

4. The exact mathematical procedure for this technique has been presented in several different ways. For fans of a graphic interpretation, we recommend starting with Cook and Weisberg (1999) or Hamilton (1992).
5. To reconstruct the transformations used in finding values for b_0 and b_1 for which RSS is at a minimum, you can do so as follows:

$$\frac{\partial \text{RSS}}{\partial b_0} = -2\sum y_i + 2nb_0 + 2nb_1 \sum x_i$$

If you set this partial derivative equal to zero and solve for b_0, you will get

$$b_0 = \overline{y} - b_1\overline{x}$$

Following the same principle, you can find the first partial derivative with respect to b_1:

$$\frac{\partial \text{RSS}}{\partial b_1} = -2\sum y_i x_i + 2b_0 \sum x_i + 2b_1 \sum x_i^2 = 0$$

Now you replace b_0 with $\overline{y} - b_1\overline{x}$. After a few transformations, you end up with

$$b_1 = \frac{\sum (x_i - \overline{x})(y_i - \overline{y})}{\sum (x_i - \overline{x})^2}$$

You can find a more detailed presentation of this derivation in Hamilton (1992, 33).

Before we continue with the mathematics, we will show you how to compute a regression with Stata, and you will see how easy and helpful it is to use statistical packages for these kinds of computations. But be careful: despite the simplicity of the computational work, you must always think carefully about what exactly you are doing. In this chapter, we will look at substantive problems caused by naively applying these regression techniques.

8.1.2 Linear regression using Stata

Here we will explain how to fit a linear regression model with Stata. In the previous subsection, we voiced a suspicion that home size is influenced by net household income. You might now be interested in a specification of this relationship. A good place to begin would be with a linear regression of home size (`sqfeet`) on net household income (`hhinc`). The Stata command you will need to perform your regression is pretty simple:

```
. use data1, clear
(SOEP'97 (Kohler/Kreuter))

. regress sqfeet hhinc
      Source |       SS       df       MS              Number of obs =    3126
-------------+------------------------------           F(  1,  3124) =  694.26
       Model |  114301950     1   114301950            Prob > F      =  0.0000
    Residual |  514327350  3124  164637.436            R-squared     =  0.1818
-------------+------------------------------           Adj R-squared =  0.1816
       Total |  628629300  3125  201161.376            Root MSE      =  405.76

------------------------------------------------------------------------------
      sqfeet |      Coef.   Std. Err.      t    P>|t|     [95% Conf. Interval]
-------------+----------------------------------------------------------------
       hhinc |   .1786114   .0067787    26.35   0.000     .1653202    .1919025
       _cons |   600.2684   14.91459    40.25   0.000      571.025    629.5118
------------------------------------------------------------------------------
```

Clearly the command consists of the `regress` statement and a list of variables. The first variable is the dependent variable, and the second is the independent variable. The output contains three different sections: the "table of ANOVA results" in the upper-left corner, the "model fit table" in the upper-right corner, and the "table of coefficients" in the bottom half of the output.

The table of coefficients

At the bottom of the table in the column labeled `Coef.`, you will find the estimated regression coefficients, that is, the values for b_0 and b_1, from (8.3).

To the right of the estimated regression coefficients are several other statistics used to evaluate the accuracy of those estimates. We discuss those after we show how to interpret the estimated coefficients and make predictions.

The value for b_0 is written in the regression-output row labeled `_cons`. b_0 is 600.2684 in this example. According to this model, every family has a home that is at least 600 ft^2, regardless of income. The value for b_1 is stored in the row that begins with `hhinc` and is about 0.1786114. According to the regression model, the home size will increase by about 0.18 ft^2 with every additional dollar of income.

Assuming that the Lopez family has a net monthly income of $1,748 at its disposal, you can use (8.1) to estimate how big the family's home might be

$$\widehat{y}_{\text{Lopez}} = 600.2684 + 0.1786114 \cdot \$1,748$$

You can calculate this amount directly within Stata using the `display` command, much as you would use a pocket calculator. Type

```
. display 600.2684 + 0.1786114 * 1748
912.48113
```

If you use the numbers displayed in the table of coefficients, you must deal with two problems: first, typing numbers by hand often leads to mistakes. Also the figures in the output have been rounded. For computations like the one above, we recommend using the results saved internally by Stata (see chapter 4). Commands that fit regression models are considered to be e-class in Stata, so you can look at the saved results with the command `ereturn list`. If you do this, you might find yourself searching in vain for the estimated regression coefficients because they are all stored in a matrix named `e(b)`. The easiest way to access the values contained in this matrix is to use the construction `_b[`*varname*`]`, where *varname* is replaced by the name of either an independent variable or the constant (`_cons`).

The computation for the Lopez family would then look like this:

```
. display _b[_cons]+_b[hhinc]*1748
912.48107
```

This number differs a bit from the number we computed above because the results saved by Stata are accurate to about the 16th decimal place. You can see the effect of raising income by $1 on home size. If you enter $1749 instead of $1748 as the value for income, you will see that the predicted value for home size increases by $b_1 = b_{\texttt{hhinc}} = 0.1786114\,\text{ft}^2$.

You might not be interested in an estimated home size for a family with a certain income but in the actual home sizes of all families in our data who have that income. To see the sizes of the homes of all the families with a net household income of $1748, you could use the following command:

```
. list sqfeet hhinc if hhinc==1748
```

As you can see here, the predicted home size of 912 ft^2 is not displayed, but various values between 474 ft^2 and 1227 ft^2 appear instead. The observed values of y_i differ from the predicted values, $\widehat{y}_i$. These differences are the residuals.

If you want to compute the predicted values for every household in your dataset, you could use the saved estimates of the regression coefficients.[6] To compute the predicted values this way, you would type[7]

```
. generate sqfeethat=_b[_cons]+_b[hhinc]*hhinc
```

This is the same principle that was used in the previous `display` command, except that the home size is predicted not only for the Lopez family but for all families. The result of this computation is stored in the `sqfeethat` variable. We use the suffix `hat` to indicate that this is a predicted variable.[8]

There is an easier and better way to get the same result: the `predict` command computes the predicted values after each regression command and stores them in a variable. If you type

```
. predict yhat1
```

Stata will store the predicted values into the new variable `yhat1`,[9] which contains the same values as the `sqfeethat` variable. If you want to convince yourself that this is the case, type `list sqfeethat yhat1`. Because it is used after estimation, the `predict` command is called a *postestimation* command.

If you have already calculated the predicted values, it is easy to calculate the values of the residuals. They are just the differences between the observed and predicted values:

```
. generate resid1=sqfeet-sqfeethat
```

This difference is nothing more than the distance you measured between each point and the line in the figure on page 186.

You can also compute the residuals by using the `predict` postestimation command with the `residuals` option and specifying a variable name (here, `resid2`):[10]

```
. predict resid2, resid
```

Let us use a graph to look at the results. We might want to draw a scatterplot with `sqfeet` against `hhinc`, overlaid by a line plot of the predicted values (`yhat1`) against `hhinc`.

---

6. You can use the saved regression coefficients anywhere Stata expects an *expression*; see section 3.1.5.
7. After entering this command, you will get a warning that some missing values have been generated. Those missing values are for all the families for whom the dataset contains no income information.
8. We placed a "hat" (circumflex) on y in the above equations to indicate a predicted value ($\widehat{y}$), as opposed to a value actually measured for a certain family (y).
9. In each subsection of this chapter, we perform a separate computation of the predicted values, meaning that we compute variables with predicted values several times. We use the name `yhat` with a running number for each of these variables. This running number has no meaning but is simply used to denote a new computation.
10. Please resist the temptation to set `e` as a name for the residuals. The name `e` may, in principle, be a valid variable name, but using it might lead to confusion if scientific notation is used for numbers. See section 5.1.1 for a list of variable names you should avoid.

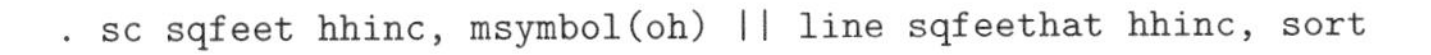

```
. sc sqfeet hhinc, msymbol(oh) || line sqfeethat hhinc, sort
```

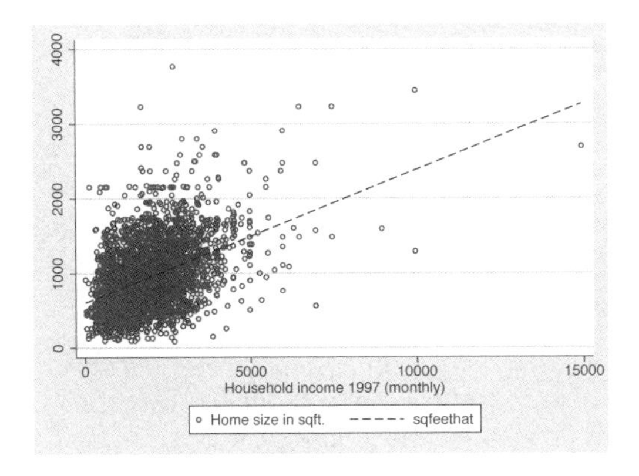

Standard errors

Thus far, we have treated the estimated coefficients of our regression model as if they have been determined without any level of uncertainty. However, your dataset is typically only a random sample from a large population. The estimated coefficients that you obtain from `regress` are therefore only estimates of the values that describe the entire population. If you could collect a second random sample from the same population, you would obtain different estimates. We therefore need a way to describe the variability that we would obtain if we were to apply our estimator to many different samples. Said slightly differently, we need a way to determine the precision with which the coefficients we obtained from our sample estimate the population parameters. Standard errors, which are really nothing more than sample standard deviations associated with estimated coefficients, are our solution.

Standard errors allow us to perform statistical inference; that is, they allow us to test hypotheses about the underlying population parameters. For example, we might want to test whether the parameter on a given independent variable is zero, which means that the variable has no impact on the dependent variable.

Techniques for statistical inference are frequently misused. These techniques are based on a series of assumptions, which tend to be fulfilled only in high-quality samples. In the context of linear regression, the default statistical inference techniques assume a simple random sample.[11] In addition the errors are assumed to be uncorrelated and homoskedastic. The confidence intervals are based on the assumption of normally distributed errors.[12] You will have to evaluate whether these assumptions hold. In section 8.3, we present several techniques that can be used to check those assumptions.

11. For complex samples see section 8.5.2.
12. An assumption that is likely to hold with moderately sized to large samples even in the nonnormal cases. Most introductory statistics textbooks will cover the central limit theorem that underlies this statement (see, for example, Moore and McCabe [2005]).

Here we give a few illustrations of the techniques for statistical inference. First, recompute the regression model we fitted earlier to analyze home size:

```
. regress sqfeet hhinc
```

On the right side of the table of coefficients, you see for each estimated coefficient its 95% confidence interval boundaries; for household income, these are 0.165 and 0.192. When you think of confidence intervals, remember that if we were to draw many random samples out of the population and for each one estimate the regression coefficient and the corresponding confidence intervals around this coefficient, then 95% of all intervals would contain the true coefficient for the population.[13]

For a quick check, you can determine whether the value zero is included in the confidence limits. If so, you can assume that the corresponding independent variable has no influence on the dependent variable in the population. Often you will see the t value used to determine the significance (statistical, not substantive) of an estimated coefficient. With the help of the t distribution, this significance test tells you how likely it is that you would observe a value at least as extreme as the particular estimated coefficient under the assumption that the "true" coefficient in the population is *zero* (null hypothesis). The probability of observing a t-value at least as large in absolute value as the one shown with a given sample size under the null hypothesis is shown in column $P > |t|$. A small value (e.g., smaller than 0.05) in this column tells you only that it is unlikely that you will observe a value like the one you would estimate if the true coefficient in the population were *zero*. This means that your test is based on the hypothesis that household income has *no* influence at all on home size—a statement that you probably would not dare to make. You should therefore keep track of the confidence intervals and the effect size itself.

To compute the 95% confidence interval, you add or subtract roughly 1.96 times the standard error to the estimated regression coefficient.[14]

The table of ANOVA results

ANOVA is short for analysis of variance. We use the term "table of ANOVA results" to describe the upper-left section of the Stata regression output, where you will find the variation in the dependent variable divided into an explained portion and an unexplained portion. For handy reference, we reproduced the table of ANOVA results that you have already seen on page 188:

13. This does *not* mean that the true value is between the interval limits with a probability of 0.95.
14. The exact value varies with the sample size; however, 1.96 is a good approximation for a sample size above 30.

```
      Source |       SS       df       MS
-------------+------------------------------
       Model |  114301950     1   114301950
    Residual |  514327350  3124  164637.436
-------------+------------------------------
       Total |  628629300  3125  201161.376
```

We can learn a bit more about the table of ANOVA results by using a fictional example. Say that you are asked to predict the size of an apartment belonging to a student named Paul. If you do not know anything about Paul, you might answer that his apartment is as big as the average student apartment. Here your guess would be reasonable because the average apartment size is the value with which you get the smallest squared error.

Table 8.1 lists the apartment sizes and household sizes of three students in a hypothetical city. The average student apartment size in that city is 590 ft^2, which we calculated using data for all the students in the city, not just the ones listed here.[15]

Table 8.1. Apartment and household size

	Apt. size	City	Diff.	HH size	Estim.	Residual
Paul	430	590	−160	1	480	−50
John	590	590	0	2	640	−50
Ringo	860	590	+270	3	800	+60

If you use 590 ft^2 to estimate the size of Paul's apartment, you end up with a number that is 160 ft^2 too high.[16] If you use the mean to estimate the other students' apartment sizes, then in one case you make a correct prediction and in the other case you underestimate the student's apartment size by 270 ft^2. If you take the squares of these differences and sum them, the result is a total squared deviation of 98,500 ft^4. This number is usually called the *total sum of squares* (TSS). In general,

$$\text{TSS} = \sum (y_i - \overline{y})^2$$

This corresponds to the expression you find in the numerator of the formula for the variance (s^2). The TSS is therefore sometimes also called the *variation*.

Maybe you should not make your prediction using only the mean. You might wish to make use of other information you have about the students. After all, it is reasonable to assume that the size of the apartment increases with the number of people living there. If all the students you know have bedrooms that are about 160 ft^2, you might think

15. We got the idea of using a table like this one from Hair et al. (1995).

16. In table 8.1, the difference between the observed value and the predicted mean is calculated as follows: 430 − 590 = −160.

this number holds true for most other students. So the apartment would have to have at least 160 ft^2 for each of the students living there, but it is likely to be even larger. An apartment usually has at least one bathroom and a kitchen, and you might think that together they take up about 320 ft^2. You might describe this hypothesis using the equation below:

$$y_i = 320 + 160x_i$$

You could use that model to compute an apartment size for each household size. If you did this, you would calculate the difference between the actual apartment size and the apartment size you predicted with your model; this is the amount displayed in the last column of the table. To compare these differences with the TSS we calculated above, you would have to square these deviations and sum them. If you did this, you would have calculated the residual sum of squares (RSS) we introduced in section 8.1.1. For your hypothesis, the value of RSS is 8,600.

If you subtract the RSS from the TSS, you get the model sum of squares (MSS), which indicates how much you have been able to improve your estimation by using your hypothesis:

$$\begin{array}{rcr} \text{TSS} & = & 98{,}500 \\ -\text{RSS} & = & 8{,}600 \\ \hline = \text{MSS} & = & 89{,}900 \end{array}$$

The squared residuals that you get when you use household size to predict apartment size are about 89,900 smaller than the ones you got without taking this knowledge into account. That means that the actual apartment sizes are much closer to your predicted values when you use household size in making your prediction.

Therefore, the MSS can be regarded as a baseline to measure the quality of our model. The higher the MSS, the better are your predictions compared with the prediction based solely on the mean. The mean can be regarded as the standard against which to judge the quality of your prediction.

In the ANOVA part of the regression output, you will find information about the MSS, RSS, and TSS in the column labeled `SS`. The first row of numbers (`Model`) describes the MSS, the second (`Residual`) describes the RSS, and the third (`Total`) describes the TSS. If you look at the output on page 193, you will see that our RSS is 514327350. The sum of the squared residuals taking the mean as the estimate (TSS) is 628629300, and the difference between these two quantities (MSS) is 114301950.

The column labeled `df` contains the number of degrees of freedom,[17] which equals the number of unknowns that can vary freely. For the MSS, the number of degrees of freedom is just the number of independent variables included in the model, that is, $k-1$, where k is the number of regression coefficients (the constant and all independent variables). The number of degrees of freedom for the RSS is $n-k$, where n is the number

17. For a well-written explanation of the concept of degrees of freedom, see Howell (1997, 53).

of observations. The number of degrees of freedom for the TSS is $n-1$. The last column contains the average sum of squares (MS). You may want to compute these numbers yourself by dividing the first column by the second column (the number of degrees of freedom).

The model fit table

Here is the model fit table from page 188:

```
Number of obs =    3126
F( 1,   3124) =  694.26
Prob > F      =  0.0000
R-squared     =  0.1818
Adj R-squared =  0.1816
Root MSE      =  405.76
```

Earlier, we showed that the MSS tells you how much the sum of the squared residuals decreases when you add independent variables to the model. If you were looking at models with different independent variables, you might want to compare the explanatory power of those models using the MSS. You could not, however, use the *absolute* value of the MSS to do so. That value depends not only on the quality of the model but also on how much variation there was in the first place as measured by TSS.

To compare models, you must look at how much the model reduces the squared residuals relative to the total amount of squared residuals. You can do this using the coefficient of determination, or R^2:

$$R^2 = \frac{\text{MSS}}{\text{TSS}} = 1 - \frac{\text{RSS}}{\text{TSS}} = 1 - \frac{\sum e_i^2}{\sum (y_i - \overline{y})^2}$$

R^2 represents the squared residuals that are explained by the model as a share of the total squared residuals. When we say that the model explains a portion of the residuals, we mean that portion of the residuals of the model without independent variables that disappears when we use a model *with* independent variables. For this reason, R^2 is called the *explained variation* or the *explained variance*. You will find this statistic in the model fit table of the Stata output, where it is called `R-squared`.

Here $R^2 = 0.1818$, meaning that household size (the independent variable in our model) explains 18% of the variation in apartment size.

R^2 is a useful indicator of a model's explanatory power, but it should not be considered in isolation. Unfortunately, people often evaluate the quality of a regression model only by looking at the size of R^2, which is not only invalid but also dangerous. In section 8.3, we will show you why.

One alternative to R^2 is the root MSE, which is the square root of the average residual of the model from the table of ANOVA results:

$$\text{root MSE} = \sqrt{\frac{\text{RSS}}{n-k}}$$

This statistic is easy to interpret, as it has the same units as the dependent variable. In our example, a root MSE of 405.76 can be interpreted as showing that we are, on average for our data, about 406 ft^2 off the mark in predicting a respondent's apartment size with our model. (This interpretation is not completely correct since it is not a literal average. After all, $\sqrt{\sum e_i^2} \neq \sum e_i$. But the above interpretation seems justified to us.)

There are two rows of the model fit table that we still haven't talked about: the rows labeled "F(1, 3124)" and "Prob > F". The values in these rows are included because we are using a sample to test our regression model and therefore want some measure of its significance.[18] The F value is calculated using the following equation:

$$F = \frac{\text{MSS}/(k-1)}{\text{RSS}/(n-k)}$$

This F statistic is the ratio of the two values in the third column of the ANOVA table. It is F distributed and forms the basis of a significance test for R^2. The value of F is used to test the hypothesis that the R^2 we estimated from our sample data is significantly different from the population value of zero.[19] That is, you want to estimate the probability of observing the reduction in RSS in the model if, in fact, the independent variables in the model have no explanatory power.[20] The value listed for "Prob > F" gives the probability that the R^2 we estimated with our sample data will be observed if the value of R^2 in the population is actually equal to zero.

8.2 Multiple regression

Load `data1.dta` into working memory:

```
. use data1, clear
```

Earlier, we introduced linear regression with one independent variable. A multiple regression is an extension of the simple linear regression presented in that section. Unlike simple regression, you can use several independent variables in a multiple regression. Analogous to (8.1), the model equation of the multiple linear regression is

$$y_i = \beta_0 + \beta_1 x_{1i} + \beta_2 x_{2i} + \cdots + \beta_{K-1} x_{K-1,i} + \epsilon_i \tag{8.6}$$

The equation for the simple linear regression has been extended with more X variables and the attendant regression coefficients. You might want to use a model like this for two reasons.

18. For more about the technical term "significance", see section 8.5.
19. A description of this relationship can be found in Gujarati (1995, 244–250).
20. This F test is often called a test of the null hypothesis—that all coefficients but the constant are zero (Gujarati 1995, 247). Incidentally, the confidence intervals might *not* contain the value zero, but the overall model may nevertheless not be significant.

In section 8.1.2, you fit a simple linear regression model of the dwelling size on household income. You were able to explain 18% of the variation in apartment size with this regression, and the average error in predicting apartment size was 406 ft^2. If you want to maximize the predictive power of our model, there is no reason to be satisfied with the performance of this simple model. You could improve the predictive power of our model by including other variables. This would be the primary reason for using a regression with more than one independent variable.

A second reason is a bit more complicated. In one of the previous sections, we used household income as an independent variable. Suppose that you also wanted to allow for the effect of household size. You have, however, reason to assume that household income is related to the size of the household, as more family members might contribute to the overall household income. At the same time, it is reasonable to assume that households with more members need more space than those with fewer members. Thus the regression coefficient that we estimate for household income may already include the effect of household size. In cases like this, the estimated regression coefficient on household income is said to be "biased". You might try to combat this bias by including more variables in the model.

We will show you how to fit a multiple linear regression model in Stata and then interpret the estimated regression coefficients. Then we will present some computations that are specific to this kind of regression. Finally, we will illustrate what is meant by the formal phrase "controlling for" when it is used for the interpretation of coefficients in multiple regression models (section 8.2.3).

8.2.1 Multiple regression using Stata

The Stata command for computing a multiple regression is the same as that for simple linear regression. You just enter more independent variables at the end of the list of variables; the order in which you enter them does not matter. You can apply the general rules for lists of variables (page 43), but remember that the dependent variable is always the first one in your list.

The output for the multiple linear regression resembles the one for a simple linear regression, except that, for each additional independent variable, you get one more row for the corresponding estimated coefficient. Finally, you obtain the predicted values using the `predict` command as you did earlier.

For example, say that you want to fit a regression model of dwelling size that contains not only household size and household income but also a location variable for the difference between East and West Germany and an ownership variable indicating owned and rented living space. To do this, you will need to recode some of the variables:[21]

```
. generate owner = renttype == 1 if renttype < .
. generate east = state >= 11 & state <= 16 if state < .
```

21. See chapter 5 if you have any problems with these commands.

Now you can fit the regression model:

```
. regress sqfeet hhinc hhsize east owner

      Source |       SS       df       MS              Number of obs =    3125
-------------+------------------------------           F(  4,  3120) =  442.09
       Model |  227333588     4    56833397            Prob > F      =  0.0000
    Residual |  401094539  3120  128555.942            R-squared     =  0.3617
-------------+------------------------------           Adj R-squared =  0.3609
       Total |  628428127  3124  201161.372            Root MSE      =  358.55

------------------------------------------------------------------------------
      sqfeet |      Coef.   Std. Err.      t    P>|t|     [95% Conf. Interval]
-------------+----------------------------------------------------------------
       hhinc |   .1168247   .0064601    18.08   0.000     .1041582    .1294912
      hhsize |   32.76697   5.185336     6.32   0.000     22.59996    42.93399
        east |  -99.99731    14.2274    -7.03   0.000    -127.8933   -72.1013
       owner |   383.6225   13.89444    27.61   0.000     356.3793    410.8657
       _cons |   524.1375   17.36074    30.19   0.000     490.0979    558.1771
------------------------------------------------------------------------------
```

The number of observations has decreased from 3,126 to 3,125 because of missing values for the state variable. Observations that have a missing value in any of the variables are dropped when you fit the model. This is called "casewise deletion" of missing values. Type `search impute` to learn about other ways of dealing with missing values.

You interpret the estimated coefficients in a multiple regression model just as you do in the simple linear regression. The only difference is that the b coefficients are now estimated *controlling for* the effect of all the other independent variables. We will discuss the meaning of that phrase in section 8.2.3. For now we will confine ourselves to once again illustrating the formal interpretation of the coefficients.

The regression coefficients reflect the average change in the size of the dwelling as the independent variable in question increases by one unit, holding all other independent variables constant. The estimated coefficient might, for example, be interpreted as saying that "with each additional dollar of household income, the predicted size of the dwelling increases by an average of about 0.117 ft^2". Similarly, the predicted dwelling size increases by an average of about 32.77 ft^2 for each additional person in the household.

The variables `east` and `owner` are dummy variables, or variables that have only two categories, denoted by the values 0 and 1.[22] In principle, you interpret these variables just as you interpret all the other variables. For example, let us look at the `owner` variable, which has a value of 0 for all renters and 1 for all owners: for each unit by which the `owner` variable increases, the dwelling increases by an average of about 384 ft^2. Since a dummy variable can be increased by one unit only once, we could also say, "Owners live in dwellings that are, on average, about 384 ft^2 larger than the ones in which renters live." Likewise, the dwellings in East Germany are, on average, around 100 ft^2 *smaller* than the dwellings in West Germany.

22. There are other possibilities for coding binary variables (Aiken and West 1991, 127–130).

The regression constant indicates how large a dwelling is whose observation has a value of 0 for all variables included in the model. This value would refer to dwelling size for western households with no household income and no household members. This is clearly useless information. A more sensible interpretation for the constant can be reached by subtracting from all values of a continuous variable the average value of that variable. In doing so that new *centered* version of the variables will have a mean of 0, and the constant term of the regression will then refer to observations with the average value of the continuous variables. If we center the two continuous variables of our last example by applying the methods described in chapter 4 and using the following commands

```
. summarize hhsize
. generate c_hhsize = hhsize - r(mean)
. summarize hhinc
. generate c_hhinc = hhinc - r(mean)
. regress sqfeet c_hhinc c_hhsize east
```

we notice that the results indicate the predicted flat size of respondents from West Germany with an average income and an average household size is 973 ft^2.

8.2.2 More computations

Adjusted R²

In adding the two dummy variables and the household size variable to our regression model, you have increased R^2 from 18% to 36%. This is an obvious improvement in the explanatory power of our model, but you need to put this improvement in perspective: R^2 almost always increases if you add variables to the model.[23] The effect of these additional variables on R^2 is offset by the effect of additional observations. Having more observations tends to result in a lower R^2 than you would obtain fitting the same model with fewer observations. You can safeguard against misleading increases in R^2 by making sure that you have enough observations to test your model. In the example above, the ratio between observations and independent variables that was used in the model is quite favorable. However, if you intend to work with only a small number of observations (e.g., if your dataset comprises country-level information for European countries) and you use many independent variables, R^2 will quickly become an unreliable measure.[24]

Perhaps it will be easier to understand why a small number of observations leads to a higher R^2 if you imagine a scatterplot with two points. These two points can be easily connected by a line, which is the regression line. Now you have explained all the variance, as there are no distances left between either of the points and the line. But does this mean that the two variables for which you made the scatterplot are really related to each other? Not necessarily. Imagine that you plotted the gross national products of Great Britain and Germany against the lengths of their coasts and drew a

23. The only situation in which R^2 does not increase is when the coefficient of the additional variable is exactly equal to zero. In practice, this case is almost never observed.
24. You will find a list of problems related to the use of R^2 in Kennedy (1997, 26–28).

regression line. You would be able to explain the difference between the gross national products of Germany and Great Britain "perfectly"; at the same time, you would be forced to leave the scientific community.

Given the effects of the number of observations and the number of independent variables on R^2, you may want a more meaningful measure of your model's explanatory power. The adjusted R^2 (`Adj R-squared`) results from a correction that accounts for the number of model parameters, k (everything on the right side of your equation), and the number of observations (Greene 2008, 35)

$$R_a^2 = 1 - \frac{n-1}{n-k}(1 - R^2)$$

where k is the number of parameters and n is the number of observations. As long as the number of observations is sufficiently large, the adjusted R^2 will be close to R^2.

Standardized regression coefficients

In our regression model, the estimated coefficient for household size is much larger than the one for household income. If you look only at the absolute size of the estimated coefficients, you might be tempted to assume that the household size has a larger influence on dwelling size than does household income. But you will recognize that the estimated coefficients reflect how much a dependent variable changes if the independent variable is changed by one unit. Thus you are comparing the change in dwelling size if household income increases by $1 with the change in dwelling size if the size of the household increases by one person!

To compare the effects of variables measured in different units, you will often use the standardized form of the estimated regression coefficients (b_k^*), which are calculated as follows

$$b_k^* = b_k \frac{s_{X_k}}{s_Y} \tag{8.7}$$

where b_k is the estimated coefficient of the kth variable, s_Y is the standard deviation of the dependent variable, and s_{X_k} is the standard deviation of the kth independent variable.

The standardized estimated regression coefficients are often called beta coefficients, which is why you use the `beta` option to look at them. If you want to reexamine the estimated coefficients of your last model in standardized form, you can redisplay the results (with no recalculation) by typing `regress, beta`. If you do this, you end up with values for `beta` in the rightmost column of the table of coefficients:[25]

25. The `noheader` option suppresses the output of the ANOVA table and the model fit table.

```
. regress, beta noheader
```

sqfeet	Coef.	Std. Err.	t	P>|t|	Beta
hhinc	.1168247	.0064601	18.08	0.000	.2789243
hhsize	32.76697	5.185336	6.32	0.000	.0942143
east	-99.99731	14.2274	-7.03	0.000	-.1009768
owner	383.6225	13.89444	27.61	0.000	.4113329
_cons	524.1375	17.36074	30.19	0.000	.

The beta coefficients are interpreted in terms of the effect of standardized units. For example, as household income increases by one standard deviation, the size of the dwelling increases by about 0.28 standard deviation. In contrast, a one-standard-deviation increase in household size leads to an increase in dwelling size of about 0.09 standard deviation. If you look at the beta coefficients, household income has a stronger effect on dwelling size than does the size of the household.

Understandably, using the standardized estimated regression coefficients to compare the effect sizes of the different variables in a regression model is quite popular. But people often overlook some important points in doing so:

- You cannot use standardized regression coefficients for binary variables. Because the standard deviation of a dichotomous variable is a function of its skewness, the standardized regression coefficient gets smaller as the skewness of the variable gets larger.[26]
- If interaction terms are used (see section 8.4.2), calculating b_k^* using (8.7) is invalid; if interactions are included in your model, you cannot interpret the beta coefficients provided by Stata. If you want to study effect sizes with beta coefficients that are appropriate for interactions, you must transform all the variables that are part of the interaction term in advance, using a z standardization (Aiken and West 1991, 28–48).
- You should not compare standardized regression coefficients estimated with different datasets, as the variances of the variables will likely differ among those datasets (Berk 2004, 28–31).

8.2.3 What does "under control" mean?

The b coefficients from any regression model show how much the predicted value of the dependent variable changes with a one-unit increase in the independent variable. In a multiple regression model, this increase is calculated *controlling for* the effects of all the

26. To make this point clear, we wrote a small do-file demonstration: `anbeta.do`. This program fits 1,000 regressions with a dichotomous independent variable that takes on the values 0 and 1. In the first regression, no observation has a value of 1 for the independent variable. In each additional regression, the number of observations where $X = 1$ increases by one, until the last regression, where all cases have the value 1 for the independent variable. A figure is drawn with the beta coefficients from each of those 1,000 regressions.

other variables. We see the effect of changing one variable by one unit while holding all other variables constant. Here we will explain this concept in greater detail by using a simpler version of the regression model used above. Here only the estimated regression coefficients are of interest to us:

```
. regress sqfeet hhsize hhinc, noheader
```

sqfeet	Coef.	Std. Err.	t	P>|t|	[95% Conf.	Interval]
hhsize	40.96258	5.809275	7.05	0.000	29.5722	52.35297
hhinc	.1650224	.0069971	23.58	0.000	.1513031	.1787418
_cons	520.5754	18.6216	27.96	0.000	484.0635	557.0872

Look for a moment at the estimated coefficient for household income, which differs from the coefficients we estimated both for the simple model (page 188) and for the multiple model (page 198): what is the reason for this change? To find an answer, you need to estimate the coefficient for household income in a slightly different way: to begin, compute the residuals of the regression of dwelling size on household size:

```
. regress sqfeet hhsize, noheader
```

sqfeet	Coef.	Std. Err.	t	P>|t|	[95% Conf.	Interval]
hhsize	79.88333	5.946378	13.43	0.000	68.22431	91.54235
_cons	738.6072	17.16877	43.02	0.000	704.9445	772.2699

```
. predict e_fs, resid
(81 missing values generated)
```

When you do this, you create a new variable that stores the residuals: `e_fs`. Before continuing, you should give some serious thought to the meaning of those residuals.

We suggest that the residuals reflect the size of the dwelling adjusted for household size. The residuals reflect that part of the dwelling size that has nothing to do with household size. You could also say that they are that part of the information about dwelling size that cannot already be found in the information about the household size.

Now compute the residuals for a regression of household income on household size:

```
. regress hhinc hhsize, noheader
```

hhinc	Coef.	Std. Err.	t	P>|t|	[95% Conf.	Interval]
hhsize	227.0473	14.14031	16.06	0.000	199.3223	254.7723
_cons	1335.758	40.74527	32.78	0.000	1255.869	1415.648

```
. predict e_hh, resid
(139 missing values generated)
```

These residuals also have a substantive interpretation. If we apply the above logic, they reflect that part of household income that has nothing to do with household size. They therefore represent household income *adjusted for* household size.

Now fit a linear regression of e_fs on e_hh.

```
. regress e_fs e_hh, noheader
```

e_fs	Coef.	Std. Err.	t	P>|t|	[95% Conf.	Interval]
e_hh	.1650187	.006996	23.59	0.000	.1513014	.178736
_cons	-1.354776	7.200201	-0.19	0.851	-15.47238	12.76283

Take a close look at the b coefficient for e_hh, which corresponds to the coefficient in the multiple regression model you estimated above.[27] If you interpreted this estimated coefficient in the same way as one from a simple linear regression, you might say that dwelling size, adjusted for household size, increases about 0.165 ft^2 with each additional dollar of household income, adjusted for household size. The same interpretation holds true for the coefficients in the multiple regression model. The regression coefficients in the multiple regression model therefore reflect the effect of the independent variable in question on the dependent variable, adjusted for the effect of all other independent variables. This is what "controlling for" means.

8.3 Regression diagnostics

It is so easy to fit a multiple regression model using modern statistical software packages that people tend to forget that there are several assumptions behind a multiple regression; if they do not hold true, these assumptions can lead to questionable results. These assumptions are called "Gauss–Markov assumptions".[28]

We will describe each of the Gauss–Markov assumptions in detail in sections 8.3.1, 8.3.2, and 8.3.3, respectively. To illustrate the importance of the underlying assumptions, open the data file anscombe.dta and fit the following regression models:[29]

```
. use anscombe, clear
. regress y1 x1
. regress y2 x2
. regress y3 x3
. regress y4 x4
```

The estimated results for each regression model are the estimated coefficients, the variance of the residuals (RSS), and the explained variance R^2. Evidently you cannot find any difference between these four models just by looking at the numbers: you got an R^2 of 0.67 in all four models. The constant (or intercept) is 3, and the slope of the regression line is 0.5. If you did not know about the regression assumptions or regression diagnostics, you would probably stop your analysis at this point, supposing that you had a good fit for all models.

27. Differences are due to rounding errors.

28. If you are already familiar with the Gauss–Markov assumptions and how to check them, you might want to get a quick overview of regression diagnostics within Stata by typing help regress postestimation.

29. The data file was created by Anscombe (1973).

Now draw a scatterplot for each of these variable combinations, and then consider which model convinces you and which one does not; you can do this by typing the commands `scatter y1 x1`, `scatter y2 x2`, etc., one after the other. We actually used `granscomb1.do` to produce the graphs.

The scatterplots in figure 8.3 show, without a doubt, that there is good reason to be cautious in interpreting regression results. Looking at just the R^2 or the estimated coefficients can be misleading!

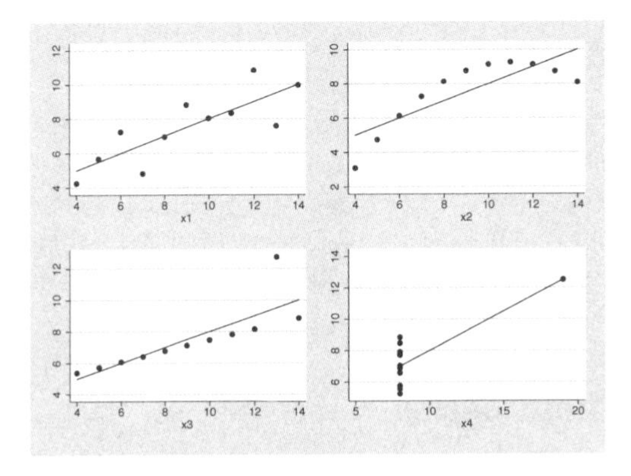

Figure 8.3. The Anscombe quartet

`granscomb1.do`

Now we want to show you how to check the Gauss–Markov conditions and correct any violations of them. Most of the diagnostic techniques we present are graphical, so you will need to understand the basics of the Stata `graph` command (see chapter 6). For an overview of various graphical diagnostic techniques, see Cook and Weisberg (1994). See Berk (2004, chapter 9) for a discussion on the limitations and potential hazards of using regression diagnostics.

8.3.1 Violation of $E(\epsilon_i) = 0$

The OLS estimation of the parameters of a regression model is based on the assumption that the "expected value" of the error terms in (8.1) and (8.6) is equal to zero, or, formally: $E(\epsilon_i) = 0$.

To understand this assumption you must first understand the meaning of an "expected value". Consider a situation in which you measure the size of the Lopez' apartment from equation (8.1) over and over again. Sure your measures will be fairly similar in each replication, but not quite identical. At the end of the day, it might be sensible for you to calculate the average across all repeated measures. This average value over an infinite number of hypothetical replications of an *experiment* is called the "expected value"; in our example, it would be the expected value of home size.

In $E(\epsilon_i) = 0$, we deal with the expected value of the error term of our regression model. As said before, the error term comprises all factors that influence the values of the dependent variable beyond the influence of the independent variables on our regression model. Hence, $E(\epsilon_i) = 0$ means that the average influence of all these factors is zero when we apply the model over and over again. Or, all influences that are not part of the model cancel out each other out in the long run.

If you estimate the parameters of a model by making such a simplifying assumption, you might ask yourself what happens when the assumption fails? The plain and simple answer is, your estimates of the regression coefficients will be biased. It is therefore important to verify that the expected value of the error term is indeed zero. All the problems that showed up in the Anscombe quartet are due to violations of this assumption.

To avoid biased estimates of the regression coefficients you should always check the underlying assumptions. Checking $E(\epsilon_i) = 0$ is of particular importance because its violation leads to biased estimators. The assumption will be violated, if

- the relationship between the dependent and independent variables is nonlinear,
- some outliers have a strong effect on the estimated regression coefficients, and/or
- some influential factors have been omitted that are in fact correlated with the included independent variables.

There are special techniques for testing each of the problems named above. You can see all three possible causes using a residual-versus-fitted plot, which is a scatterplot of the residuals of a linear regression against the predicted values. For the regression fitted last, you could build the plot by typing

```
. regress y4 x4
. predict yhat
. predict resid, resid
. scatter resid yhat
```

or more simply by using the `rvfplot` command, which generates one of the specialized statistical graphs mentioned in section 6.2.

```
. rvfplot
```

With `rvfplot`, you can use all the graphic options that are available for scatterplots. Figure 8.4 shows the residual-versus-fitted plots for all regressions in the Anscombe example.

(*Continued on next page*)

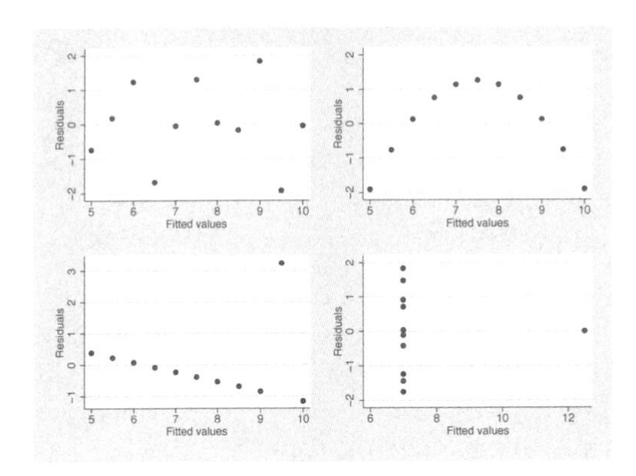

Figure 8.4. Residual-versus-fitted plots of the Anscombe quartet

granscomb2.do

In these graphs, the mean of the residuals is by definition always equal to zero. In a regression model, the regression coefficients are estimated so that the mean of the sample residuals is equal to zero. To fulfill the assumption that $E(\epsilon_i) = 0$, not only must the overall mean of the residuals be zero, but the mean of the residuals must be zero *locally*, meaning the mean of the residuals is zero for any slice of the x axis. This is true only for the first and the last regression model.

In a regression with only one independent variable, violations of the regression assumptions can be seen with a simple scatterplot of the dependent variable against the independent variable. The advantage of the residual-versus-fitted plot is that it also applies to regression models with more than just one independent variable.

In practice, a violation of $E(\epsilon_i) = 0$ is usually not as obvious as it is in the Anscombe data. Therefore, we will now introduce some special diagnostic tools for determining which of the three possibilities might be causing the violation of this assumption.

Linearity

To understand the following examples, you might want to start with a regression of home size on household income and household size using the GSOEP data:

```
. use data1, clear
. regress sqfeet hhinc hhsize
```

One of the most important requirements for a linear regression is that the dependent variable can indeed be described as a linear function of the independent variables. To examine the functional form of the relation, you should use nonparametric techniques, where you try to have as few prior assumptions as possible. A good example is a scatterplot reflecting only some general underlying assumptions derived from perception theory.

You can use a scatterplot matrix to look at the relationships between all variables of a regression model. Scatterplot matrices draw scatterplots between all variables of a specified variable list. Here is an example:

```
. graph matrix sqfeet hhinc hhsize
```

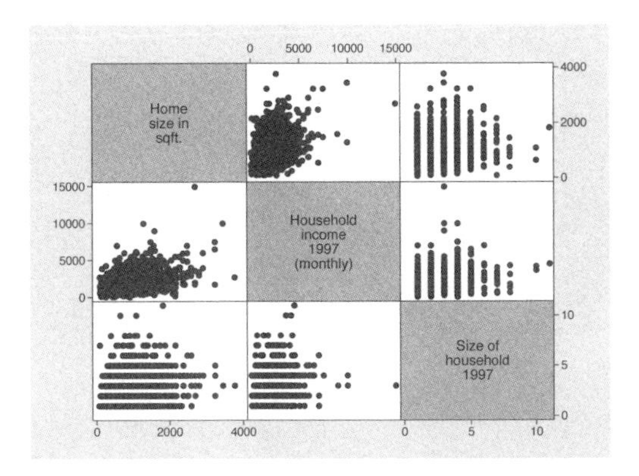

In each plot, the variable to the side of the graph is used as the Y variable, and the variable above or below the graph is used as the X variable. In the first line of the figure are scatterplots of home size against all the independent variables of the regression model.

However, scatterplots often show the functional form of a relationship only for small sample sizes. If you deal with larger sample sizes, you will need more information to improve the scatterplot. For this purpose, Stata allows you to overlay scatterplots with a scatterplot smoother (Fox 2000).

One example of a scatterplot smoother is the "median trace". To construct a median trace, you divide the variable plotted on the x axis of a twoway plot into strips and calculate the median for each strip. Then the medians are connected with straight lines. In Stata, you get the median trace as plottype `mband` of twoway graphs. The `bands(`*k*`)` option of this plottype is used to decide the number of strips into which the x axis should be divided. The smaller the number of bands, the smoother the line.

(*Continued on next page*)

```
. scatter sqfeet hhinc, ms(oh) || mband sqfeet hhinc, bands(20) clp(solid)
```

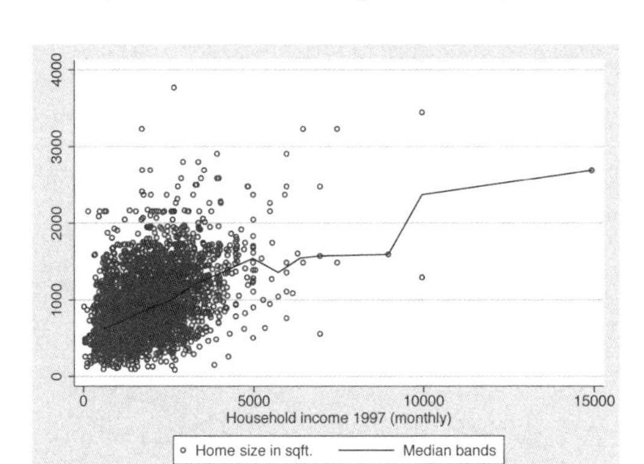

The figure shows a linear trend for most of the data and that there are many outliers on both variables: income and home size.[30] Even if you can establish a linear relationship between two variables, that relationship may change when you include other variables in the regression model. That is, the functional form of a relation between two variables may change under the influence of other variables.

One clue about the relation between one independent variable (e.g., household income) and the dependent variable (home size), if you control for other independent variables (such as household size), is given by plotting the residuals against the independent variables.[31] But plotting the residuals against one of the independent variables does not indicate the exact shape of any curvilinearity. For example, a U-shaped relation and a logarithmic relation might produce the same plot under certain circumstances (Berk and Both 1995).[32]

The component-plus-residual plots—also known as partial residual plots—are a modification of the plot just described: they allow the determination of the functional form of the relation. Within the component-plus-residual plots, instead of using the residual, the product of the residual and the linear part of the independent variable are plotted against the other independent variables. What this means is shown in the following example.

To examine the linearity between home size and household size in a multiple regression model, you can first fit the regression of home size on household income and household size and save the residuals as `e1`:

```
. regress sqfeet hhinc hhsize
. predict e1, resid
```

30. These might be observations that will heavily influence the regression result. In the next section, you will find out more about this issue.
31. When the sample size becomes large, it is reasonable to use a scatterplot smoother.
32. You must distinguish between these two kinds of relations: if there is a U-shaped relation, you must insert a quadratic term, whereas it might be sufficient to transform the dependent variable if there is a logarithmic relation (see section 8.4.3).

Then you can add the linear part of household size to the saved residuals and plot the resulting number against household size:

```
. generate e1plus = e1 + _b[hhsize]*hhsize
. scatter e1plus hhsize || mband e1plus hhsize, bands(20)
```

You would end up with the same result if you used the command `cprplot` after you fit your model using `regress`; this command will run the same procedure for any independent variable of your choice.[33] After `cprplot`, you enter the name of the independent variable for which you want to create the variable. The straight line in the resulting graph is equivalent to the regression line. We have also added a median spline, which is similar to the median trace but uses curves to connect the different medians.

```
. cprplot hhsize, mspline msopts(bands(20))
```

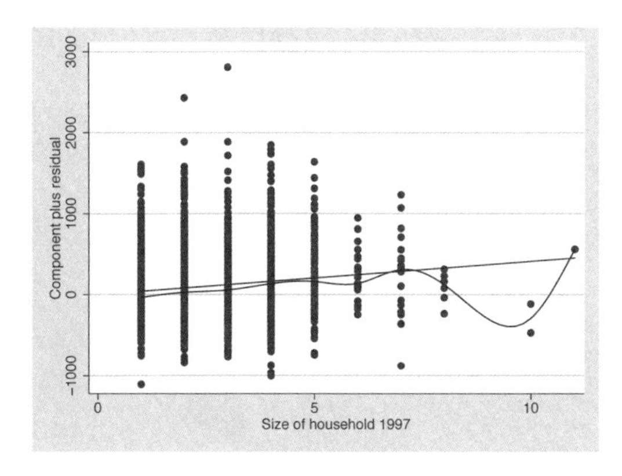

You might infer from the graph that home size decreases for a household size of seven or more. Here, however, this is probably an effect of the unstable median computation within the upper bands of household size since there are few households with seven or more members.

Potential solutions

In our example, the relations seem to be linear. In the presence of nonlinear relations, you need to transform the independent variables involved or include more quadratic terms in the equation; see section 8.4.3.

Influential cases

Influential cases are observations that heavily influence the results of a regression model. Mostly, these are observations that have unusual combinations of the regression variables

33. You will also find the augmented component-plus-residual plot from Mallows (1986): `acprplot`. Also, instead of the median trace used here, you could use the locally weighted scatterplot smoother (LOWESS) (Cleveland 1994, 168). Then you would use the option `lowess`.

included in the model (multivariate outliers). As an example, think of a person with a huge income living in a very small home.

It may not be possible to detect multivariate outliers in a bivariate scatterplot. Observations that show up as outliers in one scatterplot might in fact turn out to be normal if you controlled for other variables.

If, for example, the person mentioned had been interviewed in her secondary residence, the small home size is less surprising. Thus it is often possible to *explain* multivariate outliers. Then the solution for this problem is to include a variable in the regression model that captures the explanation. Here you would have to include in the regression model a variable that indicates whether this is the primary or secondary residence.

You can find signs of influential cases using a scatterplot matrix that is built from the variables included in the regression model. As each data point of one of these scatterplots lies on the same row or column as that of the other scatterplot, you can locate conspicuous observations over the entire set of scatterplots (Cleveland 1993, 275). Our example illustrates this with the help of one observation, which we have highlighted.

```
. gen str label = string(persnr) if hhinc == 14925
. graph matrix sqfeet hhsize hhinc, mlabel(label)
```

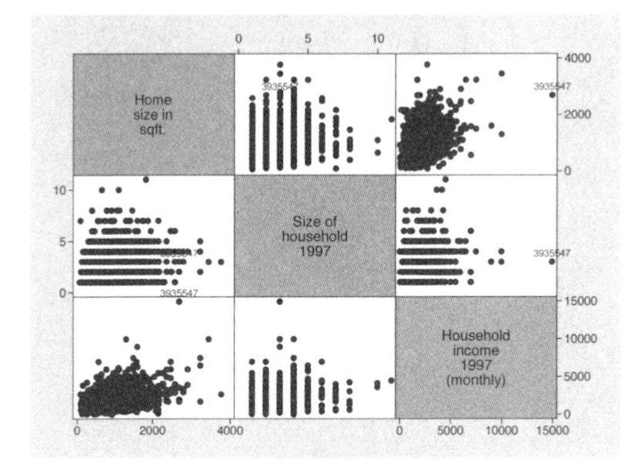

A more formal way to discover influential cases is to use DFBETAs. The computation of DFBETAs has a simple logic: first, you fit a regression model and then fit it again with one observation deleted. Then you compare the two results. If there is a big difference in the estimated coefficients, the observation that was excluded in the second computation has a big influence on the coefficient estimates. You then repeat this technique for each observation to determine its influence on the estimated regression coefficients. You compute this for each of the k regression coefficients separately. More formally, the equation for computing the influence of the ith case on the estimation of the kth regression coefficient is

$$\text{DFBETA}_{ik} = \frac{b_k - b_{k(i)}}{s_{e(i)}/\sqrt{\text{RSS}_k}}$$

where b_k is the estimated coefficient of variable k and $b_{k(i)}$ is the corresponding coefficient without observation i; $s_{e(i)}$ is the standard deviation of the residuals without observation i. The ratio in the denominator standardizes the difference so that the influences on the estimated coefficients are comparable (Hamilton 1992, 125).

In Stata, you compute values for DFBETA_{ik} using the **dfbeta** command. You enter this command after the regression command, with a variable list in which you specify the coefficients for which you want to view the change. If you do not specify a variable list, all coefficients are used. The results of the command **dfbeta** are stored in variables whose names begin with "DF".

Typing

```
. regress sqfeet hhinc hhsize
. dfbeta
```

generates two variables: DFhhinc and DFhhsize. Both variables contain, for each observation, its influence on the estimated regression coefficient. If there are indeed influential cases in your dataset, you can detect this using

```
. graph box DF*
```

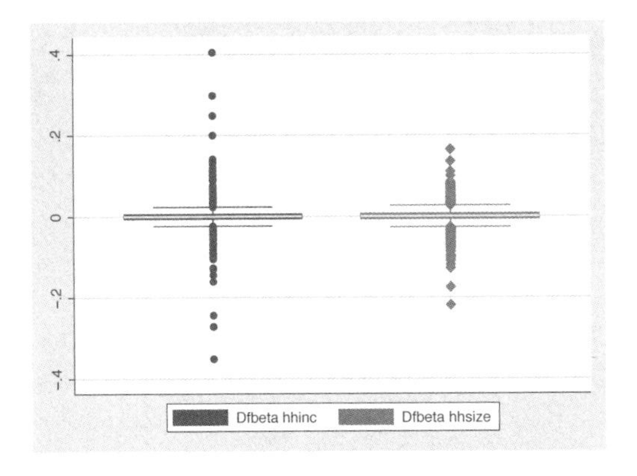

Values of $|\text{DFBETA}| > 2/\sqrt{n}$ are considered large (Belsley, Kuh, and Welsch 1980, 28).[34] In our model, several observations exceed this boundary value. With

```
. foreach var of varlist DF* {
  2. list persnr `var' if (abs(`var') > 2/sqrt(e(N))) & `var' < .
  3. }
```

you obtain a list of these observations.[35]

34. Other authors use 1 as the boundary value for DFBETA (Bollen and Jackman 1990, 267).

35. The command foreach is explained in section 3.2.2. The expression abs() is a general Stata function that returns the absolute value of the argument included in the parentheses (see section 3.1.6). Finally, e(N) is the number of observations included in the last regression model fitted (see chapter 4).

Another way to detect outliers is to use the added-variable plot (partial regression plot). To create the added-variable plot of the variable X_1, you first run a regression of Y on all independent variables besides X_1. Then you run a regression of X_1 on all remaining independent variables. You then save the residuals of both regressions and plot them against each other.[36]

In Stata, you can also create added-variable plots by using the `avplot` or `avplots` commands. `avplot` creates the added-variable plot for one explicitly named independent variable, whereas `avplots` shows all possible plots in one graph:

```
. regress sqfeet hhinc hhsize
. avplots
```

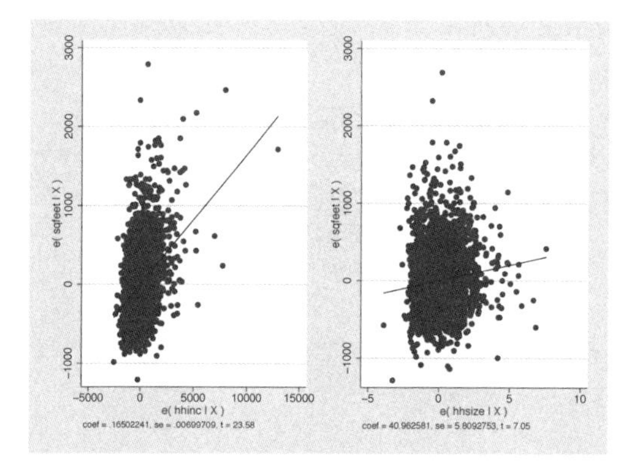

In these plots, points that are far from the regression line are “multivariate outliers”. These kinds of observations have more potential to influence the regression results. Here some observations are conspicuous in that household income is higher than you would assume by looking at the values of the remaining variables. In the plot for household income, one observation in particular is cause for concern—the one with the largest house.

You can type

```
. avplot hhinc, mlabel(persnr)
```

to identify the personal identification number of this observation:

36. The logic behind added-variable plots corresponds to the way the b coefficients are interpreted in a multiple regression model (see section 8.2.3). A scatterplot of the residuals that were created there would be an added-variable plot.

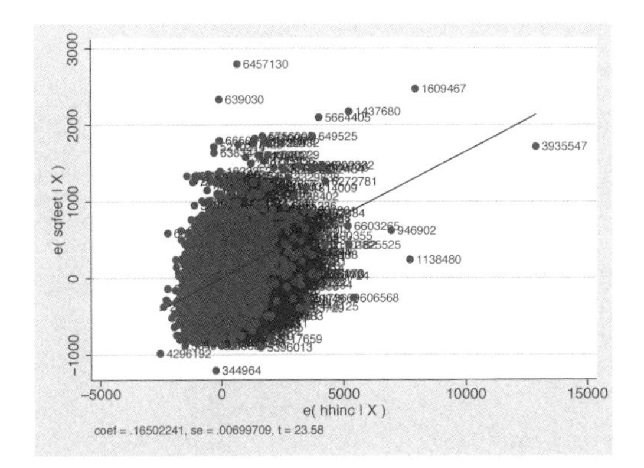

Hamilton (1992, 128–129, 141) recommends using an added-variable plot where the size of the plot symbol is proportional to DFBETA. To do this, you must create the plot yourself. In the multiple linear regression we used above, you would create such a plot for household income as follows:[37]

```
. regress sqfeet hhsize
. predict esqfeet, resid
. regress hhinc hhsize
. predict ehhinc, resid
. generate absDF = abs(DFhhinc)
. scatter esqfeet ehhinc [weight = absDF], msymbol(oh)
> || lfit esqfeet ehhinc, clpattern(solid)
```

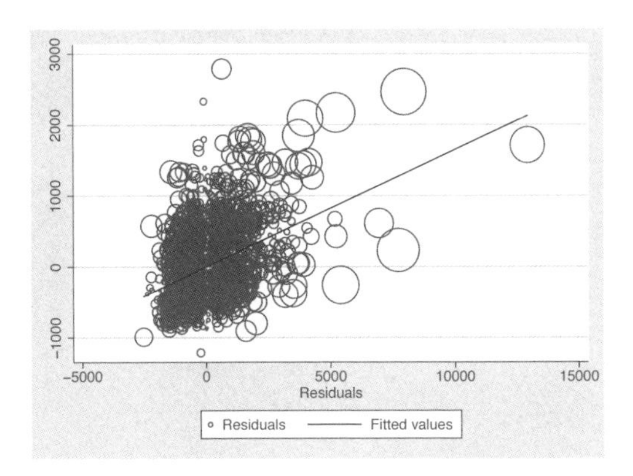

In Stata graphs, you can control the size of the plot symbol using *weights*. Here it is not important what kind of weights (`fweights` or `aweights`, for example) you use. In this example, you must pay attention to possible negative values of DFBETA, so you can compute the absolute values of DFBETA first and use these values for weighting.[38]

37. For this example, we use the variable `DFhhinc`, which we created on page 211. The axes of this graph are both labeled as "residuals" automatically by the command `predict`. If you want to change these labels, see section 6.3.4.

38. You will find some general remarks about weights in section 3.3.

The previous figure shows that the multivariate outlier has an appreciable influence on the regression line. Even stronger are two observations more to the left, but they cancel each other. Altogether, we seem to find the influential cases mainly in the upper region of income, regardless of the other variables. Those few observations with high income have a disproportionately strong influence on the regression result.

So far, the impact of single observations has been examined separately for the different coefficients. If you have many independent variables, you will find it more complicated to interpret the many DFBETA values. With "Cook's D", you have a statistic available that estimates the effect of one observation on all regression coefficients simultaneously (Fox 1991, 84) and hence the influence of one observation on the entire regression model. You get this statistic by entering `predict` after the regression command:

```
. predict cook, cooksd
```

The idea behind this statistic is that the influence of one observation on the regression model is composed of two aspects: the value of the dependent variable and the combination of independent variables. An influential case has an unusual value on Y *and* an unusual combination of values on the Xs. Only if both aspects are present will the estimated coefficients be strongly affected by this observation. The graphs in figure 8.5 clarify this. The graphs present scatterplots of home size against the income of five Englishmen in 1965, 1967, and 1971.

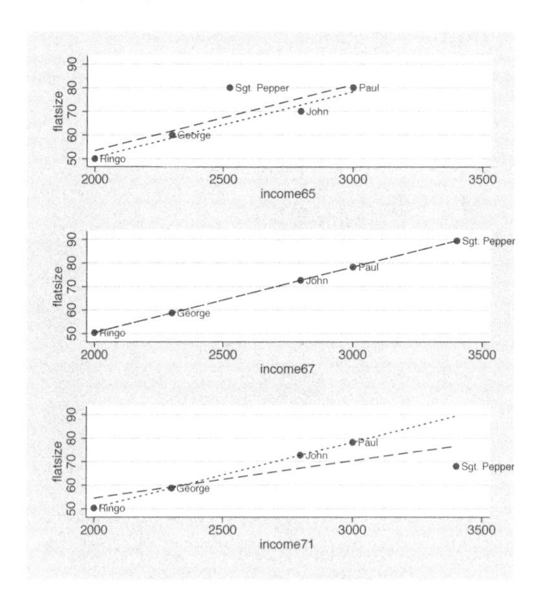

Figure 8.5. Scatterplots to picture leverage and discrepancy

`grbeatles.do`

In the first scatterplot, which shows the year 1965, Sgt. Pepper has an extraordinarily large home given his income. Sgt. Pepper's income is, however, anything but extraordinary: it is equal to the mean net income of the five Englishmen. We draw two regression lines in this picture. The dotted line is the regression line that results from a regression without Sgt. Pepper. When Sgt. Pepper is included in the regression, the regression line is shifted upward. There is no change in the slope of the line (the b coefficient of income).

In the scatterplot for 1967, Sgt. Pepper has an extraordinarily high income. The size of his home corresponds, however, exactly to the square footage we would expect from our model. Sgt. Pepper has therefore an extraordinarily large value of X but, given this value for X, a quite common Y value. The regression lines that result from the regressions with and without Sgt. Pepper are identical in this case.

In the scatterplot for 1971, Sgt. Pepper has an extraordinarily high income and, for this income, an extraordinarily small home. Here both aspects mentioned above are present. Accordingly, the regression line changes.[39]

The idea that the effect of a certain point is determined by the extreme values of X and Y can be described mathematically as

$$\text{influence} = \text{leverage} \times \text{discrepancy} \tag{8.8}$$

where the leverage signifies how extraordinary the combination of the X values is (as in the second scatterplot) and the discrepancy signifies how extraordinary the Y value is (as in the first scatterplot). As leverage and discrepancy are multiplied, the influence of any given observation is equal to 0 if one or both aspects are zero.

To compute the influence as shown in (8.8), you need some measures of the leverage and the discrepancy. First, look at a regression model with only one independent variable. Here the leverage of a specific observation increases with its distance from the mean of the independent variable. Therefore, a measure of the leverage would be the ratio of that distance to the sum of the distances of all observations.[40]

When there are several independent variables, the distance between any given observation and the centroid of the independent variables is used, controlling for the correlation and variance structure of the independent variables (see also Fox [1997, 97]). In Stata, you obtain the leverage value for every observation by using the `predict lev, leverage` command after the corresponding regression. When you type that command, Stata saves the leverage value of every observation in a variable called `lev`.

39. Think of the regression line as a seesaw, with the support at the mean of the independent variable. Points that are far away from the support and from the regression line are the most influential points.

40. Specifically,

$$h_i = \frac{1}{n} + \frac{(x_i - \overline{x})^2}{\sum_{j=1}^{n}(x_j - \overline{x})^2}$$

To measure the discrepancy, it seems at first obvious that you should use the residuals of the regression model. But this is not in fact reasonable. Points with a high leverage pull the regression line in their direction, and therefore they may have small residuals. If you used residuals as a measure of discrepancy in (8.8), you might compute small values for the influence of an observation, although the observation changed the regression results markedly.[41]

Hence, to determine the discrepancy you need a statistic that is adjusted for the leverage. The standardized residual e'_i is such a statistic. You can obtain the values of the standardized residuals by using the `predict` *varname*`, rstandard` command, which you can enter after a regression.[42]

After finding a statistic for both discrepancy and leverage, you can multiply the two statistics together in accordance with (8.8). But you should provide an appropriate weight for each value to be multiplied. We leave that task to the statisticians. Cook (1977) suggested the following computation:

$$D_i = \underbrace{\frac{h_i}{1-h_i}}_{\text{leverage}} \times \underbrace{\frac{e_i'^2}{k+1}}_{\text{discrepancy}}$$

Here e'_i is the standardized residual and h_i is the leverage of the ith observation.[43] Values of Cook's D that are higher than 1 or $4/n$ are considered large. Schnell (1994, 225) recommends using a graph to determine influential cases. In this graph, the value of Cook's D for each observation is plotted against its serial number within the dataset, and the threshold is marked by a horizontal line.

To construct this graph, you must first compute the values for Cook's D after the corresponding regression:

```
. regress sqfeet hhsize hhinc
. predict cooksd, cooksd
```

Then you save the threshold in a local macro (`max`) using the number of observations in the last regression model, which is stored by Stata as an internal result in `e(N)` (see chapter 4 and section 11.2.1):

```
. local max = 4/e(N)
```

41. This can be demonstrated with the fourth graph in the Anscombe quartet (page 204). If you computed the influence of this outlier with (8.8) and thereby used the residuals as a statistic for discrepancy, the influence of this outlier would be equal to 0.
42. You can choose any variable name for *varname*.
43. There is a useful teaching tool you can see by typing the command `regpt`, which is taken from an ado-file programmed by the Academic Technology Services of the University of California, Los Angeles. To learn more about ado-files, see chapter 11; to learn more about ado-files provided over the Internet, see chapter 12.

Now you build a variable `index`, which contains the serial observation number, and use this variable as the x axis on our graph. Next construct the graph with a logarithmic y axis:

```
. generate index = _n
. scatter cooksd index, yline(`max') msymbol(p) yscale(log)
```

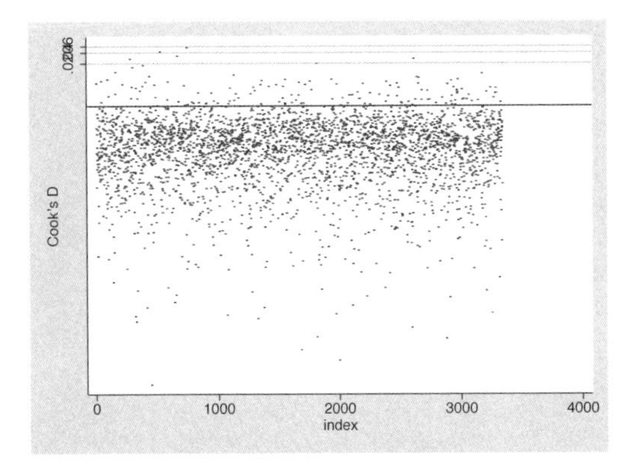

The figure shows many observations that are above the critical value, especially those with a comparatively high income:

```
. generate bigcook = cooksd > `max'
. tabulate bigcook, summarize(hhinc)
```

bigcook	Summary of Household income 1997 (monthly) Mean	Std. Dev.	Freq.
0	1860	945	2975
1	2723	1931	226
Total	1921	1068	3201

In summary, the analyses you ran in this section show a clear finding: using these diagnostic techniques, you found a few observations with high incomes to be conspicuous. The results of the model are much more strongly affected by these few observations than by all the other observations with low, medium, or high (but not very high) income. In particular, the added-variable plot on page 213 shows that these highest influential observations cancel each other after controlling for other variables in the multiple regression. High influence points therefore do not seem to be a big problem in our example. If you excluded all the observations with a very high income, the estimated coefficients would stay pretty much as they are.

Potential solutions

You may wonder what to do when influential observations are present. If an influential case can be attributed unquestionably to a measurement error, you should either correct the error or delete the observation from the file. If influential observations result from extreme values of the dependent variable, it is reasonable to use median regression (section 8.6.1).

Almost always, however, influential observations result from an incompletely specified model. Exceptional cases are in this case exceptional only because our theory explains them insufficiently. As in our example, where observations with a high income influence the regression extraordinarily, you should ask if another factor influences home size that is typically related to high (or to low) income. With *right-skewed* distributions, such as that of income, you may want to change the model to use the logarithm of household income instead of household income itself. In the current context, this means that household income is supposed to be in a logarithmic relation to home size: the higher the household income gets, the smaller the change in home size with each additional dollar of household income.

Omitted variables

Variables are called "omitted variables" or omitted factors if they influence the dependent variable and are at the same time correlated with one or more of the independent variables of the regression model. Strictly speaking, nonlinear relations and influential cases are omitted factors, too. In the first case, you may have overlooked the fact that an independent variable does not have the same influence on the dependent variable throughout the range of the dependent variable. In the second case, you may have neglected to model your theory adequately or overlooked a mechanism that would explain the outliers.

To figure out which variables have been omitted, you can begin by graphing the residuals against all variables that are not included in the model. But this is obviously possible only for those variables that are included in the data file. Even if these graphs show no distinctive features, there still may be a problem. This diagnostic tool is therefore necessary but not sufficient.

Identifying omitted factors is, first of all, a theoretical problem. Thus we warn against blindly using tools to identify omitted variables.

Multicollinearity

In trying to include all important influential factors in the model, there is another risk called *multicollinearity*. We will introduce an extreme case of multicollinearity in section 8.4.1 when we discuss how to include categorical independent variables in regression models. If there is a perfect linear relation between two variables of the regression model,[44] Stata will exclude one of them when calculating the model.

But even if the two variables are not a perfect linear combination of each other, some problems can arise: the standard errors of the estimated coefficients might increase,

44. For example, $x_1 = 2 + x_2$.

and there might be an unexpected change in the size of the estimated coefficients or their signs. You should therefore avoid including variables in the regression model haphazardly. If your model fits the data well based on R^2 but nevertheless has a few significant estimated coefficients, then multicollinearity may be a problem.

Finally, you can use the `estat vif` command to detect multicollinearity after regression. This command gives you what is called a variance inflation factor for each independent variable. See Fox (1997, 338) for an interpretation and explanation of this tool.

8.3.2 Violation of $\text{Var}(\epsilon_i) = \sigma^2$

The assumption that $\text{Var}(\epsilon_i) = \sigma^2$ requires that the variance of the errors be the same for all values of the independent variables. This assumption is called homoskedasticity, and its violation is called heteroskedasticity. Unlike the violation of $E(\epsilon_i) = 0$, heteroskedasticity does not lead to biased estimates. But when the homoskedasticity assumption is violated, the estimated coefficients of a regression model are not efficient. With inefficient estimation, there is an increasing probability that a particular estimated regression coefficient deviates from the true value for the population. That is, heteroskedasticity causes the standard errors of the coefficients to be incorrect, and that obviously has an impact on any statistical inference that you perform.

There are many possible reasons for heteroskedasticity. Frequently you find heteroskedasticity if the dependent variable of your regression model is not symmetric. To test the symmetry of variables, you will find the graphical techniques described in section 7.3.3 to be very useful.

Stata has a special technique for checking the symmetry of a distribution, called a "symmetry plot" (Chambers et al. 1983, 29). To construct a symmetry plot, you first determine the median. Then you compute the distances between the observations next in size and the median. In a symmetry plot, you plot these two quantities against each other. You do the same with the next observation, and so on. If all distances are the same, the plot symbols will lie on the diagonal. If the distances of the observations above the median are larger than those below, the distribution is right-skewed. If the reverse is true, the distribution is left-skewed.

In Stata, the `symplot` command graphs a symmetry plot of a given variable. Here we graph the symmetry plot for home size:

(Continued on next page)

```
. symplot sqfeet
```

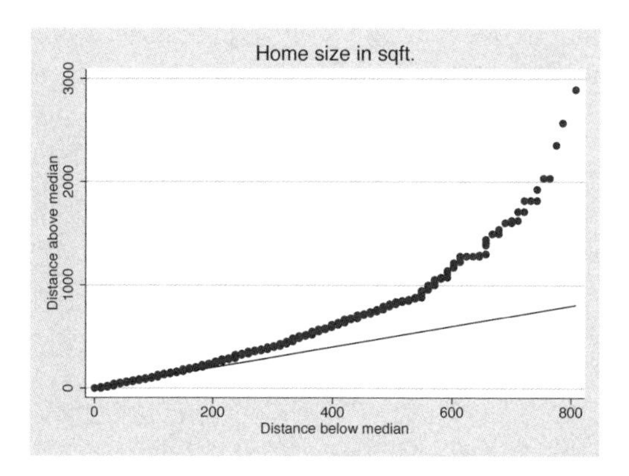

The figure shows an obviously right-skewed distribution of the variable home size. With this kind of distribution, there is risk of violating the homoskedasticity assumption.

The residual-versus-fitted plot (Cleveland 1994, 126) is the standard technique for examining the homoskedasticity assumption. We want to introduce one variation of this plot, which emphasizes the variance of the residuals. You therefore divide the x axis into k groups with the same number of observations and then draw a box plot of the studentized residuals for each group.

To do this, you again run the regression model and get the predicted values, and the studentized residuals:

```
. regress sqfeet hhinc hhsize
. predict yhat3
. predict rstud, rstud
```

For this example, we chose the number of groups used for the x axis so that each box plot contains roughly 100 observations:[45]

45. The function `round()` is a general Stata function (see section 3.1.6). The saved result `e(N)` stores the number of observations of the last regression model. The `xtile` command is described in section 7.3.1.

```
. local groups = round(e(N)/100,1)
. xtile groups = yhat3, nq(`groups')
. graph box rstud, over(groups)
```

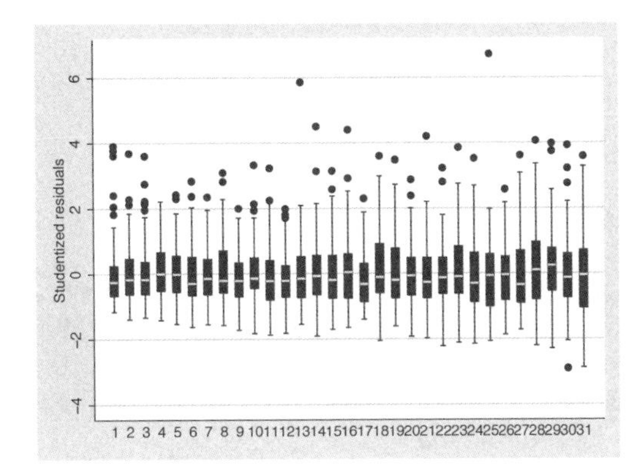

In the graph, you can see that there is a slight increase in the variance of the residuals.

Potential solutions

Often you can simply transform the dependent variable to remove heteroskedasticity. The transformation should end in a symmetric variable. For right-skewed variables, a logarithmic transformation is often sufficient. Also the `boxcox` command allows you to transform a variable so that it is as symmetric as possible. You will find more discussion of the Box–Cox transformation in section 8.4.3.

If transforming the dependent variable does not remove heteroskedasticity in the regression model, you cannot use the standard errors of the estimated coefficients (as they are given in the regression output) for a significance test. If you are nevertheless interested in a significance test, you might want to try the `vce(robust)` option in the regression command. When you use this option, the standard errors are computed so that homoskedasticity of the error terms need not be assumed.

8.3.3 Violation of Cov$(\epsilon_i, \epsilon_j) = 0,\ i \neq j$

What $\text{Cov}(\epsilon_i, \epsilon_j) = 0,\ i \neq j$, means is that the errors are not correlated. The violation of this assumption is often called "autocorrelation" or "correlated response variance", which results in inefficient estimation of the coefficients.

For example, earlier you tried to predict home size. Now suppose that you have surveyed home size by letting the interviewers estimate the size instead of asking the respondent. Here it is reasonable to assume that some of the interviewers tend to overestimate the sizes of the dwellings, whereas others tend to underestimate them. All the observations from one interviewer should be similar in over- or underestimating home size. A similar situation occurs if all people in a household are interviewed. Here, as well as in the above, there may be factors within the unobserved influences (ϵ_i) that are the same for all members of a household. The same might be true for respondents of a particular sampling unit. We showed that you can deal with a violation of the independence assumption, even with cross-sectional data.

More recently, the literature on complex samples (Lee, Forthofer, and Lorimor 1989; Lehtonen and Pahkinen 1995; Skinner, Holt, and Smith 1989) as well as in the multilevel literature (Rabe-Hesketh and Skrondal 2008), shows ways to handle the violations of the independence assumption we mentioned in the examples. We will give a more detailed discussion of this problem later.

Autocorrelation is a key concept, especially in time-series analysis, as successive observations tend to be more similar than observations separated by a large time span (serial autocorrelation). The Durbin–Watson test statistic has been developed for time-series analysis, and in Stata it is available using the `estat dwatson` command after regression. However, you must define the data as a time series beforehand.[46]

8.4 Model extensions

Here we will introduce three extensions to the linear model you have seen so far. These extensions are used for categorical independent variables, interaction terms, and modeling curvilinear relationships. Interpreting refined models can sometimes be rather difficult, so we will introduce conditional-effects plots as a graphical way to display regression results.

8.4.1 Categorical independent variables

Be careful when including a categorical variable with more than two categories in the regression model. Take, for example, marital status. The variable `marital` has six categories, namely, married, separated, unmarried, divorced, widowed, and grass-widowed (partner is living abroad):

```
. tabulate marital
```

Marital status 1997	Freq.	Percent	Cum.
married	1,860	55.69	55.69
separate	83	2.49	58.17
unmarried	800	23.95	82.13
divorce	270	8.08	90.21
widowed	312	9.34	99.55
grasswid	15	0.45	100.00
Total	3,340	100.00	

46. As we do not discuss time-series analysis in this book, we refer here to the online help `tsset` and to the manual entry [U] **26.14 Models with time-series data**.

It would not make sense to include marital status in the same way as all other independent variables since assuming that going from being married to separated has the same effect on home size as going from divorced to widowed. However, you would assume this implicitly when a categorical variable with several categories is included in a regression model without any changes. What you need instead are contrasts between the individual categories.

Let us say that you want to include a variable that differentiates between married and unmarried respondents. To do so, you can create a dichotomous variable with the response categories 0 for not married and 1 for married.

```
. generate married = marital == 1 if marital < .
```

You can interpret the resulting *b* coefficient for this variable just as for other dummy variable; accordingly, you could say that married respondents live on average in a space that is *b* square feet bigger than the one unmarried people live in. All other contrasts can be built in the same way:

```
. generate separated = marital == 2 if marital < .
. generate unmarried = marital == 3 if marital < .
. generate divorced = marital == 4 if marital < .
. generate widowed = marital == 5 if marital < .
. generate grasswid = marital == 6 if marital < .
```

Each contrast displays the difference between respondents with one particular marital status and all other respondents. Beware that Stata will automatically remove one of the dummy variables if you include all contrasts in the regression model:

```
. regress sqfeet hhinc hhsize married-grasswid, noheader
```

sqfeet	Coef.	Std. Err.	t	P>|t|	[95% Conf.	Interval]
hhinc	.1644524	.0070193	23.43	0.000	.1506894	.1782153
hhsize	39.28901	6.440657	6.10	0.000	26.66065	51.91736
married	-42.47589	47.98608	-0.89	0.376	-136.5634	51.61162
separated	(dropped)					
unmarried	-73.35993	48.77556	-1.50	0.133	-168.9954	22.27553
divorced	-107.4403	52.90216	-2.03	0.042	-211.1669	-3.713738
widowed	-20.81282	52.21819	-0.40	0.690	-123.1983	81.57269
grasswid	-329.026	120.9142	-2.72	0.007	-566.1055	-91.9465
_cons	579.1672	48.63016	11.91	0.000	483.8168	674.5176

The reason for dropping one of the contrasts is that of the six new dummy variables, only five are needed to know the marital status of each person. If five dummies indicate a person does not live with a spouse, is not divorced, is not unmarried, is not widowed, and is not grass-widowed, the person must be separated from his or her spouse. The sixth dummy variable tells you nothing that cannot be gleaned from the other five, since there are six possibilities. Computationally, it is not even possible to estimate coefficients on all six dummies in addition to the constant term because those six dummies sum to one and hence are perfectly correlated with the constant term.

The constant represents the predicted value for respondents with zero on all covariates, in this case respondents who are separated, since the separated dummy was not included in the model. The predicted home size of persons with a different family status differs by the amount of the according b coefficient. Married respondents have, therefore, on average a home size that is 42.48 ft^2 smaller than those of respondents who are separated. Even smaller are the homes of unmarried respondents—about 73.36 ft^2 smaller than those of separated respondents. All other estimated coefficients are interpreted accordingly.

The coefficients you estimated may be somewhat surprising. It would be more reasonable if married respondents have on average larger home sizes than those who are separated, other things constant; however, according to the model, the opposite is true. Upon closer inspection, though, the results are more reasonable. Separated respondents typically live alone, which means their household size is one. Whereas married couples have an average home size 42 ft^2 less than separated people holding all other factors constant, the `hhsize` variable's estimated coefficient of 39.3 implies that a married couple with no children has, in fact, a home that is an average of just 3 ft^2 less than a separated person. Also our dataset has only a few separated respondents.

Instead of creating contrast variables by hand, you can use the `tab` *varname*`,` `gen(`*newvar*`)` command, where *varname* is the name of the categorical variable you want to use. This command ensures that there will be as many variables—*newvar1* to *newvarK*—as there are categories in the categorical variable.

```
. tabulate egp, gen(egp_)
```

This command creates the variables `egp_1` to `egp_11`. You can include those in your regression model. It is usually a good idea to decide on a contrast that will be left out and used as a comparison category, but which one you use does not affect the results substantively.

The `xi` prefix is another shortcut for creating dummy variables that you can use with all kinds of model commands. You estimate the regression of home size on the categorical variable type of household by typing

```
. xi: regress sqfeet i.htyp
```

This command creates eight dummy variables (this is the number of categories in the variable `hhtype`) to be used in the regression model. By default, the smallest category is omitted and used as the reference category. The `xi` prefix is especially helpful if you intend to model interaction effects with those dummy variables. However, using `xi` multiple times for several models takes time, since all dummy variables must be created over and over for each model.

8.4.2 Interaction terms

To discuss modeling interaction effects, we will return to our analysis of income inequality between men and women from chapter 1. There we tried to explain the gross income by gender and occupational status for all respondents who have some kind of income. The analysis showed that women earn less on average than men and that this difference can be only partly explained by the difference in full-time and part-time occupation between men and women.

Let us begin by reproducing the model using the do-file `anchap1.do`:

```
. do anchap1.do
```

Assume that you hypothesize that income depends not only on the respondents' gender and occupational status but also on their educational level. A higher educational level, you believe, leads to higher income. Here you should include education in your regression model. Say that you also assume that the educational advantage is more prominent the older people are. There are three good reasons why this is a reasonable assumption:

- More highly educated people begin their occupational careers later in life. Starting salaries for highly educated people are therefore not necessarily—if at all—higher than those of their co-workers in the same age group with less education but more occupational experience. However, with increased occupational tenure, the educational advantage increases as well.
- People with less education have a higher risk of becoming unemployed. Periods of unemployment can lead to difficulties in finding a new job, with an increased likelihood of having part-time or minimum-wage jobs. This would lead in extreme cases to a loss in income as age increases for people with less education.
- Education was for a long time the most important variable for determining income. Nowadays there are professions that do not require professional training or a college degree. Income inequality between educational groups could therefore be seen as an extinguishing phenomenon that is visible only for older generations.

All three arguments could lead us to hypothesize that the effect of education on income increases with age. Such effects that vary with values of a third variable are called interaction effects, which you can include in a regression model by multiplying the relevant variables.

Here education and age are relevant variables for the interaction effect. To extend the regression model, include years of education (`yedu`). You can create the variable for age using year of birth (`ybirth`) and the date of the interview:

```
. generate age = 1997 - ybirth
```

It is advantageous to center continuous variables, such as length of education or age, before they are included in a regression model, for several reasons. This is especially true in the presence of interaction terms (Aiken and West 1991).

To center a variable, subtract the mean from each value. However, you need to compute the mean only for those respondents that will later be included in the regression model. To know which respondents these are, you need to know which respondents have valid values on all variables that will be included in the regression model. You can use the egen function rowmiss(*varlist*) to do so. This function counts the number of missing values for the specified variable list.

```
. egen miss = rowmiss(income yedu ybirth fulltime)
```

You generate the miss variable using this egen command. The variable miss takes on the value zero for all respondents that do not have missing values for any of the specified variables. This allows you to specify the correct mean, as we discussed above:[47]

```
. summarize yedu if miss == 0
. generate cyedu = yedu - r(mean) if miss==0
```

and

```
. summarize age if miss==0
. generate cage = age - r(mean) if miss==0
```

To build the interaction term, multiply both variables that are part of the interaction.

```
. generate yeduage = cyedu * cage
```

The linear regression model will be extended by the variables cage and cyedu and the interaction term yeduage that you just created:

```
. regress income men fulltime cage cyedu yeduage

      Source |       SS       df       MS              Number of obs =    1545
-------------+------------------------------           F(  5,  1539) =   50.24
       Model |   359958784     5  71991756.8           Prob > F      =  0.0000
    Residual |  2.2051e+09  1539  1432839.59           R-squared     =  0.1403
-------------+------------------------------           Adj R-squared =  0.1375
       Total |  2.5651e+09  1544  1661333.49           Root MSE      =    1197

------------------------------------------------------------------------------
      income |      Coef.   Std. Err.      t    P>|t|     [95% Conf. Interval]
-------------+----------------------------------------------------------------
         men |   444.0768   61.78785     7.19   0.000     322.8796    565.2741
    fulltime |   764.3836   90.45379     8.45   0.000     586.9579    941.8093
        cage |   -.893062   2.158071    -0.41   0.679    -5.126133    3.340009
       cyedu |   119.3288   12.50207     9.54   0.000     94.80594    143.8517
     yeduage |    1.11756   .8829518     1.27   0.206    -.6143558    2.849476
       _cons |    1007.47   86.07676    11.70   0.000     838.6304    1176.311
------------------------------------------------------------------------------
```

47. The expression r(mean) refers to the mean saved by summarize (see chapter 4).

Without going into details, we want to explain the graphical representation of the regression model results—in particular, the interpretation of the interaction terms. Even if most people use tables to display the results of a multiple linear regression, we think graphs are more reasonable for complex models.[48] Graphically displaying the results is especially useful for models with interaction terms; conditional-effects plots are commonly used.

To construct a conditional-effects plot, you draw different regression lines for different combinations of independent variables. Here you can create a graph that indicates the correlation between age and income and, according to our hypothesis about the interaction effect, the correlation for an average length of education, as well as for the lowest and highest years of education. This means that you need to fit three different regression lines.

For a simple linear regression, we showed you on page 190 how to compute the regression line according to the model results:

```
. predict yhat1
```

You could also get the predicted values by typing

```
. generate yhat2 = _b[_cons] + _b[fulltime]*fulltime + _b[men]*men +
> _b[cage]*cage + _b[cyedu]*cyedu + _b[yeduage]*yeduage
```

Unlike simple linear regression, in a multiple regression the predicted values cannot be displayed on a line in a two-dimensional graph. However, you can obtain a line if you fix the values of all variables but one. You can, for example, compute the predicted values for female respondents with part-time employment and an average education. Since you centered the variable for education, you know that 0 represents the average length of education. You can therefore compute the desired values by fixing all variables except `age` to zero:

```
. generate yh_yedu0 = _b[_cons] +  _b[cage]*cage
```

Most of the terms in the equation disappeared because you replaced the variables' values with zeros. The same is true for the interaction effect since multiplying `age` with the average value of the centered education variable is zero as well.

The predicted values for the lowest education group can be computed accordingly. You first must `summarize` the variable `cyedu` and use the saved result `r(min)`, which contains the minimum value of the variable used in the last `summarize` command. Beware that fixing the interaction term is more complicated, since it should vary with age but not with education. You therefore multiply the interaction term with the minimum value of education *and* `cage`. You can also handle the interaction term for the highest educational group in the same way:

48. In section 12.3.1, we introduce a command that is usually used to transform Stata regression tables into regression tables usually used in publications.

```
. summarize cyedu
. generate yh_yedumin = _b[_cons] + _b[cage]*cage + _b[cyedu]*r(min) +
> _b[yeduage]*cage*r(min)
. generate yh_yedumax = _b[_cons] + _b[cage]*cage + _b[cyedu]*r(max) +
> _b[yeduage]*cage*r(max)
```

You have thus created three different variables that contain values for three different educational levels for women who work part-time. Each of these three variables is itself a function of age. The values for each educational level form a line that can be displayed graphically. In this command, the star behind yh_yedu is used to include all variables beginning with yh_yedu in the graph command (see section 3.1.2):

```
. line yh_yedu* age, sort
```

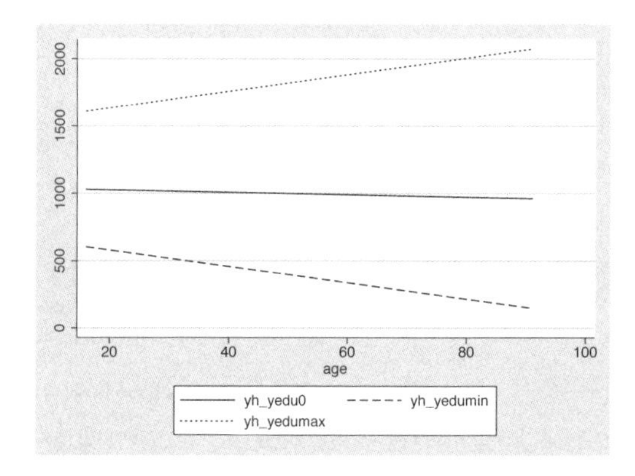

In this graph, the top line represents respondents from the highest educational group, and the bottom line represents respondents from the lowest educational group.[49] The graph shows that age has a different effect for each educational level. The higher the educational level, the greater the increase of income with increasing age. In models without interaction terms, the lines in a conditional-effects plot would always be parallel.

8.4.3 Regression models using transformed variables

There are two main reasons to use transformed variables in a regression model:

- The presence of a nonlinear relationship
- A violation of the homoskedasticity assumption

Depending on the reason you want to transform the variables, there are different ways to proceed. In the presence of a nonlinear relation, you would (normally) transform the *independent* variable, but in the presence of heteroskedasticity, you would transform the *dependent* variable. We will begin by explaining how to model nonlinear relationships and then how to deal with heteroskedasticity (see also Mosteller and Tukey [1977]).

49. To improve this legend, refer to section 6.3.4.

Nonlinear relations

We introduced regression diagnostic techniques for detecting nonlinear relationships in section 8.3.1. Often, however, theoretical considerations already provide enough reason to model a nonlinear relationship: think about the correlation between female literacy rate and birth rate. You would expect a negative correlation for these two variables, and you would also expect the birth rate not to drop linearly toward zero. Rather, you would expect birth rate to decrease with an increase in literacy rate to levels of around one or two births per woman.

Nonlinear relationships occur quite often when income is used as an independent variable. For many relationships, income changes in the lower range of income have more impact on the dependent variable than income changes in the upper part of the income distribution. Income triples with a change from $500 to $1,500, whereas the increase is only 10% for a change from $10,000 to $11,000, although the increase is in both cases $1,000.

When modeling nonlinear relationships, you first need to know or at least hypothesize a functional form of the relationship. Here you need to distinguish among three basic types of nonlinear relations: logarithmic, hyperbolic, and U-shaped. Stylized versions of these relationships can be produced with the twoway plottype `function` (see [G] **graph twoway function**):

```
. tw || function y = x^3, yaxis(1) yscale(off axis(1))
>    || function y = ln(x), yaxis(2) yscale(off axis(2))
>    || function y = (-1)* x +  x^2, yaxis(3) yscale(off axis(3))
>    || , legend(label(1 "Hyperbolic") label(2 "Logarithmic") label(3 "U-Shaped"))
```

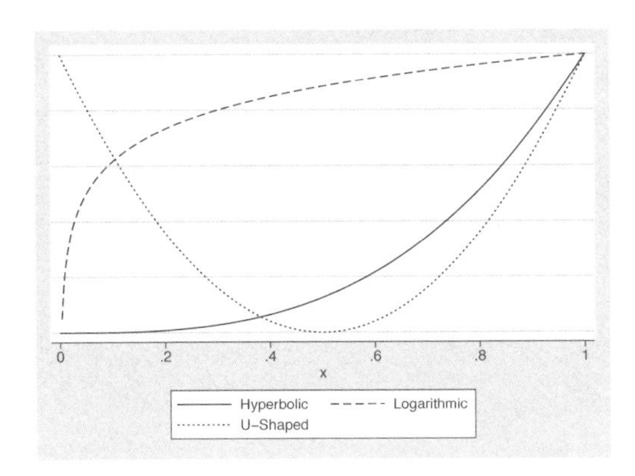

In logarithmic relationships, the dependent variable increases with increasing values of the independent variable. However, with increasing values of the independent variable, the increase in the dependent variable levels off. In hyperbolic relationships, the relation is reversed, as the dependent variable increases only moderately at the beginning and increases with increasing values of the independent variable. In U-shaped relationships, the sign of the effect of the independent variable changes. All three basic types can occur in opposite directions. For logarithmic relationships, this would mean that the values decrease rapidly at the beginning and more slowly later on. For hyperbolic relationships, the values drop slowly at the beginning and rapidly later on. For U-shaped relationships, the values first decrease and increase later on, or vice versa. In practice, logarithmic relationships occur often.

To model logarithmic relations, you first form the log of the independent variable and replace the original variable in the regression model with this new variable. A strong logarithmic relationship can be found between the countries' gross domestic product and infant mortality rate. The file `uno.dta` contains these data.[50]

```
. use uno, clear
. scatter infmort gdp
```

You can model this logarithmic relationship by first creating the log of the X variable

```
. generate loggdp = log(gdp)
```

and then using this variable instead of the original X variable:

```
. regress infmort loggdp
. predict yhat1
```

You can see the logarithmic relationship between the predicted value of the regression model `yhat1` and the untransformed independent variable:

```
. line yhat1 gdp, sort || scatter infmort gdp
```

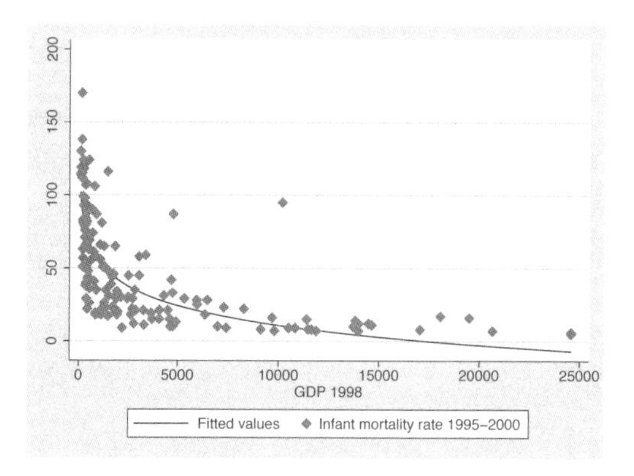

50. The original example was introduced by Fox (2000), and the values have been updated using information provided by the United Nations (http://unstats.un.org/unsd/demographic/default.htm).

You use a similar procedure to model hyperbolic relations, except that now you square the original variable instead of taking its logarithm. Here the original variable is also replaced by the newly transformed variable.[51]

The situation is different when you are modeling a U-shaped relationship. Although you still square the independent variable, the newly generated variable does not replace the original variable. Instead both variables will be used in the regression model. A U-shaped relation is one of the examples in *Anscombe's quartet* on page 204. Including the quadratic term will allow you to model this correlation perfectly:

```
. use anscombe, clear
. generate x2q = x2^2
. regress y2 x2 x2q
. predict yhat
. line yhat x2, sort || scatter y2 x2
```

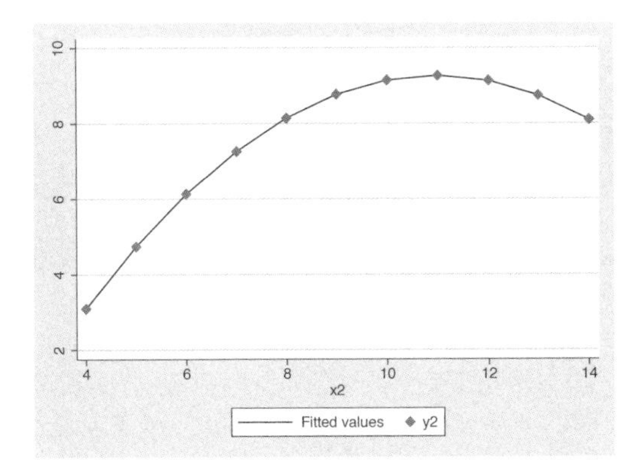

If you are thinking of using transformations of the independent variables, see Cook and Weisberg (1999, chapter 16) for some precautions.

Eliminating heteroskedasticity

In section 8.3.2, we discussed skewed dependent variables as one of the possible causes of heteroskedasticity. Here you would need to transform the dependent variable to remove heteroskedasticity. The interpretation of the regression model changes when you include a transformed variable. Transforming the dependent variable leads to a nonlinear relationship between the dependent and *all* independent variables (Hair et al. 1995, 75).

The aim in transforming a variable is to obtain a fairly symmetric or normal dependent variable. Remember the following rule of thumb: If the distribution is wide, the inverse of the variable is a useful transformation ($1/Y$). If the distribution is skewed to

51. Examples for hyperbolic relations are rare in the social sciences. The salary of a Formula 1 driver could possibly show a hyperbolic relation to the number of Grand Prix victories.

the right (such as home size in our example), taking the log is reasonable, and you can take the square root if the distribution is skewed to the left (Fox 1997, 59–82).

Besides following these rules, you can use the Stata command bcskew0, which uses a Box–Cox transformation that will lead to a (nearly) unskewed distribution.[52]

```
. use data1, clear
. bcskew0 bcsqfeet = sqfeet
```

The residual-versus-fitted plot (page 205) tells you something about the type of transformation necessary. If the spread of the residuals increases with increasing values in the predicted variable, the inverse of Y is a better dependent variable. If there is a decreasing spread of the residuals with increasing values in the predicted variable, you will want to replace the original dependent variable Y with the square root of Y (Hair et al. 1995, 70).

8.5 More on standard errors

We have mentioned that, to be valid, the standard errors reported by regress require you to make a host of assumptions. Multicollinearity, for example, typically causes standard errors to be inflated, making estimated coefficients appear insignificant. Heteroskedasticity also affects standard errors. Here we briefly introduce two Stata commands that address this point: bootstrap and svy.

These include the standard calculation based on the observed or the expected information matrix; the robust or sandwich estimator; the bootstrap; the jackknife; and in cases where the data arise from a complex sample, the linearization/robust method, balanced repeated replication, or a design-appropriate jackknife.

8.5.1 Bootstrap techniques

Confidence intervals are estimated under the assumption that the regression coefficients are normally distributed. From this assumption, you can multiply the standard errors with the critical value.[53] However, with finite samples the coefficients may not have a normal distribution. Therefore, we want to introduce a different technique to get confidence intervals. The technique is called the “bootstrap”, suggesting that we have to help ourselves using the information at hand instead of relying on distribution assumptions.[54]

---

52. Make sure that the variable used in bcskew0 does not include negative values or the value zero.

53. The critical value for a 95% confidence interval for models with more than 120 degrees of freedom is 1.96. Remember that the degrees of freedom is the number of cases minus the number of coefficients including the constant b_0.

54. See Mooney and Duval (1993, 42) for a brief overview of the different bootstrap techniques. A detailed explanation is given by Efron and Tibshirani (1993) and a pedagogical introduction is given by Stine (1990).

The bootstrap technique is based on the assumption that all the information you have about your population is contained within the sample data, meaning you use *only* the sample data to assess the population. Let us step back for a second. Assume that you could draw as many samples from the population as you wanted, and you compute confidence intervals for the estimated regression coefficient for each of these samples; about 95% of all 95% confidence intervals computed in this way would include the true value of the coefficient. Unfortunately, you cannot replicate the sample that many times, but you can use the bootstrap technique. Applying the bootstrap technique, you draw many samples out of your sample, where each sample has a similar number of observations to the original sample. In other words, you treat your original sample as if it were the population and then repeatedly draw samples from it. This may sound strange, but all that is required is sampling with replacement. Some observations will probably appear twice or three times in one of the new samples.

Now you estimate the statistic (e.g., the mean of a variable, or here the regression coefficient β) for each of the samples that has been drawn out of the original sample. The distribution of all the estimation results will let you compute confidence interval boundaries for the statistic of interest to you. According to the "percentile method", the 95% confidence interval is formed by the value above which there are 2.5% of all estimation results, and the value below which there are 2.5% of all estimation results.

In Stata, the `bootstrap` prefix command is used to perform the bootstrap.[55] All you need to do is prefix your estimation command with the keyword `bootstrap:`. You can optionally specify the number of samples to draw. If you want to interpret the confidence intervals, we suggest drawing at least 1,000 samples. This might take a while on your machine. Finally, the entire procedure is based on a random process, so unless you reset the random-number seed each time, you will get slightly different results every time you run your do-file.

```
. set seed 731
. bootstrap: regress sqfeet hhinc hhsize
```

To obtain the percentile-based confidence intervals we just discussed, use the `estat bootstrap` command afterward. `bootstrap` itself reports confidence intervals that assume the coefficients are normally distributed, which we have just argued is not always appropriate.

55. Prior to Stata 9, the `bootstrap` command was used, which follows a different syntax. Type `help bootstrap` if you are using an older version of Stata.

```
. estat bootstrap, percentile
Linear regression                                 Number of obs      =      3126
                                                  Replications       =        50

------------------------------------------------------------------------------
             |    Observed               Bootstrap
      sqfeet |       Coef.       Bias    Std. Err.  [95% Conf. Interval]
-------------+----------------------------------------------------------------
       hhinc |   .16502241  -.0015253    .0089156    .1484383   .1790049   (P)
      hhsize |   40.962581   .6106859   6.5940474    27.83905   52.51157   (P)
       _cons |   520.57536   2.741462   21.306478     489.114   574.4759   (P)
------------------------------------------------------------------------------
(P)    percentile confidence interval
```

To apply the bootstrap technique, you need to know how your sample is drawn from the population because this sampling process must be followed during the bootstrap sampling. To introduce the bootstrap, we described the simplest case of a simple random sample. Applying bootstrap techniques to most national survey data is therefore more complex than what we have introduced here since you need to replicate the sampling of, for example, a cluster sample. You can find ways to do so by typing `help bootstrap`.

8.5.2 Confidence intervals in cluster samples

Many surveys are not based on simple random sampling but on multistage clustered samples. One example for a two-stage cluster sample is the selection of hospitals and the selection of patients within each selected hospital. An example of a multistage cluster sample would be the selection of schools, the selection of classrooms within each school, and within each classroom the selection of students (Levy and Lemeshow 2008, 225). The GSOEP, which is the basis of the dataset `data1.dta` used here, is also a multistage cluster sample (Pannenberg et al. 1998). At the first stage, election districts were sampled from a nationwide register, and later respondents were drawn using what is called a random walk.

The decision to use a clustered sample (and not a simple random sample) can be made for organizational or financial reasons. For example, the absence of a general population register in many countries reduces many researchers' ability to use simple random samples. Sampling in several sampling stages, one of them at the level of small geographical clusters, allows the selection of respondents without register data. However, there is a drawback to using a clustered sample: the standard errors can be larger than they would be under simple random sampling (Kish 1965). One reason for this is the relative homogeneity of observations within a cluster. For example, respondents within the same neighborhood can probably afford similar housing and belong to a similar income group, therefore giving similar answers regarding income. Another reason for larger standard errors may be the way data were collected, where each interviewer might influence the responses (Schnell and Kreuter 2005).

Also samples are often stratified; that is, population elements are divided into exhaustive subgroups and sampling takes place within each of these subgroups. Unlike clustering, stratification can reduce the standard errors if the stratification variable is correlated with the outcome of interest. However, you will need to take sampling design information into account.

Stata provides special commands to estimate the correct standard errors for complex samples. You can find detailed information in Kreuter and Vailliant (2007). If you want to take the potentially homogenizing effects of interviewers into account for our example, you can use the following command sequence:

```
. svyset intnr

      pweight: <none>
          VCE: linearized
  Single unit: missing
     Strata 1: <one>
         SU 1: intnr
        FPC 1: <zero>

. svy: regress sqfeet hhinc hhsize
(running regress on estimation sample)

Survey: Linear regression

Number of strata   =         1                  Number of obs      =      2661
Number of PSUs     =       429                  Population size    =      2661
                                                Design df          =       428
                                                F(   2,      427)  =    216.53
                                                Prob > F           =    0.0000
                                                R-squared          =    0.1808

------------------------------------------------------------------------------
             |             Linearized
      sqfeet |      Coef.   Std. Err.      t    P>|t|     [95% Conf. Interval]
-------------+----------------------------------------------------------------
       hhinc |   .1563405   .0086396    18.10   0.000     .1393592    .1733217
      hhsize |   41.89213   6.339742     6.61   0.000     29.43122    54.35303
       _cons |   527.1673   21.69781    24.30   0.000     484.5198    569.8148
------------------------------------------------------------------------------
```

The `svyset` command specifies the name of a variable that contains identifiers for the primary sampling units (clusters). We use the variable `intnr` as our best proxy for primary sampling units. You can parse this specification to the regression by putting the `svy` prefix in front of the regression command. The same technique applies to other statistical models as well.

You now obtain results for 2,661 participants, who had been interviewed by 429 interviewers.[56] Here the difference between the correct confidence intervals and confidence intervals estimated assuming a simple random sample is negligible.[57] Make sure that your data file includes variables that contain sampling design information so that you have information (about the sampling point and about the interviewer) so you can check your results for design effects. An excellent resource for the application of advanced techniques in Stata is the book by Rabe-Hesketh and Everitt (2007).

56. The interviewer information is missing for 561 respondents in this data file.

57. This looks quite different if you look at the confidence intervals for the mean of the variables indicating respondents' concerns, for example, `np9506` with `svy: mean np9506`. With `svy: mean np9506` followed by `estat effects` you get a design effect of 1.45, which means that the confidence intervals are about 1.45 times higher than the ones estimated assuming simple random sampling.

8.6 Advanced techniques

As well as multiple linear regression, there are several related models that can be estimated in Stata. We cannot explain all of them in detail. However, a few of these models are so common that we want to describe the general ideas behind them. Each model is explained in detail in the *Stata Reference Manual*, where you will also find selected literature on the model.

8.6.1 Median regression

A median regression is quite similar to the OLS regression we talked about earlier. Whereas the sum of the squared residuals $\sum(\widehat{y}_i - y_i)^2$ is minimized in OLS regression, the sum of the absolute residuals $\sum|\widehat{y}_i - y_i|$ is minimized when applying median regression. Squaring residuals in OLS means that large residuals are more heavily weighted than small residuals. This property is lost in median regression, so it is less sensitive to outliers than OLS regression.

Median regression takes its name from its predicted values, which are estimates of the median of the dependent variable conditional on the values of the independent variables. In OLS, the predicted values are estimates of the conditional means of the dependent variable. The predicted values of both regression techniques, therefore, describe a measure of a certain property—the central tendency—of the dependent variable.

Stata treats median regression as a special case of a quantile regression. In quantile regression, the coefficient is estimated so that the sum of the weighted (i.e., multiplied by the factor w_i) absolute residuals is minimized.

$$\min \sum(|y_i - \widehat{y}_i| \times w_i)$$

Weights can be different for positive and negative residuals. If positive and negative residuals are weighted the same way, you get a median regression. If positive residuals are weighted by the factor 1.5 and negative residuals are weighted by the factor 0.5, you get a "3rd quartile regression", etc.

In Stata, you estimate quantile regressions using the `qreg` command. Just as in any other Stata model command, the dependent variable follows the command and then you specify the list of independent variables; the default is a median regression.

For this, use the dataset `data2agg.dta`, which contains the mean life satisfaction and the mean income data from the German population from 1984 to 2002.[58]

```
. use data2agg, clear
```

First, take a look at a scatterplot with the regression line of the mean life satisfaction on the mean income:

58. We used this small dataset to exemplify the effect of median regression. Be aware that working with aggregate data is prone to ecological fallacy (Freedman 2004).

```
. twoway lfit lsat inc || scatter lsat inc, mlabel(wave)
```

In this graph, the data for 1984 might influence the regression results more than any other data point. Now estimate a median regression

```
. qreg lsat inc
```

and compare the predicted values of the median regression with the standard linear fit of the OLS regression:

```
. predict medhat
. twoway lfit lsat inc || scatter lsat inc, mlabel(wave)
>                      || line medhat inc, sort
>                      ||, legend(order(1 "OLS" "Median"))
```

The regression line of the median regression is not as steep as the standard regression line because the median regression is more robust to extreme data points, such as those from 1984.

8.6.2 Regression models for panel data

Panel data, or cross-sectional time-series data, contain repeated measures of the same individuals over time. An example of panel data is the German Socio-Economic Panel (GSOEP). In the GSOEP, about 12,000 persons have been asked identical questions every year since 1984. That is, the GSOEP measures the *same* variables for the *same* respondents at *different* points in time. Panel data, however, do not arise only from such panel surveys. The same data structure is also present if you have collected certain macroeconomic indices in many different countries over time, or even data about certain features of political parties over time. Really, what defines panel data is that the *same* entities are observed at different times. In the remaining section, we will use the term "individuals" for these entities.

In Stata, all the commands that deal with panel data begin with the letters `xt`, and these commands are described in the *Longitudinal/Panel Data Reference Manual* [XT]. A list of the `xt` commands can be found by typing `help xt`. Among the `xt` commands are some of the more complex models in the statistical universe, which we will not describe here. Instead, we will help you understand the thinking behind the major approaches to analyzing panel data together with examples of how you can use these approaches in Stata.[59]

Before we describe the statistical models, we need to say a word about data management. All Stata commands for panel analysis require a panel dataset that is in long format, so the next section describes how to put your data in this format. Then we will explain fixed-effects models and error-components models.

59. For more information, see Baltagi (2008); Baum (2006); Hardin and Hilbe (2003); Diggle, Liang, and Zeger (1994); Wooldridge (2002); and the literature cited in [XT] **xtreg**.

From wide to long format

Panel data can be stored in wide format or in long format. In wide format, the observations of the dataset are the individuals observed, and the variables are their characteristics at the respective time points. For example, if we ask four specific individuals, say, John, Paul, George, and Ringo, about their life satisfaction in 1968, 1969, and 1970, we can store their answers in wide format by making a dataset with four observations, namely, John, Paul, George, and Ringo, and three variables reflecting life satisfaction in 1968, 1969, and 1970, respectively (see table 8.2). However, the same information can also be stored in long format, where the observations are the individuals *at a specific point in time* and the variables are the observed characteristics. Hence, in our example, there would be three observations for John—one for 1968, one for 1969, and one for 1970—three observations for Paul, etc. The information on life satisfaction would be in one variable. To keep the information about the timing, we would need a new variable for the year of observation.

Table 8.2. Ways to store panel data

Wide format

i	X_{1968}	X_{1969}	X_{1970}
John	7	8	5
Paul	5	2	2
George	4	3	1
Ringo	8	8	6

Long format

i	year	X
John	1968	7
John	1969	8
John	1970	5
Paul	1969	5
Paul	1969	2
Paul	1970	2
George	1968	4
⋮	⋮	⋮
Ringo	1970	6

Stata's `xt` commands generally expect panel data in long format. It is, however, more common for dataset providers to distribute panel data in wide format.[60] You will often need to reshape your dataset from wide to long.

An example of a panel data in wide format is `data2w.dta`. Please load this dataset to follow our example of changing from wide format to long format:

```
. use data2w, clear
```

60. For very large panel studies, such as the GSOEP, the American Panel Study of Income Dynamics (PSID), or the British Household Panel Study (BHPS), the situation tends to be even more complicated. These data are often distributed in more than one file. You would need to first combine these files into one file. In section 10.4, we show you how to do this using an example from the GSOEP, resulting in a dataset in wide format.

This file contains information on year of birth, gender, life satisfaction, marital status, individual labor earnings, and annual work hours of 1,761 respondents (individuals) from the German Socio-Economic Panel (GSOEP). The individuals were observed every year between 1984 and 2002. Therefore, with the exception of the time-invariate variables gender and year of birth, there are 19 variables for each observed characteristic. If you look at the file with

```
. describe
Contains data from data2w.dta
  obs:         1,761                          GSOEP 1984-2002 randomized
                                                (Kohler/Kreuter)
 vars:            80                          1 Sep 2004 09:24
 size:       294,087 (99.1% of memory free)   (_dta has notes)
-------------------------------------------------------------------------------
              storage  display     value
variable name   type   format      label      variable label
-------------------------------------------------------------------------------
hhnr            long   %12.0g                 Fix Household Number
persnr          long   %12.0g                 Person ID (n)
sex             byte   %13.0g      sex        Gender (n)
gebjahr         int    %8.0g                  Year of birth (n)
lsat1984        byte   %45.0g      sat        General Life Satisfaction
mar1984         byte   %20.0g      d1110484 * Marital Status of Individual (n)
hour1984        int    %12.0g                 Annual Work Hours of Individual
                                                (n)
inc1984         float  %9.0g                * Individual Labor Earnings (n)
lsat1985        byte   %45.0g      sat        General Life Satisfaction
mar1985         byte   %20.0g      d1110485 * Marital Status of Individual (n)
hour1985        int    %12.0g                 Annual Work Hours of Individual
                                                (n)
inc1985         float  %9.0g                * Individual Labor Earnings (n)
lsat1986        byte   %45.0g      sat        General Life Satisfaction
mar1986         byte   %20.0g      d1110486 * Marital Status of Individual (n)
  (output omitted)
-------------------------------------------------------------------------------
```

you will see that the variable names of the file have a specific structure. The first part of the variable names, namely, `lsat`, `mar`, `hour`, and `inc`, refers to the content of the variable, whereas the second part refers to the year in which the variable has been observed. Using this type of naming convention makes it easy to reshape data from wide to long.

Unfortunately, in practice variable names rarely follow this naming scheme. Even the variables in the GSOEP do not. For your convenience, we have renamed all the variables in the dataset beforehand, but generally you will need to do this on your own using the `rename` and `renpfix` commands. Renaming all the variables of panel data in wide format can be quite cumbersome. In the do-file `crdata2.do`, we have therefore constructed some loops, which use concepts explained in sections 3.2.2 and 11.2.1. If you need to rename many variables, you should review these concepts.[61]

61. The user-written Stata command `soepren` makes it easier to rename GSOEP variables. The command is available on the SSC archive; for information about the SSC archive and installing user-written commands, see chapter 12.

The command for changing data between wide and long is `reshape`. `reshape long` changes a dataset from wide to long, and `reshape wide` does the same in the other direction. Stata needs to know three pieces of information to reshape data:

- the variable that identifies the individuals in the data (i.e., the respondents),
- the characteristics that are under observation, and
- the times when the characteristics were observed.

The first piece of information is easy to obtain. In our example data, it is simply the variable `persnr`, which uniquely identifies each individual of the GSOEP. If there is no such variable, you can simply generate a variable containing the running number of each observation (see section 5.1.3).

The last two pieces of information are coded in the variable names. As we have seen, the first part of the variable names contains the characteristic under observation, and the second part contains the time of observation. We therefore need to tell Stata where the first part of the variable names ends and the second part starts. This information is passed to Stata by listing the variable name stubs that refer to the characteristic under observation. Let us show you how this works for our example:

```
. reshape long inc lsat mar hour, i(persnr) j(wave)
```

First, option `i()` is required. It is used to specify the variable for the individuals of the dataset. Second, look at what we have specified after `reshape long`. We have listed neither variable names nor a *varlist*. Instead we have specified the name stubs that refer to the characteristic under observation. The remaining part of the variable names is then interpreted by Stata as being information about the time point of the observation. When running the command, Stata strips off the year from the variables that begin with the specified name stub and stores this information in a *new* variable. Here the new variable is named `wave`, as we specified this name in the option `j()`. If we had not specified that option, Stata would have used the variable name `_j`.

Now let us take a look at the new dataset.

```
. describe
Contains data
  obs:        33,459                          GSOEP 1984-2002 randomized
                                                (Kohler/Kreuter)
 vars:             9
 size:       836,475 (97.3% of memory free)   (_dta has notes)
-------------------------------------------------------------------------------
              storage  display     value
variable name   type   format      label      variable label
-------------------------------------------------------------------------------
persnr          long   %12.0g                 Person ID (n)
wave            int    %9.0g
hhnr            long   %12.0g                 Fix Household Number
sex             byte   %13.0g      sex        Gender (n)
gebjahr         int    %8.0g                  Year of birth (n)
lsat            byte   %20.0g      sat
mar             byte   %20.0g      d1110402 *
hour            int    %12.0g
inc             float  %9.0g                *
                                            * indicated variables have notes
-------------------------------------------------------------------------------
Sorted by:  persnr  wave
     Note:  dataset has changed since last saved
. list persnr wave lsat

     +------------------------------+
     | persnr   wave           lsat |
     |------------------------------|
  1. |     76   1984              8 |
  2. |     76   1985              8 |
  3. |     76   1986              7 |
  4. |     76   1987              8 |
  5. |     76   1988              8 |
     |------------------------------|
  (output omitted)
```

The dataset now has nine variables instead of 80. Clearly there are still 1,761 *individuals* in the dataset, but since the observations made on the several occasions for each individual are stacked beneath each other, we end up with 33,459 *observations*. Hence, the data are in long format, as they must be to use the commands for panel data. And working with the `xt` commands is even more convenient if you declare the data to be panel data. You can do this with the `xtset` command by specifying the variable that identifies individuals followed by the variable that indicates time:

```
. xtset persnr wave
```

After reshaping the data once, reshaping from long to wide and vice versa is easy:

```
. reshape wide
. reshape long
```

Fixed-effects models

If the data are in long format, you can now run a simple OLS regression. For example, if you want to find out whether aging has an effect on general life satisfaction, you could run the following regression:

```
. gen age = wave - gebjahr

. replace lsat = .a if lsat < 0
(82 real changes made, 82 to missing)

. regress lsat age

      Source |       SS       df       MS              Number of obs =   33377
-------------+------------------------------           F(  1, 33375) =    3.21
       Model |  10.2027446     1  10.2027446           Prob > F      =  0.0733
    Residual |  106166.227 33375  3.18101053           R-squared     =  0.0001
-------------+------------------------------           Adj R-squared =  0.0001
       Total |  106176.429 33376  3.18122092           Root MSE      =  1.7835

------------------------------------------------------------------------------
        lsat |      Coef.   Std. Err.      t    P>|t|     [95% Conf. Interval]
-------------+----------------------------------------------------------------
         age |  -.0006269     .00035    -1.79   0.073     -.001313    .0000592
       _cons |   7.195336   .0192932   372.95   0.000      7.15752    7.233151
------------------------------------------------------------------------------
```

From this regression model, you learn that with age, life satisfaction tends to decrease, but since the estimated coefficient is not significant, you might also say that age does not affect life satisfaction. However, after having read this chapter, you probably do not want to trust this regression model, particularly because of omitted variables. Should you control the relationship for quantities like gender, education, and the historical time in which the respondents grew up?

Now let us imagine that you include a dummy variable for each GSOEP respondent. As there are 1,761 individuals in the dataset, this would require a regression model with 1,760 dummy variables, which might be overwhelming to work with. But for small datasets like those presented in table 8.2, this is not a problem. So let us deal with these data for now.

```
. preserve

. use beatles, clear

. describe

Contains data from beatles.dta
  obs:            12
 vars:             4                          7 Jul 2004 15:36
 size:           108 (99.9% of memory free)
-------------------------------------------------------------------------------
              storage  display     value
variable name   type   format      label      variable label
-------------------------------------------------------------------------------
persnr          byte   %9.0g       name       Person
time            int    %9.0g                  Year of observation
lsat            byte   %9.0g                  Life Satisfaction (fictive)
age             byte   %9.0g                  Age in Years
-------------------------------------------------------------------------------
Sorted by:
```

This dataset contains the age and (artificial) life satisfaction of four Englishmen at three points in time in long format. The command

```
. regress lsat age

      Source |       SS       df       MS              Number of obs =      12
-------------+------------------------------           F(  1,    10) =    1.97
       Model |   13.460177     1   13.460177           Prob > F      =  0.1904
    Residual |  68.2064897    10  6.82064897           R-squared     =  0.1648
-------------+------------------------------           Adj R-squared =  0.0813
       Total |  81.6666667    11  7.42424242           Root MSE      =  2.6116

------------------------------------------------------------------------------
        lsat |      Coef.   Std. Err.      t    P>|t|     [95% Conf. Interval]
-------------+----------------------------------------------------------------
         age |   .6902655   .4913643     1.40   0.190    -.4045625    1.785093
       _cons |  -14.32153   13.65619    -1.05   0.319    -44.74941    16.10635
------------------------------------------------------------------------------
```

mirrors the regression analysis from above, showing a slight insignificant, positive effect of age on life satisfaction. Incorporating dummy variables for each individual of the dataset into this regression is straightforward.

```
. tabulate persnr, generate(d)

     Person |      Freq.     Percent        Cum.
------------+-----------------------------------
       John |          3       25.00       25.00
       Paul |          3       25.00       50.00
     George |          3       25.00       75.00
      Ringo |          3       25.00      100.00
------------+-----------------------------------
      Total |         12      100.00
. regress lsat age d2-d4

      Source |       SS       df       MS              Number of obs =      12
-------------+------------------------------           F(  4,     7) =   90.95
       Model |      80.125     4    20.03125           Prob > F      =  0.0000
    Residual |  1.54166667     7  .220238095           R-squared     =  0.9811
-------------+------------------------------           Adj R-squared =  0.9703
       Total |  81.6666667    11  7.42424242           Root MSE      =   .4693

------------------------------------------------------------------------------
        lsat |      Coef.   Std. Err.      t    P>|t|     [95% Conf. Interval]
-------------+----------------------------------------------------------------
         age |     -1.625    .165921    -9.79   0.000    -2.017341   -1.232659
          d2 |  -6.916667   .5068969   -13.65   0.000    -8.115287   -5.718046
          d3 |  -8.541667   .6281666   -13.60   0.000    -10.02704   -7.056289
          d4 |   1.333333    .383178     3.48   0.010     .4272613    2.239405
       _cons |   53.45833    4.81933    11.09   0.000     42.06243    64.85424
------------------------------------------------------------------------------
```

Now it appears that age has a strong negative effect on life satisfaction. The sign of the age effect has reversed, and we will soon see why. But let us first say something about the individual dummies. The estimated coefficients of the individual dummies reflect how strongly the life satisfaction of the four Englishmen differs. You can see that persons 1 and 4 have a much higher life satisfaction than persons 2 and 3. You do not know *why* these people differ in their life satisfaction; the differences are not surprising since different people perceive life differently. Maybe they live in different neighborhoods, have different family backgrounds, grew up under different circumstances, or just have different habits about answering odd questions in population surveys. What is important here is that, since you put individual dummies into the regression model, you have reliably controlled for any differences between the persons. In this sense, the estimated coefficient for age cannot be biased because we omitted stable characteristics of these persons. It is a pure aging effect, which rests solely on the development of the life satisfaction during the aging process of these four men.

This interpretation of the estimated age coefficient can be illustrated with the following plot.

```
. predict yhat
. separate lsat, by(persnr)
. separate yhat, by(persnr)
. line yhat1-yhat4 age, clstyle(p1 p1 p1 p1)
> || lfit lsat age, clstyle(p2)
> || scatter lsat1-lsat4 age
> || , legend(order(1 2 3 4 5 9) label(5 "Fixed Effects Pred.")
>            label(9 "Standard OLS Pred."))
```

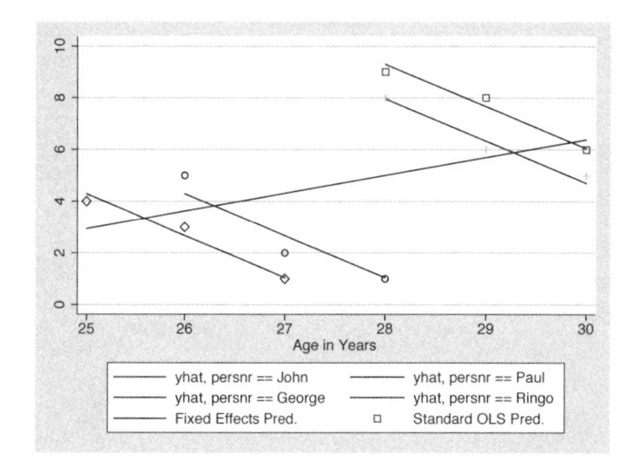

The plot is an overlay of a standard scatterplot with different markers for each person in the dataset (`scatter lsat1-lsat4 age`), a conditional-effects plot of the regression model with the person dummies (`line yhat1-yhat4 age`), and a simple regression line for all the data (`lfit lsat age`). If you look at the markers for each person separately, you will find that the life satisfaction decreases as the person gets older. At the same time, however, Ringo and John, the two oldest people in the dataset, have a higher life satisfaction than Paul and George. If we do not control for this, differences between people contribute to the age effect. The age effect of the simple OLS regression just shows that the older people have a higher life satisfaction than the younger ones. After we control for the personal differences, the only variation left is that within each person, and the age effect reflects the change in life satisfaction as each person gets older.

As the regression model with person dummies restricts itself on the variation within each person, the model is sometimes called the *within estimator*, covariance model, individual dummy-variable model, or *fixed-effects model*.

Whereas the derivation of the fixed-effects model is straightforward, the technical calculation of the model in huge datasets is not. The problem arises because the number of independent variables in regression models is restricted to 800 in Stata/IC and to 11,000 in Stata/MP and Stata/SE. Therefore, you cannot estimate a fixed-effects model by incorporating individual dummies in datasets with more than 800 or 11,000 individuals, respectively.

Fortunately, in the linear regression model, you can use an algebraic trick to estimate the fixed-effects model, anyway. And even more fortunately, Stata has a command that does this algebraic trick for you: `xtreg` with the option `fe`. Here you can use the command for our small example data

```
. xtreg lsat age, fe i(persnr)
```

which reproduces the estimated age coefficient of the model with dummy variables exactly. You do not need to list the dummy variables in this command. Instead you either `xtset` your data or else specify the name of the variable, which identifies the individuals in the option `i()`.

The same logic applies if you want to estimate the fixed-effects model for larger datasets. Therefore, you can also use the same command with our previously constructed dataset. As you have already used the command `xtset` above (see page 241), you do not need to specify the `i()` option.

```
. restore
. xtreg lsat age, fe
```

The values of the estimated coefficients for the 1,761 dummy variables are not shown in the output and were not estimated. But the coefficient for age in the model is estimated as if the dummy variables were present. The fixed-effects model controls for all time-invariant differences between the individuals, so the estimated coefficients of the fixed-effects models cannot be biased because of omitted time-invariant characteristics. This feature makes the fixed-effects model particularly attractive.

One side effect of the features of fixed-effects models is that they cannot be used to investigate time-invariant causes of the dependent variables. Technically, time-invariant characteristics of the individuals are perfectly collinear with the person dummies. Substantively, fixed-effect models are designed to study the causes of changes within a person. A time-invariant characteristic cannot cause such a change, because it is constant for each person.

8.6.3 Error-components models

Let us begin our description of error-components models with the simple ordinary least-squares regression:

```
. regress lsat age
```

This model ignores the panel structure of the data and treats data as cross-sectional. From a statistical point of view, this model violates an underlying assumption of ordinary least-squares regression, namely the assumption that all observations are independent of each other. In panel data, you can generally assume that observations from the same individual are more similar to each other than observations from different individuals.

In observing the similarity of the observations from one individual, you might say that the residuals of the above regression are correlated. That is, an individual with a high positive residual at the first time of observation should also have a high positive residual at the second time point, etc.

Let us show you that the residuals of the above regression model are in fact correlated. First, calculate the residuals from the above regression model:

```
. predict res, resid
```

and then we change the dataset to the wide format. Since you have generated a new variable in the long format since last using `reshape`, you cannot just type `reshape wide`; instead, you need to use the full syntax:

```
. reshape wide lsat mar hour inc age res, i(persnr) j(wave)
```

You end up with 19 variables containing the residuals for each individual for every year. These variables can be used to construct a correlation matrix of the residuals. We will display this correlation matrix only for the residuals from the eighties:

```
. corr res198?
(obs=1741)

             |  res1984  res1985  res1986  res1987  res1988  res1989
-------------+------------------------------------------------------
     res1984 |   1.0000
     res1985 |   0.4406   1.0000
     res1986 |   0.3861   0.4570   1.0000
     res1987 |   0.3705   0.4200   0.5006   1.0000
     res1988 |   0.3310   0.3590   0.4276   0.5241   1.0000
     res1989 |   0.3159   0.3411   0.4164   0.4755   0.5098   1.0000
```

The residuals are in fact highly correlated. Let us now define this correlation matrix as $\mathbf{R}_{t,s}$

$$\mathbf{R}_{t,s} = \begin{pmatrix} 1 & & & \\ r_{e_{i2},e_{i1}} & 1 & & \\ \vdots & \vdots & \ddots & \\ r_{e_{iT},e_{i1}} & r_{e_{iT},e_{i2}} & \cdots & 1 \end{pmatrix}$$

As we said, in computing the simple OLS regression on panel data, you assume, among other things, that all correlations of this correlation matrix are 0, or more formally

$$\mathbf{R}_{t,s} = \begin{cases} 1 & \text{for } t = s \\ 0 & \text{otherwise} \end{cases}$$

As we have seen, this assumption is not fulfilled in our example regression. Hence, the model is not correctly specified. This is almost always the case for panel data. With panel data, you should expect correlated errors. In error-components models, you can therefore hypothesize about the structure of $\mathbf{R}_{t,s}$. Probably the simplest model after the simple regression model is the random-effects model:

$$\mathbf{R}_{t,s} = \begin{cases} 1 & \text{for } t = s \\ \rho & \text{otherwise} \end{cases}$$

Here the hypothetical structure of $\mathbf{R}_t, s$ is that observational units are more similar to each other over time than observations across observational units. The Stata command for random-effects models is `xtreg` with the option `re`.

```
. reshape long
. xtreg lsat age, re
```

Another reasonable assumption for the correlation structure of the residuals might be that the similarity between observations within each observational units is greater the shorter the elapsed time between the observations. This structure can be imposed using an AR(1) correlation matrix:[62]

$$\mathbf{R}_{t,s} = \begin{cases} 1 & \text{for } t = s \\ \rho^{|t-s|} & \text{otherwise} \end{cases}$$

Different structures for the correlation matrix allow for a nearly infinite number of model variations. All these variations can be estimated using the `xtgee` command with the `corr()` option for specifying predefined or customized correlation structures. Typing

```
. xtgee lsat age, corr(exchangeable)
```

specifies the random-effects model and produces results similar to those from `xtreg, re`.[63]

```
. xtgee lsat age, corr(ar1)
```

which produces a model with an AR(1) correlation matrix. Typing

```
. xtgee lsat age, corr(independent)
```

produces the standard OLS regression model described at the beginning of this section.

62. AR is short for autoregression.

63. `xtgee, corr(exchangeable)` and `xtreg, re` produce slightly different results because of implementation details that are too technical to discuss here. In practice, the results are usually quite similar.

You can interpret the estimated coefficients of error-components models just like the estimated coefficients of a simple OLS regression model. But unlike in the simple OLS model, in an error-components model, if the error structure is correctly specified, the estimates are more accurate. As the estimated coefficients are based on variations within and between the individuals, you should have no problem investigating the effects of time-invariant independent variables on the dependent variable. Unlike in the fixed-effects model, the estimated coefficients can be biased because of omitted time-invariant covariates.

8.7 Exercises

1. Set up a regression model with data of `data1.dta` to test the hypothesis that household income (`hhinc`) increases general life satisfaction (`np11701`). How many observations are used in your model? What is the estimated life satisfaction for respondents with a household income of $1,000? How does the predicted life satisfaction change when household income increases by $500?
2. Check whether the correlation between general life satisfaction and household income is due to factors that are influences of life satisfaction or correlated with income (like gender, age, education, employment status). How does the effect of household income change when you take these other variables into account?
3. Check your model for influential cases. List the three observations with the highest absolute value of the DFBETA for household income.
4. Check your model for nonlinearities. Discuss your findings. Go through the following three steps and see how your model improves.
    - Recalculate your regression model with log household income instead of household income, and add a squared age variable to the regression model.
    - Reiterate the regression diagnosis for the new model. What would you further do to improve your model?
    - Reiterate the process until you are satisfied.
5. Check the homoskedasticity assumption for your final model. What do you conclude? Recalculate the model with robust standard errors, if necessary.
6. Sociologist Ronald Inglehart has proposed the theory that younger generations are less materialistic than older generations.
    - Which testable implication has this hypothesis for your regression model?
    - Change your regression model to test this implication. Is Inglehart correct?
    - Create a conditional effects plot to illustrate the results of your analysis of Inglehart's hypotheses.

9 Regression models for categorical dependent variables

Researchers in the social sciences often deal with categorical dependent variables, whose values may be dichotomous (e.g., rented apartment, yes or no), nominal (party identification: CDU, SPD, or Green Party), or ordinal (no concerns, some concerns, strong concerns). Here we will present several procedures used to model variables such as these by describing a procedure for dealing with dichotomous dependent variables: logistic regression.

Logistic regression is most similar to linear regression, so we will explain it as an analogy to the previous chapter. If you have no experience or knowledge of linear regression, first read chapter 8 up to page 191.

As in linear regression, in logistic regression a dependent variable is predicted by a linear combination of independent variables. A linear combination of independent variables can look like this:

$$\beta_0 + \beta_1 x_{1i} + \beta_2 x_{2i} + \cdots + \beta_{K-1} x_{K-1,i}$$

Here x_{1i} is the value of the first independent variable for interviewee i, x_{2i} is the respective value of the second independent variable, and so on. The regression parameters $\beta_1, \beta_2, \ldots, \beta_{K-1}$ represent the weights assigned to the variables.

We did not say, however, that the mean of the dependent variable y_i is equal to that linear combination. In contrast to linear regression, in logistic regression you must consider a particular transformation of the dependent statistic. Why such a transformation is required and why linear regression is inappropriate are explained in section 9.1, whereas the transformation itself is explained in section 9.2.1.

Section 9.2.2 explains the method by which the logistic regression coefficients are estimated. As this explanation is slightly more difficult and is not required to understand logistic regression, you can skip it for now.

Estimating a logistic regression with Stata is explained in section 9.3. Then we discuss methods of verifying the basic assumptions of the model in section 9.4. The procedure for verifying the joint significance of the estimated coefficients is discussed in section 9.5, and section 9.6 demonstrates a few possibilities for refining the modeling of correlations.

For an overview of further procedures, in particular procedures for categorical variables with more than two values, see section 9.7.

As with linear regression (chapter 8), you will need to do some additional reading if you want to understand the techniques we describe fully. We suggest that you read Hosmer and Lemeshow (2000) and Long (1997).

9.1 The linear probability model

Why is linear regression not suitable for categorical dependent variables? Imagine that you are employed by an international ship safety regulatory agency and are assigned to take a closer look at the sinking of the *Titanic*. You are supposed to find out whether the seafaring principle of "women and children first" was put into practice or if there is any truth in the assumption made by the film *Titanic*, in which the first-class gentlemen took the places in the lifeboats at the expense of the third-class women and children.

For this investigation, we have provided you with data on the sinking of the *Titanic*.[1] Open the file by typing[2]

```
. use titanic2
```

and before you continue to read, make yourself familiar with the contents of the dataset by using the commands

```
. describe
. tab1 _all
```

You will discover that the file contains details on the age (`age2`), gender (`sex`), and passenger class (`class`) of the *Titanic*'s passengers, as well as whether they survived the catastrophe (`survived`).

To clarify the disadvantages of using linear regression with categorical dependent variables, we will go through such a model. First, we will investigate whether children really were rescued more often than adults. What would a scatterplot look like where the Y variable represents the variable for survival and the X variable represents age? You may want to sketch this scatterplot yourself.

The points can be entered only on two horizontal lines: either at the value 0 (did not survive) or at 1 (survived). If children were actually rescued more often than adults, the number of points on the 0-line should increase in relation to those on the 1-line the farther to the right you go. To check whether your chart is correct, type

1. The data provided are real. The dataset and its exact description can be found at http://amstat.org/publications/jse/archive.htm. For teaching purposes, we have changed the original dataset in that we have divided adults and children into further fictional age groups, as the original set differentiates merely between adults and children. Among the files you installed in the steps discussed in the *Preface* is the do-file we used to change the dataset, (`crtitanic2.do`), as well as the original dataset (`titanic.dta`).
2. Please make sure that your working directory is `c:\data\kk2`; see page 3.

```
. scatter survived age2
```

This diagram is not particularly informative, as the plot symbols are often directly marked on top of each other, hiding the number of data points.

With the help of the `scatter` option `jitter()`, you can produce a more informative diagram. `jitter()` adds a small random number to each data point, thus showing points that were previously hidden under other points. Within the brackets is a number between 1 and 30 that controls the size of the random number; you should generally use small numbers if possible.

```
. scatter survived age2, jitter(2)
```

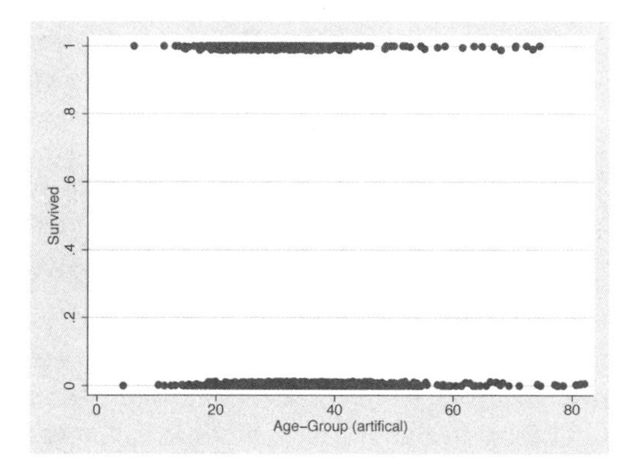

On examining the chart, you might get the impression that there is a negative correlation between ages and survival of the *Titanic* disaster. This impression is confirmed when you draw the regression line on the chart (see also section 8.1.2):

```
. regress survived age
. predict yhat
. scatter survived age2, jitter(2) || line yhat age2, sort
```

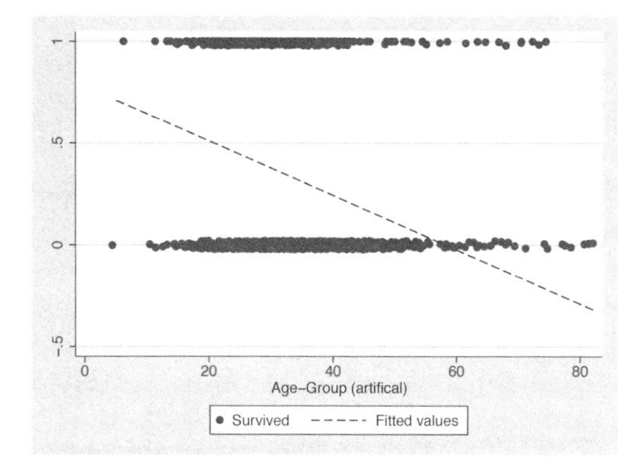

The chart reveals one central problem of linear regression for dichotomous dependent variables: the regression line in the illustration shows predicted values of less than 0 from around the age of 60 onward. What does this mean with regard to the content? Remind yourself of how the predicted values of dichotomous dependent variables are generally interpreted. Until now, we have understood the predicted values to be the estimated average extent of the dependent variables for the respective combination of independent variables. In this sense, you might say, for example, that the survival of a 5-year-old averages around 0.7. This is an invalid interpretation if you consider that passengers can only survive or not survive; they cannot survive just a little bit.

However, the predicted value of the dichotomous dependent variable can also be interpreted in a different way. You need to understand what the arithmetic mean of a dichotomous variable with the values of 0 and 1 signifies. The mean of the variable `survived`, for example, is 0.3230. This reflects the share of passengers who survived.[3] So, we see that the share of survivors in the dataset amounts to around 32%, or in other words, the probability that you will find a survivor in the dataset is 0.32. In general, the predicted values of the linear regression are estimates of the conditional mean of the dependent variable. Thus you can use the probability interpretation for every value of the independent variable: the predicted value of around 0.7 for a 5-year-old means a predicted probability of survival of 0.7. From this alternative interpretation, the linear regression model for dichotomous dependent variables is often called the linear probability model or LPM (Aldrich and Nelson 1984).

How can you interpret the negative predicted values for passengers over 60 with the help of the probability interpretation? In fact, you cannot, as according to the mathematical definition of probabilities, they must be between 0 and 1. Given sufficiently small or large values of the X variable, a model that uses a straight line to represent probabilities will, however, inevitably produce values of more than 1 or less than 0. This is the first problem that affects OLS regression of dichotomous variables.[4]

The second problem affects the homoskedastic assumption of linear regression that we introduced in section 8.3.2. According to this assumption, the variance of errors for all values of X (and therefore all values of $\widehat{Y}$) should be constant. We suggested that the scatterplot of the residuals against the predicted values indicated a possible violation of this assumption. You can achieve a graph such as this for our linear probability model by typing

```
. predict r, resid
. scatter r yhat, yline(-.6 .4) ylab(-.6 .4 1.4) xline(.6)
```

3. You can confirm this for yourself by typing `tab survived`.
4. In practice, this is a problem of little importance when predicted values of more than 1 or less than 0 do not appear for real values of the independent variables. However, using a model that would prevent such impossible probabilities from the start seems sensible.

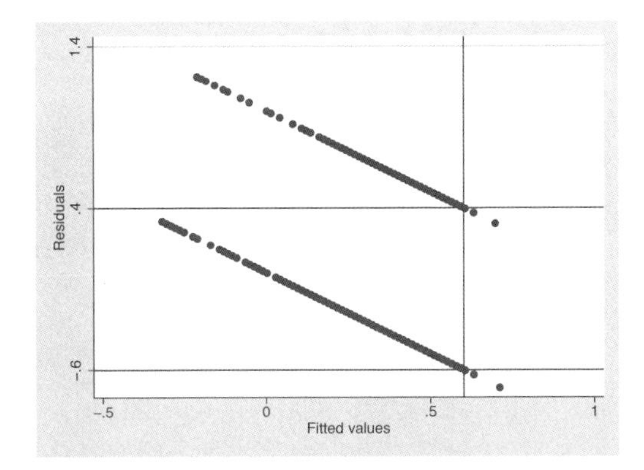

Here you can observe that only two possible residuals can appear for every predicted value. Less apparent is that both of these residuals result directly from the predicted values. If a survivor (`survived` = 1) has a predicted value of 0.6 due to her age, she will have a residual of $1 - 0.6 = 0.4$. If you predict a value of 0.6 for an individual who did not survive (`survived` = 0), you will receive a value of $0 - 0.6 = -0.6$.

Thus the residuals are either $1-\widehat{y}_i$ or $-\widehat{y}_i$. The variance of the residuals is $\widehat{y}_i \times (1-\widehat{y}_i)$ and is therefore larger as the predicted values approach 0.5. The residuals of the linear probability model are therefore by definition heteroskedastic, so that the standard errors of the estimated coefficients will be wrong.

In conclusion, although a linear regression with a dichotomous dependent variable is possible, it leads to two problems. First, not all predicted values can be interpreted, and second, this model does not allow for correct statistical inference. To avoid these problems, you need a model that produces probabilities only between 0 and 1 and relies on assumptions that are maintained by the model. Both are fulfilled by logistic regression, the basic principles of which we will introduce now.

9.2 Basic concepts

9.2.1 Odds, log odds, and odds ratios

Earlier we found that the linear OLS regression of dichotomous dependent variables can produce unwanted predicted values. This is clearly because we attempted to represent values between 0 and 1 with a line. The values estimated with a linear regression[5] are basically not subject to any restrictions.

This means that, theoretically, values between $-\infty$ and $+\infty$ may emerge. Therefore, regression models that are based on a linear combination should use only dependent variables whose range of values are equally infinite.

5. We showed you a linear combination like this on page 249.

As the range of values for probabilities lies between 0 and 1, they are not suited to be estimated with a linear combination. An alternative is the logarithmic chance, which we will explain using the *Titanic* data from the previous section.

We previously received indications that children had a higher chance of survival than adults did. Now we want to investigate whether women were more likely to survive than men. You can obtain an initial indication of the chance of survival for women and men through a two-way table between `sex` and `survived`:

```
. tabulate sex survived, row
```

Key
frequency *row percentage*

Gender	Survived no	yes	Total
women	126 26.81	344 73.19	470 100.00
man	1,364 78.80	367 21.20	1,731 100.00
Total	1,490 67.70	711 32.30	2,201 100.00

In section 7.2.1, we interpreted tables like this using row or column percentages. By using the available row percentages, we determine that the overall share of survivors was around 32%, whereas that of the women was about 50 percentage points higher than that of the men (73% compared with 21%). You can do a similar comparison by dividing the number of survivors by the number of dead. For the women, this ratio would be 344:126.

```
. display 344/126
2.7301587
```

You will get the same number[6] if you divide the proportional values (in this case, the row percentages)

```
. display .7319/.2681
2.7299515
```

You can interpret these ratios as follows: for women, the estimated probability of surviving is almost three times as high as the estimated probability of dying. The estimated probability of dying is around one-third ($1 : 2.73 = 0.366$) the estimated probability of surviving. In practice, we would say that the estimated odds of surviving are generally around 2.73 to 1, whereas the estimated odds of dying are around 1 to 2.73.

6. The deviations are due to roundoff error.

In general, this relationship can be written as

$$\text{odds}_{\text{surviving}} = \frac{\text{Probability}_{\text{surviving}}}{\text{Probability}_{\text{dying}}}$$

or slightly shorter by using symbols instead of text:

$$\text{odds} = \frac{\Pr(Y=1)}{1-\Pr(Y=1)} \tag{9.1}$$

The probabilities of survival $\Pr(Y = 1)$ and dying $\Pr(Y = 0)$ can be found, respectively, in the numerator and the denominator. Since the only two alternatives are surviving or dying, their probabilities sum to one, so we replace $\Pr(Y = 0)$ with $1 - \Pr(Y = 1)$.

You can also estimate the chance of survival for men: their odds of survival are considerably lower than those of the women: $367/1364 = 0.269$. This means that for men the estimated probability of survival stands at 0.269 : 1; men are 3.72 times more likely to be among the victims.

Of course, you can compare the odds of survival for men and women using a measured value. For instance, you can compare the estimated chance of survival for men with that of women by dividing the odds for men by the odds for women:

```
. display .269/2.73
.0985348
```

This relationship is called the *odds ratio*.

Here we would say that the odds of survival for men are 0.099 times, or a 10th of the odds for women. Apparently, the principle of "women and children first" appears to have been adhered to. Whether this *appearance* actually holds is something that we will investigate in more detail on page 282.

However, first we should consider the suitability of using odds for our statistical model. Earlier we looked at the probabilities of surviving the *Titanic* catastrophe by passenger age. We found that predicting these probabilities with a linear combination could result in values outside the definition range of probabilities. What would happen if we were to draw upon odds instead of probabilities?

(*Continued on next page*)

Table 9.1. Probabilities, odds, and logits

$\Pr(Y=1)$	odds = $\frac{\Pr(Y=1)}{1-\Pr(Y=1)}$		ln(odds)
0.01	1/99 =	0.01	−4.60
0.03	3/97 =	0.03	−3.48
0.05	5/95 =	0.05	−2.94
0.20	20/80 =	0.25	−1.39
0.30	30/70 =	0.43	−0.85
0.40	40/60 =	0.67	−0.41
0.50	50/50 =	1.00	0
0.60	60/40 =	1.50	0.41
0.70	70/30 =	2.33	0.85
0.80	80/20 =	4.00	1.39
0.95	95/5 =	19.00	2.94
0.97	97/3 =	32.33	3.48
0.99	99/1 =	99.00	4.60

In the first column of table 9.1, we list several selected probability values. You will see that at first the values increase slowly, then rapidly, and finally slowly again. The values are between 0 and 1. If we presume that the values represent the estimated chance of survival for passengers on the *Titanic* of different ages, the first row would contain the group of the oldest passengers with the lowest chance of survival, and the bottom row would contain the group of the youngest passengers with the highest chance of survival. Using (9.1), you can calculate the odds that an individual within each of these groups survived the *Titanic* catastrophe. Furthermore, imagine that each of these groups contains 100 people. As the first group has a probability of 0.01, one person out of 100 should have survived: a ratio of 1 to 99 (1:99). If you calculate 1/99, you get 0.010101. You can perform this calculation for each row in the table. The values of the odds lie between 0 and $+\infty$; odds of 0 occur if there are no survivors within a specific group, whereas odds $+\infty$ occur when everyone in a large group survives. If the number of survivors is equal to the number of victims, we get odds of 1.

Odds are therefore *slightly* better suited than probabilities to be estimated with a linear combination. No matter how high the absolute value is when predicting with a linear combination, it will not be outside the definition range of the odds. However, a linear combination also allows for negative values, but negative odds do not exist. You can avoid this problem by using the natural logarithm of the odds. These values, called *logits*, are displayed in the last column of the table.

Now look at the values of the logits more closely: although the odds have a minimum boundary, the logarithmic values have no lower or upper boundaries. The logarithm of 1 is 0. The logarithm of numbers less than 1 results in lower figures that stretch to $-\infty$

as you approach 0. The logarithm of numbers over 1 stretch toward $+\infty$. Note also the symmetry of the values. At a probability of 0.5 the odds lie at 1:1 or 50:50. The logarithmic value lies at 0. If you look at the probabilities above and below 0.5, you will see that at equal intervals of probabilities of the odds' logarithm, only the algebraic sign changes.

The logit is not restricted and has a symmetric origin. It can therefore be represented by a linear combination of variables and hence is better suited for use in a regression model. Unfortunately, the logit is not always easy to interpret. Your employers are unlikely to understand you if you tell them that the logarithmic chance of survival of a male *Titanic* passenger is -1.31, whereas that of a female passenger is $+1.00$. However, by simply transforming (9.1), you can convert the values of the logits back into probabilities

$$\Pr(Y=1) = \frac{e^L}{1+e^L} \tag{9.2}$$

where L is logit and e is Euler's constant ($e \approx 2.718$). A functional graph of this transformation can be drawn as follows:

```
. twoway function y=exp(x)/(1+exp(x)), range(-10 10)
```

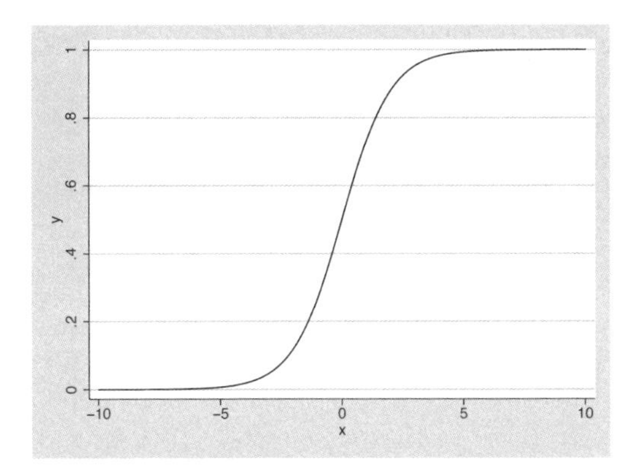

In this graph, we see another interesting characteristic of logits: although the range of values of the logits has no upper or lower boundaries, the values of the probabilities calculated from the logits remain between 0 and 1. For logits between around -2.5 and 2.5, the probabilities increase relatively rapidly; however, the closer you get to the boundary values of the probabilities, the less the probabilities change. The probabilities asymptotically approach the values 0 and 1, but they *never* go over the boundaries. From this we can deduce that on the basis of a linear combination, predicted logits can always be converted into probabilities within the permitted boundaries of 0 and 1.

To summarize, the logarithmic chance is well suited to be estimated with a linear combination and can therefore be used in a regression model. The equation for such a model could be

$$L_i = \beta_0 + \beta_1 x_{1i} + \beta_2 x_{2i} + \cdots + \beta_{K-1} x_{K-1,i} \tag{9.3}$$

This is called the logistic regression model, or logit model. The formal interpretation of the β coefficients of this model is identical to that of the linear regression (OLS): when an X variable increases by one unit, the predicted values (the logarithmic odds) increase by β units.

Before we use logistic regression, let us examine the procedure for estimating the β coefficients of (9.3). For the linear regression, we used the OLS procedure for estimation. For the logistic regression, we instead use the process of maximum likelihood. The logic of this process is somewhat more complex than that of OLS, even though the basic principle is similar: you look for the β coefficients that are optimal in a certain respect. We will explain this process in detail in the following section. However, you do not need to work through the example to understand the section that follows it!

9.2.2 Excursion: The maximum likelihood principle

In discussing linear regression, we explained the OLS process used to determine the b coefficients, which are estimators of the β coefficients. In principle, you could calculate the logarithmic odds for each combination of the independent variables and use these in an OLS regression model. Nevertheless, for reasons we will not explain here, a procedure such as this is not as *efficient* as the process of estimation applied in logistic regression: the maximum likelihood principle.[7] Using this technique, you can determine the b coefficients so that the proportionate values you observed become maximally probable. What does this mean? Before we can answer this question, we need to make a little detour:

On page 252, we informed you that 32.3% of the *Titanic* passengers survived. Suppose that you had determined this figure from a sample of the passengers. Here you could ask yourself how likely such a percentage may be, when the *true* number of the survivors amounts to, say, 60%, of the passengers? To answer this question, imagine that you have selected one passenger from the population. If 60% of the passengers survived, the estimated probability that this passenger will be a survivor is 0.6 and the estimated probability that he or she will be a victim is 0.4. Now select a second person from the population. Whether this person is a survivor or a victim, the estimated probabilities remain the same (sampling with replacement).

In figure 9.1, we have conducted all possible samples with three observations. We obtained $2^n = 2^3 = 8$ samples with $n = 3$. In the first sample, we observed only survivors (S). The probability that a sample randomly selects three survivors is $0.6 \times 0.6 \times 0.6 = 0.6^3 = 0.216$. In the second, third, and fifth samples, we observed two survivors and one victim (V). Each of these samples has probability $0.6 \times 0.6 \times 0.4 = 0.6^2 \times 0.4^1 = 0.144$. In total, the probability of such a sample is $0.144 \times 3 = 0.432$. The probabilities of samples

7. Andreß, Hagenaars, and Kühnel's (1997, 40–45) introduction to the maximum likelihood principle served as a model for the following section.

4, 6, and 7 are each $0.6 \times 0.4 \times 0.4 = 0.6 \times 0.4^2 = 0.096$. In total, the probability of these samples is therefore $0.096 \times 3 = 0.288$. Finally, there is sample 8, where the probability lies at $0.4 \times 0.4 \times 0.4 = 0.4^3 = 0.064$. If, from the samples given in the mapping, we ask how likely it is that one of three survives, the answer is that it is as likely as samples 4, 6, and 7 together, i.e., 0.288.

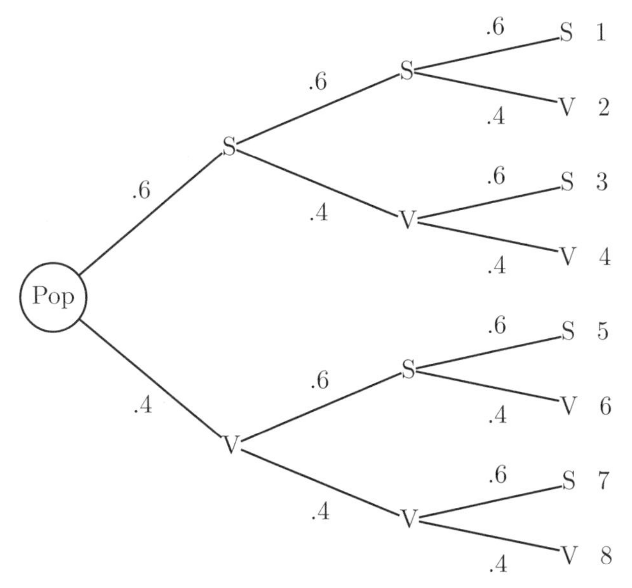

Figure 9.1. Sample of a dichotomous characteristic with the size of 3

Generally, the probability of observing h successes in a sample of size n is

$$\Pr(h|\pi, n) = \binom{n}{h} \pi^h (1-\pi)^{n-h} \tag{9.4}$$

where π defines the probability of a positive outcome in the population. The term $\binom{n}{h}$ stands for $n!/h!(n-h)!$. It enables us to calculate the number of potential samples in which the dichotomous characteristic appears n times. In Stata, the probability of samples 4, 6, and 7 in our mapping can be calculated with this command:

```
. display comb(3,1) * .6^1 * .4^2
.288
```

In practice, we are usually not interested in this figure; instead our attention is on π, the characteristic's share in the population. Although π is unknown, we can consider what value of π would make the given sample most probable. For this, we can use various values for π in (9.4) and then select the value that results in the highest probability. Formally, this means that we are searching for the value of π for which the likelihood

$$\mathcal{L}(\pi|h,n) = \binom{n}{h}\pi^h(1-\pi)^{n-h} \tag{9.5}$$

is maximized. We can forgo a calculation of $\binom{n}{h}$, as this term remains constant for all values of π. The likelihood is calculated with the same formula as in (9.4). If (9.4) is evaluated for all possible values of h, the probabilities sum to 1, but this is not the case for the values of $\mathcal{L}$ and all possible values of π. Therefore, we must differentiate between likelihood and probability.

You can do this for sample 2 from figure 9.1 (two survivors and one victim) by creating an artificial dataset with 100 observations:

```
. clear
. set obs 100
obs was 0, now 100
```

Now generate the variable `pi` by rendering a series of possible values for π:

```
. generate pi = _n/100
```

As h and n are known from the sample, you can calculate the likelihood for the various values of π:

```
. generate L = pi^2 * (1 - pi)^(3-2)
```

With the help of a graph, you can then analyze which π results in a maximal likelihood:

```
. line L pi, sort
```

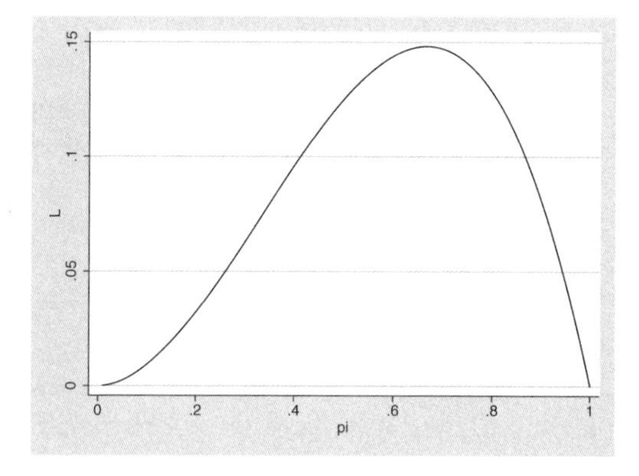

The maximum of the likelihood lies around $\pi = 0.66$. This is the maximum likelihood estimate of the share of survivors from the population, given the sample contains two survivors and one victim.

How can you estimate the β coefficients of our regression model with the maximum likelihood principle from (9.3)? The answer is simple. Instead of directly inserting the values for π, you can calculate π with the help of our regression model. Now insert (9.3) in (9.2)

$$\Pr(Y=1)=\frac{e^{\beta_0+\beta_1 x_{1i}\cdots+\beta_{K-1}x_{K-1,i}}}{1+e^{\beta_0+\beta_1 x_{1i}\cdots+\beta_{K-1}x_{K-1,i}}}$$

and again in (9.5):

$$\mathcal{L}(\beta_k|f,n,m)=\Pr(Y=1)^h\times\{1-\Pr(Y=1)\}^{n-h}=$$
$$\left(\frac{e^{\beta_0+\beta_1 x_{1i}+\cdots+\beta_{K-1}x_{K-1,i}}}{1+e^{\beta_0+\beta_1 x_{1i}+\cdots+\beta_{K-1}x_{K-1,i}+}}\right)^h\times\left(1-\frac{e^{\beta_0+\beta_1 x_{1i}+\cdots+\beta_{K-1}x_{K-1,i}}}{1+e^{\beta_0+\beta_1 x_{1i}+\cdots+\beta_{K-1}x_{K-1,i}}}\right)^{n-h}$$

After doing this, you can attempt to maximize this function by trying out different values of β_k. However, as is the case with OLS regression, it is better to reproduce the first derivative from β_k and to set the resulting standard equation as zero. The mathematical process is made easier when the log likelihood; i.e., $\ln\mathcal{L}$, is used. You will not find an analytical solution with this model, unlike linear OLS regression. For this reason, iterative algorithms are used to maximize the log likelihood.

We have introduced the maximum likelihood principle for logistic regression with a dichotomous dependent variable. In principle, we can apply it to many different models by adapting (9.5) to reflect the distributional assumptions we wish to make. The resulting likelihood function is then maximized using a mathematical algorithm. Stata has a command called `ml` to do this, which is described in detail in Gould, Pitblado, and Sribney (2003).

9.3 Logistic regression with Stata

Let us briefly set aside our *Titanic* example for an alternative. Say that you assumed that when the age and household income of a surveyed individual increases, the probability of living in an apartment or house they own also increases. Also you expect that the share of individuals who own their own residence[8] to be higher in West Germany than it is in East Germany.

Now let us load our dataset `data1.dta`.

```
. use data1, clear
```

8. In the following, we will refer to living in an apartment or house that the individual owns as *residence ownership*. In this respect, children may also be considered to "own" housing. For household income, we will use the word "income".

To check your assumption, you can calculate a logistic regression model of residence ownership against the independent variables of age, household income, and an East–West variable.

Stata has two commands for fitting logistic regression models, `logit` and `logistic`. The commands differ in how they report the estimated coefficients. `logit` reports the actual b's in (9.3), whereas `logistic` reports the odds ratios discussed previously. Because we have emphasized using a linear combination of variables to explain the dependent variable, we will focus on `logit` and show you how to obtain odds ratios after estimation. Some researchers, particularly biostatisticians and others in the medical field, focus almost exclusively on odds ratios and therefore typically use `logistic` instead. Regardless of how the estimated coefficients are reported, both commands fit the same underlying statistical model.

At least one category of the dependent variable must be 0, as `logit` takes a value of 0 to represent failure and any other value as representing success. Normally you use a dependent variable with the values 0 and 1, where the category assigned the value of 1 means success. Here the variable `owner` should be generated with the values of 1 for house owner and 0 for tenant as follows:[9]

```
. generate owner = renttype == 1 if renttype < .
```

We generate the East–West variable analogously as we did previously for our linear regression model (page 197):

```
. generate east = state>=11 & state<=16 if state<.
```

It would also be sensible to generate an age variable for our regression model from the year-of-birth variable available in our dataset.

```
. generate age = 1997-ybirth
```

Furthermore, we recommend that you center the two continuous independent variables `age` and `hhinc`, i.e., deduct the mean of the variable from each value. The mean of centered variables is zero, making it easier to interpret regression models at various points (Aiken and West 1991). You can center `age` and `hhinc` with the following commands:[10]

```
. summarize age if hhinc < . & owner < . & east < .
. generate cage = age - r(mean) if hhinc < . & owner < . & east < .
. summarize hhinc if age < . & owner < . & east < .
. generate chhinc = hhinc - r(mean) if age < . & owner < . & east < .
```

Now we are ready to fit the logistic regression model:

9. For details on this command, see page 76. The command `label list` determines the assignment of the values to labels (section 5.5).
10. The Stata commands used here are explained in chapter 4.

```
. logit owner cage chhinc east
Iteration 0:   log likelihood = -2091.5129
Iteration 1:   log likelihood = -1930.1356
Iteration 2:   log likelihood = -1927.6015
Iteration 3:   log likelihood = -1927.5979
Logistic regression                               Number of obs   =       3200
                                                  LR chi2(3)      =     327.83
                                                  Prob > chi2     =     0.0000
Log likelihood = -1927.5979                       Pseudo R2       =     0.0784

------------------------------------------------------------------------------
       owner |      Coef.   Std. Err.      z    P>|z|     [95% Conf. Interval]
-------------+----------------------------------------------------------------
        cage |   .0189758   .0021862     8.68   0.000     .0146909    .0232608
      chhinc |   .0006504   .0000418    15.54   0.000     .0005684    .0007324
        east |  -.0583019   .0864511    -0.67   0.500    -.2277431    .1111392
       _cons |  -.6023514   .0462412   -13.03   0.000    -.6929826   -.5117202
------------------------------------------------------------------------------
```

The results table is similar to the one from linear regression. At the bottom of the output is the coefficient table, which contains the estimated coefficients for the dependent variables and the constants along with their standard errors, significance tests, and confidence intervals. At the top left is the iterations block with some results that are related to the maximum likelihood algorithm, and at the top right we see block describing the model fit. We will discuss each of these blocks along the lines of our explanation of linear regression.

9.3.1 The coefficient table

The b coefficients can be found in the first column of the coefficient table.[11] The b coefficients formally indicate how the predicted values change when the corresponding independent variables increase by one unit, just like in linear regression, although here the predicted values are the logarithmic odds of success, not the mean of the dependent variable. For example, you would interpret the estimated regression coefficient of `cage` as follows: the logarithmic odds of residence ownership rise on average by 0.0189758 if age increases by 1 year. You would interpret the estimated regression coefficient of `chhinc` the same way. From the estimated regression coefficient of `east`, we can say that with every one-unit increase of the variable `east`, the estimated logarithmic chance of residence ownership falls on average by 0.0583019. As `east` can increase by one unit only once, we might instead say that East Germans have, on average, a 0.06 smaller estimated logarithmic chance of residence ownership than West Germans. The estimated regression constants provides the predicted value for those individuals surveyed for whom all the other independent variables show the value 0. Because of centering, this means that the predicted logarithmic chance of residence ownership for West German individuals with a mean age and mean income lies at -0.6023514.

11. In the second column, you will find the standard errors of the regression coefficients, which will help you calculate significance tests, as well as confidence interval limits. You can interpret these figures the same way you did the figures in the linear regression (section 8.5). In logistic regression, you usually evaluate the significance of the coefficients using a likelihood-ratio test (section 9.5).

Because changes in the logarithm of the odds of a positive outcome are hard to interpret, we will discuss their interpretation in more detail.

Sign interpretation

As a first step, consider just the signs and relative sizes of the estimated coefficients. A positive sign means that the estimated probability or chance of residence ownership increases with the respective independent variable, whereas a negative sign means that it decreases. Here the estimated probability of house ownership increases with age and with income. The estimated probability of home ownership is lower in the East than it is in the West.

Interpretation with odds ratios

Using the model equation, we want to calculate the predicted logarithmic chance of a West German with mean income and mean age. For the centered income variable (`chhinc`), the individuals surveyed whose income matched the mean have the value 0. This also is true for the centered age variable (`cage`), where individuals with the mean age have the value 0. Finally, West Germans surveyed are assigned 0 for the variable `east`. Thus the predicted logit for a West German of mean age and mean income is simply equal to the constant term.

By calculating the exponential of the regression function, you can convert the logarithmic odds to odds:[12]

```
. display exp(_b[_cons])
.54752267
```

Similarly, you can calculate the predicted odds for those who are exactly 1 year older than the average:

```
. display exp(_b[_cons] + _b[cage]*1)
.55801156
```

An older person's predicted odds of owning their residence is therefore slightly larger than that of those with average age. We can use the *odds ratio* (page 255) to compare the outcomes of the two ages. Here it amounts to:

```
. display exp(_b[_cons] + _b[cage])/exp(_b[_cons])
1.019157
```

This means that if the age increases by 1 year, a person is 1.02 times as likely to own her residence. Increasing age by 2 years increases the odds of owning a residence by $1.02 \times 1.02 \approx 1.04$. Odds ratios work in a multiplicative fashion.

12. We covered working with the saved coefficients in detail in section 9.1.

You can reduce the complexity of calculating odds ratios if you consider that, to determine the odds ratios, you must first calculate the odds for a particular value of X and then for the value $X+1$. After that, you divide both results by each other, which can be presented as follows:

$$\widehat{\text{odds ratio}} = \frac{e^{b_0+b_1(X+1)}}{e^{b_0+b_1X}} = \frac{e^{b_0+b_1X}e^{b_1}}{e^{b_0+b_1X}} = e^{b_1}$$

You can therefore obtain the estimated odds ratio simply by computing the exponential of the corresponding b coefficient.

Many logistic regression users prefer the interpretation of results in the form of the odds ratios. For this reason, Stata also has the command `logistic` that directly reports the estimated odds ratios. If you have already fit your model using `logit`, typing in `logistic` will redisplay the output in terms of odds ratios.

Probability interpretation

The third possibility for interpreting the coefficient is provided by (9.2), which we used to show you how to convert predicted logits into predicted probabilities. For example, to compute the estimated probability of residence ownership for West Germans with mean ages and incomes, you could type

```
. display exp(_b[_cons])/(1 + exp(_b[_cons]))
.35380591
```

The `predict` command lets you generate a new variable that contains the predicted probability for every observation in the sample. You just enter the `predict` command along with the name of the variable you want to contain the predicted probabilities.

```
. predict Phat
```

Here the name `Phat` indicates that this variable deals with predicted probabilities. You can also calculate the predicted logits with the `xb` option of the `predict` command.

One difficulty in interpreting probabilities is that they do not increase at the same rate for each unit increase in an independent variable. For example, consider the following three probabilities where we have increased the age by 10 years at a time:

```
. display exp(_b[_cons] + _b[cage]*10)/(1 + exp(_b[_cons] + _b[cage]*10))
.39829047

. display exp(_b[_cons] + _b[cage]*20)/(1 + exp(_b[_cons] + _b[cage]*20))
.44452061

. display exp(_b[_cons] + _b[cage]*30)/(1 + exp(_b[_cons] + _b[cage]*30))
.49173152
```

Comparing West Germans of the mean age with those 10 years older, the predicted probability of residence ownership *increases* by around $0.3983 - 0.3538 = 0.0445$. If age increases by 10 more years, the predicted probability increases by $0.4445 - 0.3983 = 0.0462$ and then by $0.4917 - 0.4445 = 0.0472$. We see that an increase in age of 10 years at a time does not lead to a constant change in the predicted probability.

One solution would be to show the probabilities in a conditional-effects plot. Similar to the graphs shown in section 8.4.2, this plot graphs the predicted values for various characteristics of the independent variables. Thus you could, for example, generate a variable with income-dependent predicted probabilities of West Germans with the mean age:[13]

```
. generate Phat2 = exp(_b[_cons]+_b[chhinc]*chhinc)/
> (1+exp(_b[_cons]+_b[chhinc]*chhinc))
```

and display this in a graph:

```
. line Phat2 hhinc, sort
```

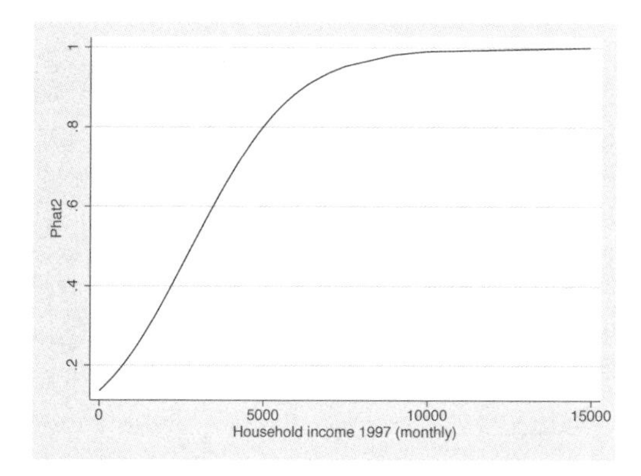

The graph shows that the increase in predicted probabilities is not constant over all income values. Depending on income, the predicted probability of home ownership will rise either rapidly or slowly.

9.3.2 The iteration block

In the upper-left part of the `logit` output (see page 263) are several rows beginning with the word *iteration*. This sort of output is typical for models whose coefficients are estimated by maximum likelihood. As we mentioned in our discussion of this procedure, when you use the maximum likelihood principle, there is typically no closed-form mathematical equation that can be solved to obtain the b coefficients. Instead, an iterative procedure must be used that tries a sequence of different coefficient values. As the algorithm gets closer to the solution, the value of the likelihood function changes by less and less.

13. At the mean age, the value of the centered age variable is 0, so you can omit age from the equation. The variable `east` is zero for West Germans, so you can omit it from the calculations as well.

The first and last figures of the iteration block are in some respects similar to the figures given in the ANOVA block of the linear regression (see section 8.1.2), which contained figures for TSS, RSS, and MSS. TSS was the sum of the squared residuals from predicting all the values of the dependent variables through arithmetic means. RSS was the sum of the squared residuals from the regression model, and MSS was the difference between TSS and RSS. MSS thus represents how many fewer errors we make when using the regression model instead of the mean for predicting the dependent variable.

In the logistic regression model, the residuals used to estimate the regression coefficients cannot be interpreted in the same way as with linear regression. Two values of the likelihood function are of particular interest, namely, the first and the last. The first likelihood shows how probable it is that all β coefficients of the logistic regression apart from the constant term equal 0 ($\mathcal{L}_0$).[14] The last likelihood represents the maximized value. The larger the difference between the first and last values of the log likelihood, the stronger is the advantage of the model with independent variables compared with the null model. In this sense, you can consider TSS analogous to $\mathcal{L}_0$, RSS to $\mathcal{L}_K$, and MSS to $\mathcal{L}_0 - \mathcal{L}_K$.

Other than the first and last log likelihoods, the rest of the figures in the iteration block are of little interest, with one exception. Sometimes the maximum likelihood process delivers a solution for the coefficients that is not optimal. This may occur if the domain where you are searching for the coefficients is not concave or flat. This is a somewhat technical issue, and we do not wish to delve any further. However, many iterations may indicate a difficult function to maximize, though it is difficult to say how many iterations are too many. You should generally expect more iterations the more independent variables in your model.

In general, the logistic regression model's likelihood function is "well behaved", meaning that it is relatively easy to maximize. However, if you do have problems obtaining convergence, you may want to remove a few independent variables from your specification and try again.

9.3.3 The model fit block

R^2 was used to assess the fit of a linear regression model. R^2 is so commonly used because it has, on one hand, clear boundaries of 0 and 1, and on the other, a clear interpretation of the *share of explained variance*. There is no comparable generally accepted measured value for logistic regression. Instead, many different statistics have been suggested, some of which we will introduce here.

14. This is true for Stata's `logit` and `logistic` commands. However, other maximum likelihood commands use alternative starting values, and then the iteration-zero log likelihood is *not* the value obtained when all the slope parameters are set to zero.

One such measure of fit is reported in the model fit block of `logit`: the pseudo-R^2 (p^2). Nevertheless, it is already a mistake to speak of *the* pseudo-R^2. There are various definitions for pseudo-R^2 (Veall and Zimmermann 1994; Long and Freese 2006). Therefore, you should always indicate which pseudo-R^2 you are referring to. The one reported by Stata is that suggested by McFadden (1973), which is why we refer to it as p^2_{MF}.

McFadden's p^2_{MF} is defined in a way that is clearly analogous to the R^2 in linear regression (recall that $R^2 = \text{MSS}/\text{TSS} = 1-\ \text{RSS}/\text{TSS}$). p^2_{MF} is defined as

$$p^2_{\mathrm{MF}} = \frac{\ln \mathcal{L}_0 - \ln \mathcal{L}_K}{\ln \mathcal{L}_0} = 1 - \frac{\ln \mathcal{L}_K}{\ln \mathcal{L}_0}$$

where $\mathcal{L}_0$ is the likelihood from the model with just a constant term and $\mathcal{L}_K$ is the likelihood of the full model. As is the case in R^2, p^2_{MF} lies within the boundaries of 0 and 1; however, interpreting the content is disproportionately more difficult. "The higher, the better" is really all you can say of p^2_{MF}. In our example (page 263), the value of p^2_{MF} around 0.08 is what most people would say is rather small.

Besides McFadden's pseudo-R^2, the likelihood-ratio χ^2 value ($\chi^2_{\mathcal{L}}$) is another indicator of the quality of the overall model. It, too, is based on the difference between the likelihood functions for the full and constant-only models. However, unlike p^2_{MF}, this difference is not standardized to lie between 0 and 1. It is defined as

$$\chi^2_{\mathcal{L}} = -2(\ln \mathcal{L}_0 - \ln \mathcal{L}_K)$$

$\chi^2_{\mathcal{L}}$ follows a χ^2 distribution, and as with the F value in linear regression, you can use $\chi^2_{\mathcal{L}}$ to investigate the hypothesis that the independent variables have no explanatory power or, equivalently, that all the coefficients other than the constant are all zero. The probability of this hypothesis being true is reported in the line that reads "Prob > chi2". Here it is practically 0. Therefore, we can assume that at least one of the two β coefficients in the population is not 0. As is the case of the linear regression F test, rejection of this null hypothesis is not sufficient for us to be satisfied with the results.

As with linear regression, you should not judge a model's suitability purely by the measured values within the model fit block, especially in logistic regression, as there is no single generally accepted measured value for doing so. Therefore, we will discuss other measures that are not reported in the output.

Classification tables

The fit of the linear regression model was assessed primarily on the basis of the residuals $(y - \widehat{y})$. In logistic regression, one way to assess fit is with a classification table, in which every observation is assigned one of the two outcomes of the dependent variable. The positive outcome is normally assigned when the model predicts a probability of more than 0.5, whereas an observation is assigned a negative outcome if a probability of less than 0.5 is predicted. For example, you could do this manually assuming you have already created the variable Phat containing the predicted probabilities by typing

```
. generate ownerhat = Phat >= .5 if Phat < .
```

The classified values so generated are typically presented in a classification table, a simple cross-classified table containing the classified values and the original values:

```
. tabulate ownerhat owner, cell column

+-------------------+
| Key               |
|-------------------|
|     frequency     |
| column percentage |
|  cell percentage  |
+-------------------+

           |         owner
  ownerhat |         0          1 |     Total
-----------+----------------------+----------
         0 |     1,858        829 |     2,687
           |     90.77      71.90 |     83.97
           |     58.06      25.91 |     83.97
-----------+----------------------+----------
         1 |       189        324 |       513
           |      9.23      28.10 |     16.03
           |      5.91      10.12 |     16.03
-----------+----------------------+----------
     Total |     2,047      1,153 |     3,200
           |    100.00     100.00 |    100.00
           |     63.97      36.03 |    100.00
```

The sensitivity and the specificity of the model are commonly used by people in the medical profession. Sensitivity is the share of observations classified as residence owners within the observations who actually do own their residences. Specificity is the share of observations classified as tenants among those who are actual tenants. Here the sensitivity is 28.10%, and the specificity is 90.77%.

The count R^2 is commonly used in the social sciences. It deals with the share of overall correctly predicted observations, which you can determine by adding the overall shares in the main diagonal of the above-generated table. However, it is easier to use the estat classification command, which you can use to generate the table in a different order, as well as derive the sensitivity, specificity, count, R^2, and other figures:

(Continued on next page)

```
. estat classification

Logistic model for owner

              -------- True --------
Classified |         D            ~D  |      Total
-----------+--------------------------+-----------
     +     |       324           189  |        513
     -     |       829          1858  |       2687
-----------+--------------------------+-----------
   Total   |      1153          2047  |       3200

Classified + if predicted Pr(D) >= .5
True D defined as owner != 0
--------------------------------------------------
Sensitivity                     Pr( +| D)   28.10%
Specificity                     Pr( -|~D)   90.77%
Positive predictive value       Pr( D| +)   63.16%
Negative predictive value       Pr(~D| -)   69.15%
--------------------------------------------------
False + rate for true ~D        Pr( +|~D)    9.23%
False - rate for true D         Pr( -| D)   71.90%
False + rate for classified +   Pr(~D| +)   36.84%
False - rate for classified -   Pr( D| -)   30.85%
--------------------------------------------------
Correctly classified                        68.19%
--------------------------------------------------
```

The classification table shows that we have a total of 513 observations classified as 1. For 324 observations, this corresponds to the true value, but for 189 it does not. We have assigned the value 0 to 2,687 observations, which turned out to be correct for 1,858 of the observations. In total, we correctly classified $R^2_{\text{count}} = (324 + 1858)/3200 = 68.19\%$ of the observation, as shown in the final line of the table.

Overall, we might think that our model looks good. However, you can also correctly classify some cases without knowing about the independent variable. If you know only the distribution of the dependent variable, you will make fewer errors if you assign all observations to the most-frequent category. If you were to predict that all the observations are tenants, you would already be correct in $2047/3200 = 63.97\%$ of the cases. By comparing the correct classification obtained from the marginal distribution with the correct classification obtained with the knowledge of the independent variable, you can calculate the adjusted count R^2 (Long 1997, 108)

$$R^2_{\text{AdjCount}} = \frac{\sum_j n_{jj} - \max_c(n_{+c})}{n - \max_c(n_{+c})}$$

where n_{+c} is the sum of column c, and $\max_c(n_{+c})$ is the column with the higher value of n_{+c}. $\sum_j n_{jj}$ is the sum of cases in the main diagonal of the classification table, i.e., the amount of correctly classified cases. Here we receive an R^2_{AdjCount} of

```
. display ((324 + 1858) - 2047)/(3200 - 2047)
.11708586
```

This means that when predicting with a model that includes independent variables, our error rate drops by 12% compared with prediction based solely on the marginal distribution of the dependent variable. You can receive the adjusted count R^2, as well as other model fit–measured values through Scott Long and Jeremy Freese's ado package `fitstat`, available from the SSC archive (see section 12.3.2). Two further common fit statistics, the AIC and BIC, are available with the `estat ic` command.

Pearson chi-squared

A second group of fit statistics is based on the Pearson residuals. For you to understand these, we must explain the term *covariate pattern*, which is defined as every possible combination of a model's independent variables. In our example, this is every possible combination of the values of household income, age, and region. Every covariate pattern occurs m_j times, where j indexes each covariate pattern that occurs. By typing

```
. predict cpatt, number
. list cpatt
```

you can view the index number representing the covariate pattern of each observation.

The Pearson residuals are obtained by comparing the number of successes having covariate pattern j (y_j) with the predicted number of successes having that covariate pattern ($m_j\widehat{P}_j$, where $\widehat{P}_j$ is the predicted probability of success for covariate pattern j). The Pearson residual is defined as

$$r_{P(j)} = \frac{(y_j - m_j\widehat{P}_j)}{\sqrt{m_j\widehat{P}_j(1-\widehat{P}_j)}}$$

(Multiplying $\widehat{P}_j$ by the number of cases with that covariate pattern results in the predicted number of successes in pattern j.) Unlike residuals in linear regression, which are in general different for each observation, the Pearson residuals for two observations differ only if those observations do not have the same covariate pattern. Typing

```
. predict pres, resid
```

generates a variable containing the Pearson residuals. The sum of the square of this variable over all covariate patterns produces the Pearson chi-squared statistic. You can obtain this statistic by typing

```
. estat gof
Logistic model for owner, goodness-of-fit test
       number of observations =       3200
 number of covariate patterns =       3183
          Pearson chi2(3179) =       3218.80
                 Prob > chi2 =          0.3066
```

This test is for the hypothesis of the conformity of predicted and observed frequencies across covariate patterns. A small χ^2 value (high p-value) indicates small differences between the observed and the estimated frequencies, whereas a large χ^2 value (low p-value) suggests that the difference between observed and estimated values cannot be explained by a random process. Be careful when interpreting the p-value as a true "significance" level: a p-value less than 0.05 may indicate that the model does not represent reality, but values greater than 5% do not necessarily mean that the model fits the data well. A p-value of, say, 6% is still fairly small, even though you cannot formally reject the null hypothesis that the difference between observed and estimated values is completely random at significance levels below 6%.

The χ^2 test is unsuitable when the number of covariate patterns (here, 3,183) is close to the number of observations in the model (here, 3,200). Hosmer and Lemeshow (2000, 140–145) have therefore suggested modifying the test by sorting the data by the predicted probabilities and dividing them into g approximately equal-sized groups. They then suggest comparing the frequency of the observed successes in each group with the frequency estimated by the model. A large p-value indicates a small difference between the observed and the estimated frequencies.

You can obtain the Hosmer–Lemeshow test by using `estat gof` together with the `group()` option. Enter the number of groups into which you want to divide the data between the brackets; $g = 10$ is often used.

```
. estat gof, group(10)
```

9.4 Logistic regression diagnostics

We now discuss two methods to test the specification of a logistic regression model. First, logistic regression assumes a linear relationship between the logarithmic odds of success and the independent variables. Thus you should test the validity of this assumption before interpreting the results.

Second, you need to deal with influential observations, meaning observations that have a strong influence on the results of a statistical procedure. Occasionally these outliers, as they are also known, turn out to be the result of incorrectly entered data, but usually they indicate that variables are missing from the model.

9.4.1 Linearity

We used graphical analyses to discover nonlinear relationships in the linear regression model, and we used smoothing techniques to make the relationship more visible. You can also use certain scatterplots for logistic regression, but you should consider two issues. First, the median trace used in linear regression as a scatterplot smoother does not work for dichotomous variables because the median can take values of only 0 and 1.[15]

15. The value 0.5 can occur if there is an equal number of 0 and 1 values.

Second, the functional form of the scatterplot does not have to be linear, as linearity is assumed only with respect to the logits. The functional form between the probabilities and the independent variable has the shape of an S (see the graph on page 257).

You may use a local mean regression as the scatterplot smoother instead of the median trace. Here the X variable is divided into bands the same way as for the median trace, and the arithmetic mean of the dependent variable is calculated for each band. These means are then plotted against the respective independent variable.

Ideally, the graph should show the local mean regression to have an S-shaped curve like the illustration on page 257. However, the graph often only depicts a small section of the S-shape, so if the band means range only from about 0.2 to 0.8, the mean regression should be almost linear. U-shaped, reverse U-shaped, and other noncontinuous curves represent potential problems.

Stata does not have a specific command for simple local mean regression, but you can do it easily nonetheless:[16]

```
. generate groupage = autocode(age,15,16,90)
. egen mowner = mean(owner), by(groupage)
. scatter owner age, jitter(2) || line mowner age, sort
```

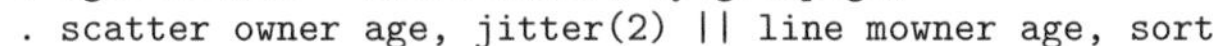

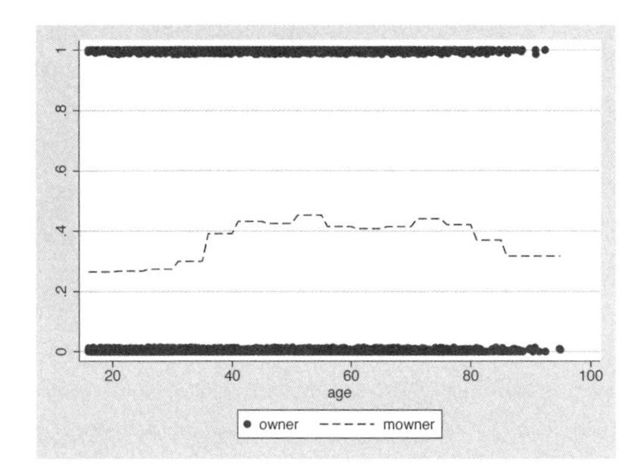

In this graph, the mean of residence ownership first increases with age and then remains constant until dropping with the oldest individuals surveyed. This is referred to as a reverse U-shaped correlation, and it certainly does not match the pattern assumed by logistic regression.

Cleveland's (1979) locally weighted scatterplot smoother (LOWESS)[17] is an alternative that is often better for investigating functional forms. You can use this smoother with the `twoway` plottype `lowess` or the statistical graph command `lowess`. We will not discuss the calculation of this smoother but refer you to the excellent explanation of the logic behind LOWESS in Cleveland (1994). You can adjust the level of smoothing

16. For the function `autocode()`, see page 157. For the command `egen`, see section 5.2.2 on page 84.
17. The process has recently also become to be known as *loess*. We use the older term, as it corresponds to the name of the Stata plottype.

by specifying a value between 0 and 1 in the `bwidth()` option, with higher numbers specifying increased smoothing. And LOWESS is a computationally intensive process, so it may take some time to display the following graph on your screen:

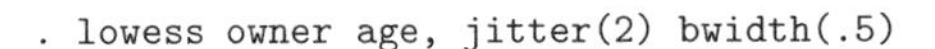

```
. lowess owner age, jitter(2) bwidth(.5)
```

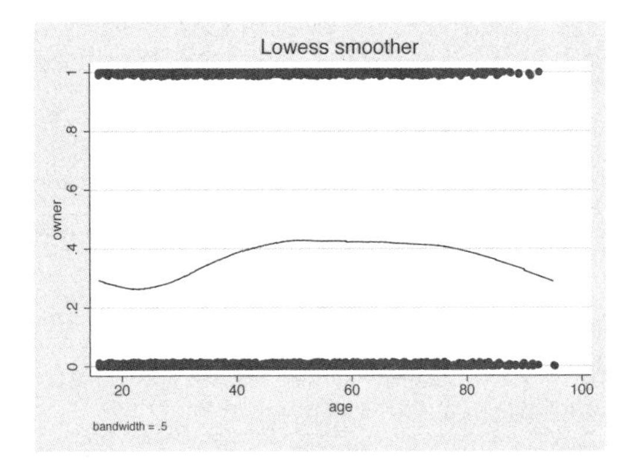

This graph also displays a reverse U-shaped correlation between residence ownership and age. The middle-age groups have a higher predicted probability of residence ownership than the upper- and lower-age groups. The youngest individuals surveyed, who presumably still live with their parents, are likely to live in their own houses or apartments.

Both graphs show a correlation that contradicts the S-shaped correlation required by logistic regression. As with linear regression, U-shaped relationships can be modeled through the generation of polynomials. Nevertheless, before you do this, check if the U-shaped relationship is still visible after controlling for household income. You can do this using a technique similar to the local mean regression discussed in the previous chapter for linear regression.[18] In this process, you replace the age variable of your regression model with a set of dummy variables (see page 282 and section 8.4.1.) for the grouped version of the age variable on page 273:

```
. tabulate groupage, gen(aged)
. logit owner aged2-aged15 chhinc east
```

Fitting this model yields a total of 14 b coefficients for the age variables. Each b coefficient indicates how much higher the predicted logarithmic chance of residence ownership is for the respective age group compared with the youngest surveyed individual. When the correlation between age and (the predicted logarithmic chance of) residence ownership is linear, the age b coefficients should increase continuously and steadily.

18. For the following process, see Hosmer and Lemeshow (2000, 90). Also scatterplots with smoothers of the dependent variables can be made against one independent variable controlling for various combinations of the other independent variables (Schnell 1994, 253). Fox (1997) demonstrates a process related to the component-plus-residual plot (page 208).

This does not appear to be the case for the estimated coefficients in front of us. To graphically evaluate the rise of the b coefficients, start with the following commands:

```
. matrix b = e(b)'
. svmat b, names(b)
```

The symbol "'" in the `matrix` command is a single quotation mark and can be found next to the ";" key on English keyboards. Here it stands for the transpose of a matrix, which we will explain later.

Explanation: Stata stores the estimated coefficients of statistical models in a row vector called `e(b)`. These are nothing more than the stored results to which we have repeatedly referred. Matrices and vectors are of particular interest, as they contain many saved results. The vector `e(b)`, for example, contains the estimated coefficients of the regression model. By typing

```
. matrix list e(b)
```

you can take a closer look at `e(b)`.

As with results saved in `r()` macros, you can also perform calculations on matrices saved in `e()`-class macros using `matrix` commands.[19] In the commands above, we used a `matrix` command to transpose the row vector `e(b)` into the column vector `b`; i.e., we turned rows into columns and columns into rows. This is important, as the estimated coefficients are next to each other in row vector `e(b)`. By contrast, in the newly generated column vector, the estimated coefficients are under one other.

The column vector is therefore nothing more than a list of numbers. The `svmat` command writes these numbers as a variable in our dataset; specifying the `names()` option gives the variable a name. Stata automatically adds the character 1 onto the name you choose. Stata does this because the command can also save matrices with several columns as variables, whereby every column in the matrix becomes a variable.

After you use the `svmat` command, your dataset will contain the new variable `b1`, which contains the 14 age b coefficients, the b coefficient for household income, and the b coefficient for the constants. The first 14 numbers in the variable $b1$ are the estimated age coefficients. Here the estimated coefficients for the age dummies will appear first in the dataset, and so we can graph them by typing

19. For an overview of the `matrix` commands, type `help matrix`.

```
. generate index = _n
. line b1 index in 1/14, sort
```

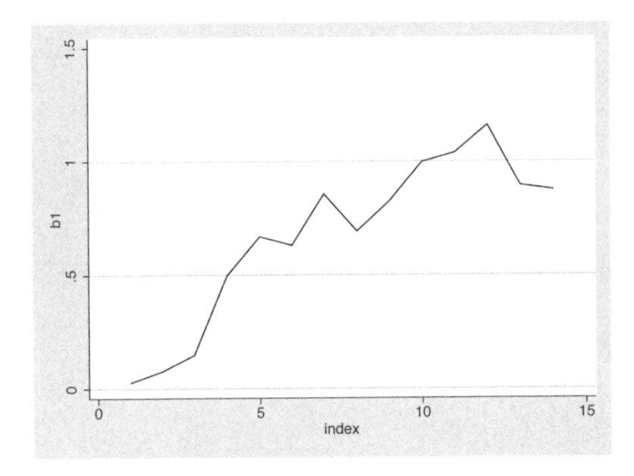

The graph shows a falling predicted logarithmic chance of residence ownership for the last two age groups. In this respect, the slight reverse U-shaped correlation remains. Including a quadratic term for age within a regression model results in a slight (albeit significant) improvement in the model fit. We will discuss this further in section 9.5 on page 279 and on page 282.

9.4.2 Influential cases

Influential data points are observations that heavily influence the b coefficients of a regression model. That is, if we were to remove an influential data point and then refit our model, our coefficient estimates would change by more than a trivial amount. As explained on page 214, influential observations are observations that exhibit an unusual combination of values for the X variable (leverage), as well as an unusual characteristic (given the X values) of the Y variable (discrepancy). Correspondingly, the measured value of Cook's D is calculated by multiplying leverage and discrepancy.

This concept is somewhat more problematic in logistic regression than it is in linear regression, as you can measure only the approximate leverage and discrepancy (Fox 1997, 459). In Stata, you can approximate the leverage values by typing

```
. logit owner cage chhinc east
. predict leverage, hat
```

You must refit the original model with `cage` and `chhinc` as independent variables, since `predict` always refers to the last model fitted, and we just fit a model including dummy variables to control for age. After getting the predicted leverage values, you can obtain the standardized residuals as an approximation to discrepancy by typing

```
. predict spres, rstandard
```

In logistic regression, the standardized residuals for observations having the same covariate pattern are identical. The same also applies for the leverage values. To isolate those covariate patterns having high leverage and discrepancy values, you can produce a graph that compares the standardized residuals with the leverage values. Fox (1997, 461) uses a diagram with vertical lines at the mean of the leverage values and at two and three times the mean. To produce this graph, we first calculate the mean of the variable leverage. We save the mean, as well as its doubled and tripled values, in the local macros ‘a’, ‘b’, and ‘c’ (see chapter 4) to later use them as vertical lines in the graph.

```
. summarize leverage
. local a = r(mean)
. local b = 2 * r(mean)
. local c = 3 * r(mean)
```

Next we generate the graph with the standardized residuals against the leverage values. To generate vertical lines, we use the `xline()` option. We use the number of covariate patterns as the plot symbol. These patterns are found in the variable `cpatt`, which we generated on page 271:

```
. scatter spres leverage, xline(‘a’ ‘b’ ‘c’) yline(-2 0 2) mlabel(cpatt)
> mlabpos(0) ms(i)
```

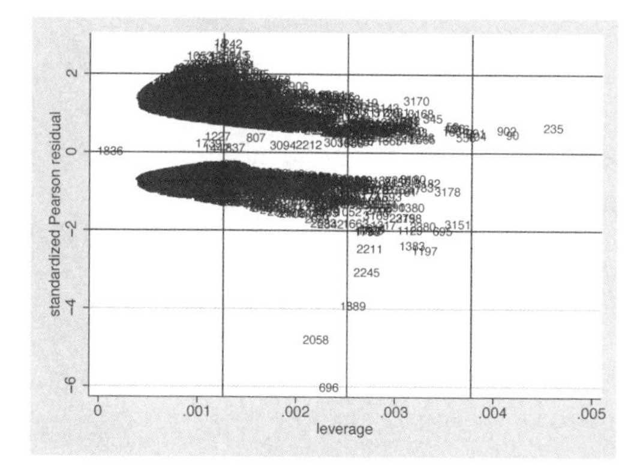

Eight covariate patterns in the graph are particularly conspicuous: both patterns having the lowest standardized residuals, and the six patterns having standardized residuals less than −2 and leverage values more than twice the average. The following command shows that invariably the latter consist of observations from West Germany with comparatively high income but not residence ownership:

(*Continued on next page*)

```
. list cpatt owner age hhinc east if leverage > 'b' & spres < -2
```

	cpatt	owner	age	hhinc	east
279.	1889	0	48	6965	0
421.	2245	0	56	5970	0
1127.	1197	0	35	5970	0
2382.	2211	0	55	5326	0
2526.	1795	0	46	4977	0
3005.	1383	0	38	5721	0

The model therefore appears to be unsuitable for explaining cases such as these.

In linear regression, the influence of individual observations on the regression result is determined by Cook's D (see section 8.3.1), which involves multiplying the leverage and discrepancy. An analogous measured value for logistic regression is

$$\Delta\beta = \underbrace{\frac{r^2_{P(j)}}{(1-h_j)^2}}_{\text{Discrepancy}} \times \underbrace{h_j}_{\text{Leverage}}$$

where h_j is the value for the leverage. In Stata, you can obtain this value with

```
. predict db, dbeta
```

as a variable under the name `db`. A scatterplot of $\Delta\beta$ against the predicted probabilities is often used, in which observations with success as the outcome are displayed in a different color or symbol than those of failure. The `separate` command is particularly useful for the latter:[20]

```
. separate db, by(owner)
. scatter db0 db1 Phat
```

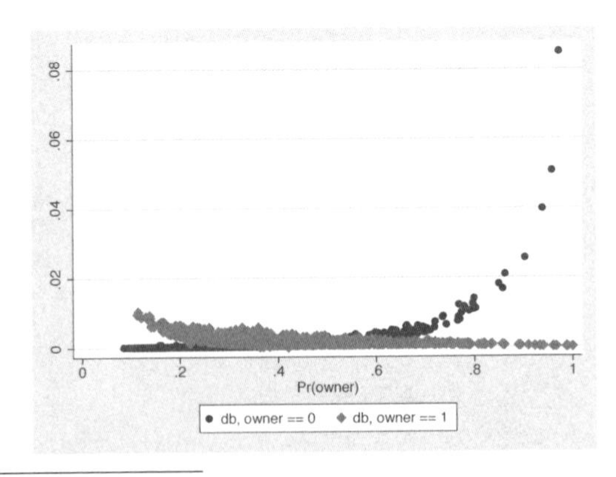

20. For an explanation of `separate`, type `help separate`. The variable `Phat` was generated on page 265 with `predict Phat`.

The curve from the bottom left to the top right consists of all tenants, whereas the curve that slopes downward from left to the bottom right consists of all residence owners. Several covariate patterns for tenants are noteworthy, namely those which have a high predicted probability of residence ownership. If you enter the number of covariate patterns into the graph instead of the symbols, you will see that these patterns are the same ones our previous analysis detected.

The Pearson residuals allow for a further test statistic for influential observations. As shown in section 9.3.3, the sum of the squared Pearson residuals is a measure of the deviation of the predicted values from the observed values. The contribution of each covariate pattern to this measure matches the square of the Pearson residual. If you divide this contribution by $1 - h_j$, you get $\Delta\chi^2_{P(j)}$, which indicates the change in the Pearson chi-squared statistic when the covariate pattern j is removed from the dataset. The scatterplot of $\Delta\chi^2_{P(j)}$ against the predicted probabilities is well suited to the discovery of covariate patterns that are hard to predict through the model. Here it would be useful to enter Hosmer and Lemeshow's raw threshold value of $\Delta\chi^2_{P(j)}$ of four into the graph (Hosmer and Lemeshow 2000, 163):

```
. predict dx2, dx2
. separate dx2, by(owner)
. scatter dx20  dx21 Phat, yline(4) mlabel(cpatt cpatt)
```

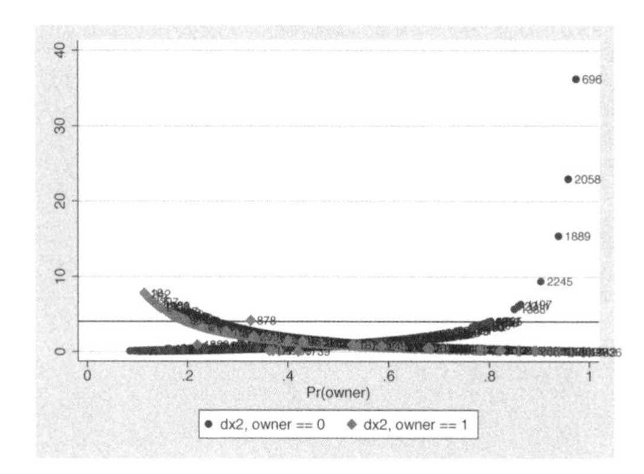

Once again, several covariate patterns stand out, and again they are the usual suspects: patterns for which residence ownership was incorrectly predicted. If you can eliminate data errors, you should determine if a variable important to the model was left out. This could be a subgroup for which the assumed correlation between age, household income, region, and residence ownership does not hold.

9.5 Likelihood-ratio test

In section 9.3.3, we showed you how to calculate $\chi^2_{\mathcal{L}}$. That statistic compares the likelihood of the fitted model with that of a model in which all the coefficients other than the constant are set to 0. A large value of $\chi^2_{\mathcal{L}}$ indicates that the full model does significantly better at explaining the dependent variable than the constant-only model.

You can apply the same principle to determine whether the addition of more independent variables achieves a significant increase in the explanatory power of our model compared with a null model with fewer independent variables. For example, you can ask whether the fit of a model on residence ownership against household income increases if we include an age variable. To answer this question, you can carry out a calculation that is analogous to the test of the overall model by again using -2 times the difference between the log likelihood of the model without age ($\ln \mathcal{L}_{\text{without}}$) and the log likelihood of the model with age ($\ln \mathcal{L}_{\text{with}}$):

$$\chi^2_{\mathcal{L}(\text{Diff})} = -2(\ln \mathcal{L}_{\text{without}} - \ln \mathcal{L}_{\text{with}})$$

Like $\chi^2_{\mathcal{L}}$, this test statistic also follows a χ^2 distribution, in which the degrees of freedom is the difference in the number of parameters between the two models.

You can easily calculate $\chi^2_{\mathcal{L}(\text{Diff})}$ in Stata using the `lrtest` command. Here we want to investigate the significance of the age effects. First, we calculate the model with the variable we want to investigate:

```
. logit owner cage chhinc east
```

We store this model internally using the command `estimates store`, and we name the model `full`:

```
. estimates store full
```

Now we calculate the reduced model.

```
. logit owner chhinc east
```

Then you can use `lrtest` to test the difference between this model and the previously stored model. You can simply list the name of the stored model (`full`) and, optionally, the name of the model against which it should be compared. If you do not specify a second name, the most recent model is used:

```
. lrtest full
Likelihood-ratio test                                 LR chi2(1)  =     76.89
(Assumption: . nested in full)                        Prob > chi2 =    0.0000
```

The probability of receiving a $\chi^2_{\mathcal{L}(\text{Diff})}$ value of 76.89 or higher in our sample is very small when the age coefficient in the population is 0. You can therefore be fairly certain that the age coefficient is not zero. However, this statistic reveals nothing about the degree of influence of age on residence ownership; for that you need to consider the estimated coefficient on age.

When using the likelihood-ratio test, only models that are nested can be compared with one another. This means that the full model must contain all the variables of the reduced model. Furthermore, both models must be calculated using the same set of observations. The latter may be problematic if, for example, some observations in your full model must be excluded due to missing values, while they may be included in the reduced model if you leave out a variable. In such cases, Stata displays a warning message ("observations differ").

If you wish to compare models not fitted to the same sets of observations, an alternative is to use "information criteria" that are based on the log-likelihood function and are valid even when comparing nonnested models. Two of the most common information criteria are the BIC (Bayesian information criterion) and AIC (Akaike's information criterion), which are obtained through the `estat ic` command mentioned earlier (page 271). An excellent introduction to the statistical foundations of these indices is provided by Raftery (1995).

9.6 Refined models

As with the linear regression model, the logistic regression model can also be expanded in various ways to investigate complicated causal hypotheses, particularly, in three ways: specifying nonlinear relationships, comparing subgroups (categorical variables), and investigating varying correlations between subgroups (interaction effects). Because these procedures for expanding the model are similar to those for linear regression, we will use a few examples from our discussion of linear regression.

9.6.1 Nonlinear relationships

During the diagnosis of our regression model, we saw signs of a U-shaped correlation between age and the logarithmic chance of residence ownership (section 9.4.1). In this respect, U-shaped correlations are only one form of nonlinear relationships. Logarithmic or hyperbolic relationships can also occur. The model assumption of logistic regression is violated only if these relationships appear between the logits and the independent variables. With respect to probabilities, logarithmic or hyperbolic relationships are to a certain extent already taken into account by the S-shaped distribution of the logit transformation.

There are many ways to account for nonlinear relationships. If you have an assumption as to why older people are less likely to own a residence than middle-aged people, you should incorporate the variable in question into the regression model. If, for example, you suspect that the observed decline is a consequence of older people moving into nursing homes to receive full-time care, you might want to introduce some type of variable indicating whether a person is in poor health.

Another way of controlling for a nonlinear relationship is to categorize the independent variable into several groups and use a set of dummy variables instead of the continuous variable. We discussed a strategy like this on page 274. A more common strategy is to use transformations or polynomials of the independent variables. The rules for linear regression apply here, too: for hyperbolic relationships, the X variable is squared, and for logarithmic relationships, the logarithm of the X variable is used. For U-shaped relationships, we use the squared X variable in addition to the original X variable.

To model the U-shaped relationship between residence ownership and age, you could proceed as follows:

```
. generate cage2 = cage^2
(140 missing values generated)

. logit owner cage cage2 chhinc east, nolog

Logistic regression                               Number of obs   =       3200
                                                  LR chi2(4)      =     332.12
                                                  Prob > chi2     =     0.0000
Log likelihood = -1925.4516                       Pseudo R2       =     0.0794

------------------------------------------------------------------------------
       owner |      Coef.   Std. Err.      z    P>|z|     [95% Conf. Interval]
-------------+----------------------------------------------------------------
        cage |   .0209879   .0024164     8.69   0.000      .016252    .0257239
       cage2 |  -.0002469   .0001196    -2.06   0.039    -.0004814   -.0000124
      chhinc |   .0006378   .0000422    15.10   0.000      .000555    .0007206
        east |  -.0675665   .0866436    -0.78   0.435    -.2373849    .1022519
       _cons |  -.5198484   .0608729    -8.54   0.000    -.6391571   -.4005397
------------------------------------------------------------------------------
```

It would be best to display the results of this regression model in a conditional-effects plot (see section 8.4.2).

9.6.2 Categorical independent variables

Categorical variables are used in logistic regression the same way they are used in linear regression (section 8.4.1). This means that a set of dummy variables is generated from a categorical variable and is then introduced into the model with the omission of a *reference category*.

Let us continue our investigation into the *Titanic* catastrophe (see section 9.1). You want to see whether the seafaring principle of women and children first was put into practice or whether, as shown in the film *Titanic*, the first-class gentlemen took their places in the lifeboats at the expense of the third-class women and children.

You have previously established that women and children evidently really did have better chances of survival than men did (and adults, respectively). To look into this more closely, load the original dataset:

```
. use titanic, clear
```

This dataset contains dichotomous variables for survival, age (children are coded zero and adults are coded one), and sex (females are coded as zero and males as one), as well as a categorical variable for first-class passengers (1), second-class passengers (2), third-class passengers (3), and crew (4).

The film *Titanic* assumed that, besides sex and age, passenger class was also a criterion for a place in the lifeboats. You can verify this assumption by using a logistic regression model of survival against age, sex, and class. To use the independent variable `class` in the regression model, first transform it into a set of dummy variables. You can do this with the following command:[21]

```
. tabulate class, gen(class)
```

This command generates four dummy variables named `class1` to `class4`, which you can now use in your regression model. As in linear regression, you must choose one of the variables as the reference category and omit it from the model. Here you will want to use the first-class passengers as the reference category.

```
. logit survived age sex class2-class4, nolog
Logistic regression                               Number of obs   =       2201
                                                  LR chi2(5)      =     559.40
                                                  Prob > chi2     =     0.0000
Log likelihood = -1105.0306                       Pseudo R2       =     0.2020
```

survived	Coef.	Std. Err.	z	P>|z|	[95% Conf.	Interval]
age	-1.061542	.2440257	-4.35	0.000	-1.539824	-.5832608
sex	-2.42006	.1404101	-17.24	0.000	-2.695259	-2.144862
class2	-1.018095	.1959976	-5.19	0.000	-1.402243	-.6339468
class3	-1.777762	.1715666	-10.36	0.000	-2.114027	-1.441498
class4	-.8576762	.1573389	-5.45	0.000	-1.166055	-.5492976
_cons	3.10538	.2981829	10.41	0.000	2.520952	3.689808

According to the signs on the age dummy, it appears that the survival chance for adults was smaller than that for children and that the survival chance for men was smaller than that for women. So far, this supports the principle of women and children first. However, it also becomes apparent that the first-class passengers have the largest estimated chance of survival compared with the rest. The third-class passengers had the smallest chances of survival; in fact, their chances of survival were even smaller than those of the crew. In conclusion, you can state that women and children were indeed favored for rescue, but apparently passenger class also played a role.

To test formally whether class played a role in determining survival, you need to test whether the estimated coefficients on the class dummies are *jointly* significantly different from zero, and for that you can use the likelihood-ratio test mentioned above. First, we save the full model you just calculated:

```
. estimates store full
```

and then you calculate the model without the class dummies:

```
. logit survived age sex
```

21. An alternative would be the command `xi`; see section 8.4.1.

A comparison of both models with the likelihood-ratio test,

```
. lrtest full
Likelihood-ratio test                                 LR chi2(3)  =    119.03
(Assumption: . nested in full)                        Prob > chi2 =    0.0000
```

shows that it is highly unlikely that the class variable has no influence whatsoever in the population. Thus passenger class did have an impact on survival.

9.6.3 Interaction effects

The logistic regression model calculated earlier shows one more weakness. It assumes that a person's sex plays the same role for adults and children. However, with the principle of women and children first, children should be preferentially treated, regardless of their sex. Sex should be used primarily for adults as a criterion for a place in the lifeboats.

If both boys and girls are treated equally, the estimated coefficient for sex should be smaller for children than it is for adults; in fact, it should be zero. In other words, the effect of sex on survival varies with age. Effects of independent variables that vary between subgroups are called interaction effects.

You can model interaction effects in logistic regressions the same way as in linear regression models. Multiplying the variables involved in the interaction effect generates interaction terms. Here we do not recenter any variables, because both age and sex are dichotomous.

```
. use titanic, clear
. tabulate class, generate(class)
. generate menage = sex * age
```

You can then incorporate the interaction terms into the model:

```
. logit survived sex age menage class2-class4, nolog
Logistic regression                               Number of obs   =       2201
                                                  LR chi2(6)      =     577.41
                                                  Prob > chi2     =     0.0000
Log likelihood = -1096.0213                       Pseudo R2       =     0.2085
```

survived	Coef.	Std. Err.	z	P>|z|	[95% Conf.	Interval]
sex	-.7150863	.406223	-1.76	0.078	-1.511269	.0810961
age	.1099979	.335319	0.33	0.743	-.5472153	.7672111
menage	-1.902104	.4330925	-4.39	0.000	-2.750949	-1.053258
class2	-1.033786	.1998153	-5.17	0.000	-1.425417	-.6421551
class3	-1.810499	.1759416	-10.29	0.000	-2.155338	-1.46566
class4	-.8033246	.1598088	-5.03	0.000	-1.116544	-.4901051
_cons	2.071621	.3528719	5.87	0.000	1.380005	2.763237

Consider first the case in which the interaction term is zero, which happens for observations on children or females. The estimated coefficient on `sex` then indicates how much lower the predicted logarithmic chance of survival is for male children than for female children. Here the estimated coefficient on age indicates that adult women had a greater predicted chance of surviving than girls, though the coefficient is not significantly different from zero.

The interaction effect indicates how much the influence of sex changes when one considers adults instead of children. If male children already had a -0.72 smaller predicted logarithmic chance of survival than female children, this would yield a $-0.72 + (-1.90) = -2.62$ smaller predicted log-chance of survival for a male adult than for a female adult. Therefore, the predicted survival chance of men was only around $1/14$ ($e^{-2.62} = 0.07 = 1/14$) that of the women.

9.7 Advanced techniques

Stata allows you to fit many related models in addition to the logistic regression we have described above. Unfortunately, we cannot show them in detail. However, we will describe the fundamental ideas behind some of the most important procedures. For further information, we will specifically refer you to the entry in the *Stata Reference Manual* corresponding to each command. There you will also find references to the literature.

9.7.1 Probit models

In the logistic regression model, we attempted to predict the probability of a success through a linear combination of one or more independent variables. To ensure that the predicted probabilities remained between the limits of 0 and 1, the probability of the success underwent a logit transformation. However, using the logit transformation is not the only way to achieve this. An alternative is the probit transformation used in probit models.

To get some idea of this transformation, think of the density function of the standard normal distribution:

(*Continued on next page*)

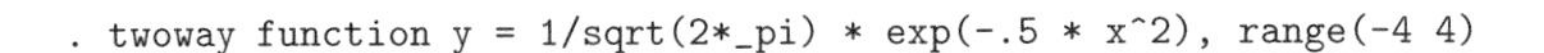

```
. twoway function y = 1/sqrt(2*_pi) * exp(-.5 * x^2), range(-4 4)
```

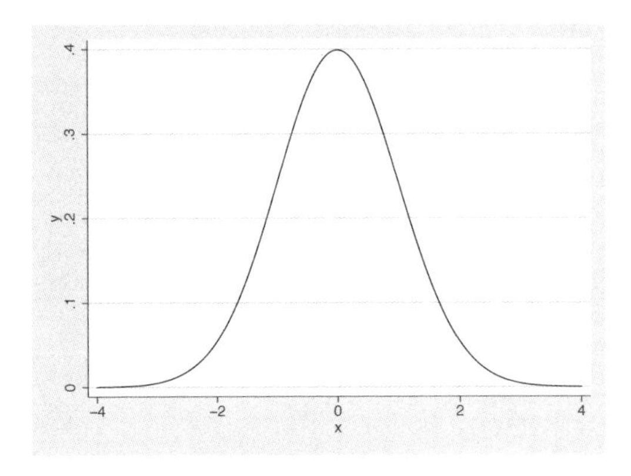

You can interpret this graph in the same way as a histogram or a kernel density estimator (see section 7.3.3); i.e., for this variable, the values around 0 occur most often, and the larger or smaller they become, the more rarely they occur.

Suppose that you randomly selected an observation from the variable X. How large would the probability be of selecting an observation that had a value of less than -2? As values under -2 do not occur very often in the X variable, the intuitive answer is, not very large. If you want to know the exact answer, you can determine the probability through distribution-function tables for standard normal distribution or through the Stata command

```
. display normal(-2)
.02275013
```

The probability of selecting an observation with a value of less than or equal to -2 from a standard normal variate is therefore 0.023. You can repeat the same calculation for any value of X and then enter the calculated probabilities against the values of X in a scatterplot. This results in the cumulative density function for the standard normal distribution Φ depicted in the following graph:

```
. twoway function y = normal(x), range(-4 4)
> || function y = exp(x)/(1+exp(x)), range(-4 4)
> || , legend(order(1 "Probit-Transformation" 2 "Logit-Transformation"))
```

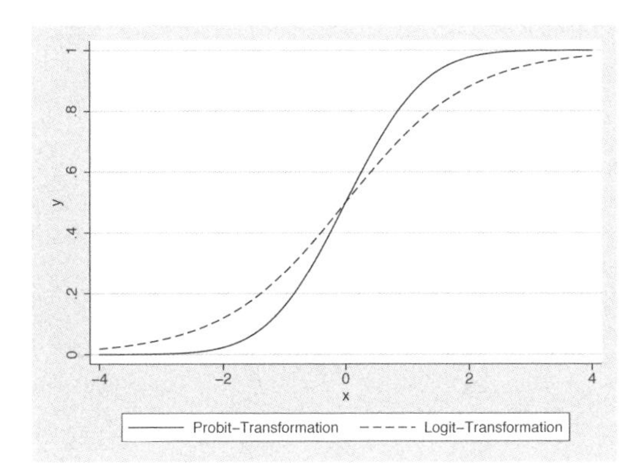

The function shows a S-shaped curve, similar to the probabilities assigned to the logits, which we have also included in the graph.

As with the logit transform you used with logistic regression, the normal distribution function can also be used to transform values from $-\infty$ and $+\infty$ into values between 0 and 1. Correspondingly, the inverse of the distribution function for the standard normal distribution (Φ^{-1}) converts probabilities between 0 and 1 for a success [$\Pr(Y=1)$] into values between $-\infty$ and $+\infty$. The values of this probit transformation are thus also suited to be estimated with a linear model. This yields the probit model:

$$\Phi^{-1}\{\Pr(Y=1)\} = \beta_0 + \beta_1 x_{1i} + \beta_2 x_{2i} + \cdots + \beta_{K-1} x_{K-1,i} \tag{9.6}$$

You can estimate the b coefficients of this model through maximum likelihood. You can interpret the estimated coefficients the same way as in logistic regression, except that now the value of the inverse distribution function of the standard normal distribution increases by b units instead of the log-odds ratio increasing by b units for each one-unit change in the corresponding independent variable. Using the distribution function for the standard normal distribution, you can then calculate probabilities of success. Usually, the predicted probabilities of probit models are nearly identical to those of logistic models, and the estimated coefficients are often about 0.58 times the value of those of the logit models (Long 1997, 49).

The Stata command used to calculate probit models is `probit`. For example, you can refit the previous model (see page 284) using `probit` instead of `logit`:

```
. probit survived sex age menage class2-class4, nolog
Probit regression                                 Number of obs   =       2201
                                                  LR chi2(6)      =     573.48
                                                  Prob > chi2     =     0.0000
Log likelihood =  -1097.988                       Pseudo R2       =     0.2071

------------------------------------------------------------------------------
    survived |      Coef.   Std. Err.      z    P>|z|     [95% Conf. Interval]
-------------+----------------------------------------------------------------
         sex |  -.4554407   .2533275    -1.80   0.072    -.9519534     .041072
         age |    .084439   .2076534     0.41   0.684    -.3225543    .4914322
      menage |  -1.102542   .2673721    -4.12   0.000    -1.626582   -.5785024
      class2 |  -.6327645   .1193993    -5.30   0.000    -.8667829   -.3987462
      class3 |   -1.02884   .0999088   -10.30   0.000    -1.224657   -.8330219
      class4 |  -.5055305   .0962533    -5.25   0.000    -.6941836   -.3168775
       _cons |   1.223367   .2160143     5.66   0.000     .7999871    1.646748
------------------------------------------------------------------------------
```

See [R] **probit** for more information on this model.

9.7.2 Multinomial logistic regression

Multinomial logistic regression is used when the dependent variable exhibits more than two categories that cannot be ranked. An example for this would be party preference with values for the German parties CDU, SPD, and all other parties.

The main problem with using multinomial logistic regression is in the interpretation of the estimated coefficients, so this will be the focus point of this section. Nevertheless, to understand this problem, you must at least intuitively grasp the statistical fundamentals of the process. These fundamentals will be discussed shortly (cf. Long 1997).

In multinomial logistic regression, you predict the probability for every value of the dependent variable. You could initially calculate a binary[22] logistic regression for every value of the dependent variable. Here you could calculate three separate logistic regressions: one with the dependent variable CDU against non-CDU, one with the dependent variable SPD against non-SPD, and one with the dependent variable for the other parties against the CDU and SPD together:

$$\begin{aligned}
\ln \frac{\Pr(Y=\text{CDU})}{\Pr(Y=\text{not-CDU})} &= \beta_0^{(1)} + \beta_1^{(1)} x_{1i} + \beta_2^{(1)} x_{2i} + \cdots + \beta_{K-1}^{(1)} x_{K-1,i} \\
\ln \frac{\Pr(Y=\text{SPD})}{\text{P}(Y=\text{not-SPD})} &= \beta_0^{(2)} + \beta_1^{(2)} x_{1i} + \beta_2^{(2)} x_{2i} + \cdots + \beta_{K-1}^{(2)} x_{K-1,i} \\
\ln \frac{\Pr(Y=\text{Sonst.})}{\text{P}(Y=\text{not-Sonst.})} &= \beta_0^{(3)} + \beta_1^{(3)} x_{1i} + \beta_2^{(3)} x_{2i} + \cdots + \beta_{K-1}^{(3)} x_{K-1,i} \qquad (9.7)
\end{aligned}$$

22. To differentiate it from multinomial logistic regression, we call the logistic regression of a dichotomous dependent variable a binary logistic regression.

The superscript in parentheses means that the β coefficients differ between the individual regression equations: $\beta_k^{(1)} \neq \beta_k^{(2)} \neq \beta_k^{(3)}$. To simplify the notation, we refer to $\beta_1^{(1)} \ldots \beta_{K-1}^{(1)}$ as $\mathbf{b}^{(1)}$ and refer to the sets of b coefficients from the other two equations as $\mathbf{b}^{(2)}$ and $\mathbf{b}^{(3)}$, respectively.

Each of the unconnected regressions allows for a calculation of the predicted probability of every value of the dependent variable. These predicted probabilities do not all add up to 1. However, they should, as one of the three possibilities—SPD, CDU or other—must[23] occur.

Therefore, it is sensible to jointly estimate $\mathbf{b}^{(1)}$, $\mathbf{b}^{(2)}$, and $\mathbf{b}^{(3)}$ and to adhere to the rule that the predicted probabilities must add up to 1. However, it is not possible to estimate all three sets of coefficients. To do so, you must constrain one of the coefficient vectors to be equal to a fixed value, zero being by far the most common choice. After making such a normalization, you can estimate the remaining coefficients using the maximum likelihood principle. Which one of the three sets of coefficients you constrain to be zero does not matter. By default, Stata's `mlogit` command constrains the coefficient vector corresponding to the most frequent outcome.

Let us show you an example of interpreting the estimated coefficients. Please load `data1.dta`:

```
. use data1, clear
```

Now generate a new variable for party choice with values for the CDU, the SPD, and the other parties from the original variable for party preferences (`np9402`). One way of doing this is

```
. recode np9402 2 3=1 1=2 4/8=3, generate(party)
. label define party 1 "CDU" 2 "SPD" 3 "Other"
. label values party party
```

This creates the variable `party` with the values 1 for the CDU/CSU, 2 for the SPD, and 3 for the other parties. Respondents without a party preference have a missing value.

The Stata command for multinomial logistic regression is `mlogit`. The syntax for the command is the same as for all estimation commands; i.e., the dependent variable follows the command and is in turn followed by the list of independent variables. With the `baseoutcome()` option, you can select the equation for which the b coefficients are set to 0.

Let us calculate a multinomial logistic regression for party preference against education (in years of education) and year of birth. Here the b coefficients of the equation for the CDU are set at 0:

23. Here you disregard the possibility of no party preference. If you did not, you would have to calculate a further regression model for this alternative. The predicted probabilities for the four regression should then add up to 1.

```
. mlogit party yedu ybirth, baseoutcome(1) nolog
Multinomial logistic regression                   Number of obs   =       1360
                                                  LR chi2(4)      =      92.69
                                                  Prob > chi2     =     0.0000
Log likelihood = -1379.4301                       Pseudo R2       =     0.0325

------------------------------------------------------------------------------
       party |      Coef.   Std. Err.      z    P>|z|     [95% Conf. Interval]
-------------+----------------------------------------------------------------
SPD          |
        yedu |  -.0039571   .0255271    -0.16   0.877    -.0539892    .0460751
      ybirth |   .0126934   .0034126     3.72   0.000     .0060047     .019382
       _cons |  -24.54483   6.627974    -3.70   0.000    -37.53542   -11.55424
-------------+----------------------------------------------------------------
Other        |
        yedu |   .1305466   .0295313     4.42   0.000     .0726663     .188427
      ybirth |   .0352889   .0046591     7.57   0.000     .0261573    .0444206
       _cons |  -71.04625   9.092251    -7.81   0.000    -88.86673   -53.22576
------------------------------------------------------------------------------
(party==CDU is the base outcome)
```

In contrast to binary logistic regression, the coefficient table is split into two parts. The upper part contains the estimated coefficients of the equation for the SPD, whereas the lower part contains the estimated coefficients of the equation for the other parties. The coefficients of the equation of the CDU were set at 0 and are therefore not displayed.

As a result of setting $\mathbf{b}^{(\text{CDU})} = 0$, you can interpret the estimated coefficients of the other two equations in relation to the CDU supporters. By this, we mean that estimated coefficients in the equation for the SPD indicate how much the predicted logarithmic chance of preferring the SPD and not the CDU changes when the independent variables increase by one unit. The equation for the other parties indicates changes in the predicted logarithmic chance of preferring the other parties and not the CDU.

Interpreting the estimated coefficients for a multinomial logistic regression is not as easy as for binary logistic regression, as the sign interpretation cannot be used. The negative sign for length of education in the SPD equation does not necessarily mean that the predicted probability of a preference for the SPD declines with education. In our regression model, we can demonstrate this with the estimated coefficient for the variable `ybirth` from the equation for the SPD. Writing the probability of preferring the SPD as P_{SPD} and the probability of preferring the CDU as P_{CDU}, the b coefficient for `ybirth` in the SPD equation can be written as

$$
\begin{aligned}
b_{\text{ybirth}}^{(\text{SPD})} &= \ln\left(\frac{\widehat{P}_{\text{SPD}|\text{ybirth}+1}}{\widehat{P}_{\text{CDU}|\text{ybirth}+1}}\right) - \ln\left(\frac{\widehat{P}_{\text{SPD}|\text{ybirth}}}{\widehat{P}_{\text{CDU}|\text{ybirth}}}\right) \\
&= \ln\left(\frac{\widehat{P}_{\text{SPD}|\text{ybirth}+1}}{\widehat{P}_{\text{SPD}|\text{ybirth}}} \times \frac{\widehat{P}_{\text{CDU}|\text{ybirth}}}{\widehat{P}_{\text{CDU}|\text{ybirth}+1}}\right)
\end{aligned}
$$

The b coefficient for year of birth in the equation for the SPD, on one hand, depends on the change in the predicted probability of SPD preference with the year of birth. On the other hand, it also depends on the respective change in predicted probability for choosing the CDU. In contrast to the binary logit model, in the multinomial logit model, the change in the predicted probability of CDU preference does not completely depend on the change in the probability of SPD preference. In this respect, the b coefficient can be solely, mainly, or partly dependent on the probability relationship in the base category.

To avoid misinterpreting the multinomial logistic regression, we recommend that you use the conditional-effects plot for the predicted probabilities.[24] To create this plot, first generate the predicted probabilities of the model with `predict`. As there is a predicted probability for every value of the dependent variable, you will have to provide three variable names for the predicted probabilities.

```
. predict PCDU PSPD POther
```

To illustrate the effect of the year of birth, we plot these variables against the year of birth, holding the length of education at a particular value.

If you fix length of education at the highest value (18 years), you can establish that the predicted probability of preferring the SPD declines with the year of birth, despite the regression model indicating a significant, positive estimated coefficient for the year of birth.[25]

```
. line PCDU PSPD POther ybirth if yedu==18, sort
```

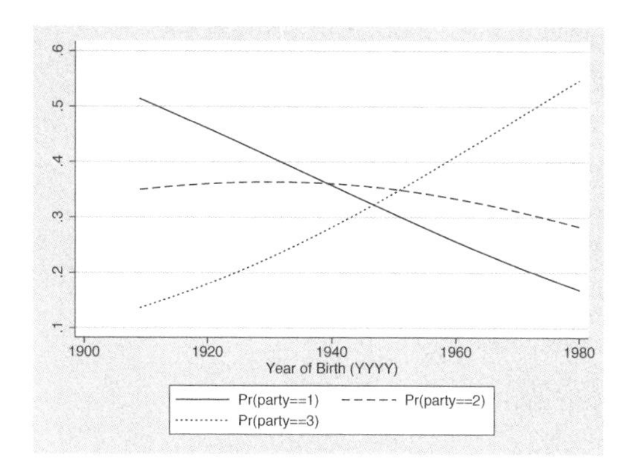

Conditional-effects plots cannot always be produced as easily as above. Problems occur when too many variables are included in the model, so fixing the values of the independent variables may mean that there are too few observations remaining for a

24. One alternative is the *method of recycled predictions*, which is described in [R] **mlogit**. A further alternative is calculating marginal effects with the command `mfx compute`.

25. In the legend, `Pr(party==1)` stands for the CDU, `Pr(party==2)` stands for the SPD, and `Pr(party==3)` stands for the other parties.

sensible plot. The method of recycled predictions described in [R] **mlogit** appears to be good solution to this problem. An even more powerful possibility for producing conditional-effects plots is the `prgen` command, which you can download from the SSC archive (see section 12.3.2) and which is more fully described by Long and Freese (2006).

9.7.3 Models for ordinal data

Models for ordinal data are used when the dependent variable has more than two values that can be ordered. An example would be the question regarding concerns about the increase of crime, which respondents could answer with "no concerns", "moderate concerns", or "strong concerns". In a dataset, these could be assigned respective values of 0, 1, and 2, or equivalently, 0, 10, and 12. The difference between two consecutive categories is immaterial—all that matters is that the outcomes can be ordered.

In principle, there are two strategies available for modeling ordinal dependent variables. The first uses multinomial logistic regression, whereby certain constraints are imposed upon the coefficients (stereotype model). The second strategy generalizes binary logistic regression for variables with more than two values (proportional-odds model). Anderson (1984) discusses the implementation prerequisites for both models.

The logic behind the stereotype model is simple. In multinomial logistic regression, every value of the dependent variable has its own set of coefficients. The length of education in the regression model on page 290 had a negative effect on the predicted chance of preferring the SPD (and not the CDU) and at the same time has a positive effect on the predicted chance of preferring another party (and not the CDU). If the dependent variable indicates the presence of ranking, you would normally not expect a directional change in the effects. For example, consider the variable for concerns about increasing crime (`np9506`), which contains the values 1 for no concerns, 2 for moderate concerns, and 3 for strong concerns. First, calculate a multinomial logistic regression for this variable against the length of education. Before you do this, you should, however, mirror the variable `np9506` so that large values stand for strong concerns and vice versa:

```
. generate worries = 4 - np9506
. mlogit worries yedu, base(1)
```

You will get an estimated coefficient of around -0.05 in the equation for moderate concerns and -0.11 in the equation for strong concerns. The direction of the effects does not change here. This should come as little surprise, since if education reduces the chance of having moderate concerns (and not of having no concerns), it should also reduce the chance of having strong concerns (and not of having no concerns). However, if you calculate a multinomial logistic regression, this assumption is ignored. Nevertheless, you can include such assumptions in the model by imposing constraints on the b coefficients.

Using constraints, you can impose certain structures for the b coefficients before calculating a model. You could, for example, require that education reduces the chance of having moderate concerns (and not of having no concerns) to the same extent that it does for having strong concerns (and not of having moderate concerns). Here the coefficient of education for strong concerns would have to be exactly twice as large as the coefficient of education for moderate concerns. With the `constraint` command, you can set this structure for the `mlogit` command. With

```
. constraint define 1 [3]yedu = 2*[2]yedu
```

you define constraint number 1, which states that the coefficient of the variable `yedu` in the third equation must be twice as large as the coefficient of the variable `yedu` in the second equation. You impose the constraint by specifying the `constraints()` option of the `mlogit` command. Here you would enter the number of the constraint you wish to use in the parentheses.

```
. mlogit worries yedu, base(1) constraints(1)
```

If you calculate this model, you will discover that it is almost identical to the previous model. However, it is far more economical, as in principle only one education coefficient has to be estimated. The other estimate is derived from the ordinal structure of the dependent variable and our assumption that education proportionately increases concerns.

Establishing specific constraints that take into account the ordinal structure of the dependent variable is one way of modeling the ordinal dependent variable. Nevertheless, the constraint is just one example of many alternatives. See [R] **slogit** for more information about this model.

The proportional-odds model follows a different approach: the value of the ordinal variable is understood as the result of categorizing an underlying metric variable. Here you could assume that answers in the `worries` variable provide only a rough indication of the attitudes toward the increase in crime. The attitudes of people probably vary between having infinitely many concerns and no concerns whatsoever, so they might take any value between; that is, attitude is actually a continuous variable, E. Instead of observing E, however, all you see are the answers reported on the survey—no concerns, moderate concerns, or strong concerns. Since you have three outcomes in the model, there must also exist two points, κ_1 and κ_2, that partition the range of E into the three reported answers. That is, if $E < \kappa_1$, then the person reported no concerns if $\kappa_1 \leq E \geq \kappa_2$, the person reported moderate concerns and if $E > \kappa_2$ the person reported strong concerns.

Remember the predicted values $(\widehat{L})$ of the binary logistic regression. These values can take on any values from $-\infty$ to $+\infty$. In this respect, you could interpret these predicted values as the unknown metric attitude, E. If you knew the value of κ_1 and κ_2 by assuming a specific distribution for the difference between E and $\widehat{L}$, you could determine the probability that each person reported each of the three levels of concern. The proportional-odds model estimates the b's in the linear combination of independent variables as well as the cutpoints needed to partition the range of E into discrete categories.

An example may clarify this concept. The command for the *proportional odds* model in Stata is `ologit`. The syntax of the command is the same as that for all other model commands: the dependent variable follows the command and is in turn followed by the list of independent variables. We will calculate the same model as above:

```
. ologit worries yedu
Iteration 0:   log likelihood = -1962.7767
Iteration 1:   log likelihood = -1955.4719
Iteration 2:   log likelihood = -1955.4689
Ordered logistic regression                       Number of obs   =       1907
                                                  LR chi2(1)      =      14.62
                                                  Prob > chi2     =     0.0001
Log likelihood = -1955.4689                       Pseudo R2       =     0.0037

------------------------------------------------------------------------------
     worries |      Coef.   Std. Err.      z    P>|z|     [95% Conf. Interval]
-------------+----------------------------------------------------------------
        yedu |  -.0684365   .0179959    -3.80   0.000    -.1037078   -.0331652
-------------+----------------------------------------------------------------
       /cut1 |  -1.195727   .2154466                     -1.617995   -.7734599
       /cut2 |    .809241   .2151594                      .3875364    1.230946
------------------------------------------------------------------------------
```

The predicted value of this model for respondents with 10 years of education is $S_{10} = -0.068 \times 10 = -0.68$. The value for κ_1 and κ_2 are provided beneath the coefficient table. The predicted probability that respondents with a predicted value of -0.68 are classified as individuals with no concerns matches the probability of $-0.68 + u_j \leq -1.196$, or, in other words, the probability that $u_j \leq -0.516$. If you assume that the error term follows the logistic distribution, the predicted probability is $1/(1 + e^{-0.516}) = 0.37$.

For more information on ordered logistic regression in Stata, see [R] **ologit**.

9.8 Exercises

1. Download a subset of the 1988 National Longitudinal Survey by typing

```
. webuse nlsw88, clear
```

2. Create a regression model where `union` is the dependent variable. Create and use the following independent variables for your model:

   - `tenure` (centered)
   - `age` (centered)
   - `collgrad`

3. Calculate the predicted values for all cases.
4. Request the display of:

   a. the predicted odds of being unionized for workers with mean age and mean tenure without a college diploma.

b. the predicted odds of being unionized for white workers with mean age and mean tenure with a college diploma.
c. the estimated odds ratio of being unionized for college graduates versus non-college graduates.
d. the odds ratio for all covariates using the `logistic` command.
e. the probability of being unionized for workers with mean age, mean tenure, and a college diploma.

5. Predict the probability of being unionized as a function of tenure for workers with mean age and without a college diploma. Display these probabilities graphically.
6. Predict the respective probability for mean aged workers with a college diploma. Add a line for these probabilities to the graph from the last problem.
7. Investigate the functional form of the effect of tenure. Does the relationship appear linear?
8. Generate a classification table manually and then using the built-in function.
9. Perform a likelihood-ratio test between the full model including age and a reduced model that excludes age. What do you conclude from this test?
10. Produce a plot of $\delta\chi^2$ by predicted probabilities. Label the marker symbol with the covariate pattern. Describe the problem your models suffers most from.
11. Investigate the correlation of high influence points with the industrial branch. What do you conclude? See how your results change if you reformulate your model by including a variable for an industrial branch with three categories.

10 Reading and writing data

In previous chapters, we asked you to load some data into Stata. Usually you just needed to type the command `use` followed by a filename, for example,[1]

```
. use data1
```

In practice, however, reading data into Stata is not always that easy—either because the data you want to use are not in Stata format but another format, such as SAS, SPSS, or Excel, or because they are not available as a machine-readable dataset.

After defining a rectangular dataset in section 10.1, we will explain how to *import* different forms of machine-readable data into Stata in section 10.2. Machine-readable data are data stored on a hard drive, a CD-ROM, a memory stick, a web site, or any other medium that can be read by a machine. However, most types of machine-readable data, such as information in statistical yearbooks or questionnaires, are not in Stata format and therefore cannot be read into Stata with `use`. In section 10.3, we will discuss how to deal with data that are not yet machine readable. Then we will move on to data that are already in Stata format but are spread across several data files. Section 10.4 shows you how to combine the information from different files into one rectangular dataset using data from the GSOEP database. In section 10.5, we will discuss importing and managing data and then show you how to save the Stata files you created. Section 10.6 gives general tips for dealing with large and even *too* large datasets.

10.1 The goal: the data matrix

Before you start, look again at the dataset `data1.dta`, which we used in previous chapters:

```
. describe
```

As you can see, the dataset consists of 3,340 observations and 47 variables. To get an impression of the data, type

```
. browse
```

1. Make sure that your working directory is `c:\data\kk2` before invoking the command (see page 3 for details).

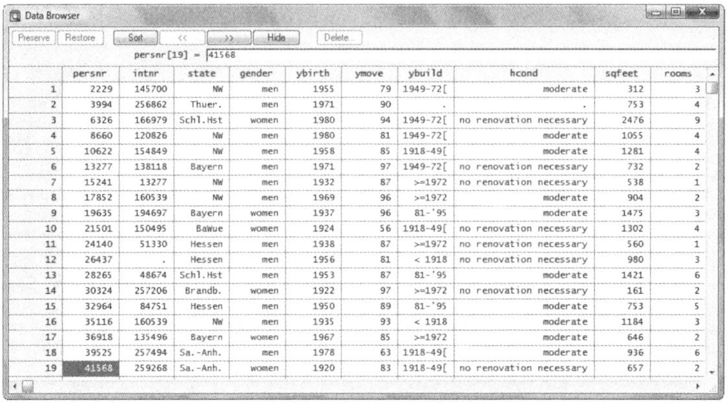

Data Browser

Preserve Restore Sort << >> Hide Delete...

persnr[19] = 41568

	persnr	intnr	state	gender	ybirth	ymove	ybuild	hcond	sqfeet	rooms
1	2229	145700	NW	men	1955	79	1949-72[	moderate	312	3
2	3994	256862	Thuer.	men	1971	90	.	.	753	4
3	6326	166979	Schl.Hst	women	1980	94	1949-72[	no renovation necessary	2476	9
4	8660	120826	NW	men	1980	81	1949-72[	moderate	1055	4
5	10622	154849	NW	men	1958	85	1918-49[	moderate	1281	4
6	13277	138118	Bayern	men	1971	97	1949-72[	no renovation necessary	732	2
7	15241	13277	NW	men	1932	87	>=1972	no renovation necessary	538	1
8	17852	160539	NW	men	1969	96	>=1972	moderate	904	2
9	19635	194697	Bayern	women	1937	96	81-'95	moderate	1475	3
10	21501	150495	BaWue	women	1924	56	1918-49[	no renovation necessary	1302	4
11	24140	51330	Hessen	men	1938	87	>=1972	no renovation necessary	560	1
12	26437	.	Hessen	men	1956	81	< 1918	no renovation necessary	980	3
13	28265	48674	Schl.Hst	men	1953	87	81-'95	moderate	1421	6
14	30324	257206	Brandb.	women	1922	97	>=1972	no renovation necessary	161	2
15	32964	84751	Hessen	men	1950	89	81-'95	moderate	753	5
16	35116	160539	NW	men	1935	93	< 1918	moderate	1184	3
17	36918	135496	Bayern	women	1967	85	>=1972	moderate	646	2
18	39525	257494	Sa.-Anh.	men	1978	63	1918-49[	moderate	936	6
19	41568	259268	Sa.-Anh.	women	1920	83	1918-49[	no renovation necessary	657	2

Figure 10.1. The Data Browser in Stata for Windows

This command opens the Data Browser (figure 10.1), which is a separate window with a table containing text or numerical cells. The gray headline (the first row in the table) shows the names of the variables. The first column, which is also gray, displays a serial number for each observation. Correspondingly, each row in this table contains information on one of the 3,340 interviewed persons (cases, observations). Each column in the table contains an entry for each of the 47 variables for all 3,340 persons. We call such a table a "data matrix".

The first white column displays the personal identification number, which is unique for each interviewed person. If you look in the fifth column of the row for the person identified by the number 3994, you will find that this person was born in 1971. In row 12 (identification number 26,437), the information on the interviewer (in the second column) is missing. Instead of a specific value, the Data Browser displays a dot representing a missing value. Since dots are used in all empty cells ("missings"), the data matrix is rectangular; that is, the matrix has the same number of observations for all variables and the same number of variables for all observations.

Now you have seen what the structure of the data should resemble. In the following sections, you will learn how to put your data into such a structure. At this point, you can close the Data Browser: press *Alt+F4*, or use the mouse to close the window as appropriate in your operating system.

10.2 Importing machine-readable data

Statistical offices and other research institutions produce large amounts of data on machine-readable media (e.g., CD-ROM) or on the Internet.[2] You can use these data for secondary data analyses.[3]

2. A list of data archives can be found at http://www.ifdo.org/network/network_archive.html.
3. Analyses are called secondary if they are carried out using data collected by others.

However, such datasets, or data that you might get from a collaborator, are often not available in Stata format. Often you will be confronted with data from other statistical packages (e.g., SAS, SPSS, and R), databases (e.g., Access, Oracle, and MySQL), spreadsheet programs (e.g., Excel and Gnumeric), or plain American Standard Code for Information Interchange (ASCII) files. ASCII files can be read directly into Stata, whereas files from other statistical packages must be translated into Stata format with the help of specialized software packages.

Within Stata, the easiest way to determine the format of a new file is to use the Stata command `type filename`. With this command, the contents of the specified file are displayed in the Results window, without loading the file into memory. You can see the difference between Stata datasets and ASCII files by typing the following commands:

```
. type popst2.dta
. type popst2.sav
. type popst2.raw
```

The output of the first two commands is rather confusing. Only the third command yields output that you can read, but all three files contain the same information.

The file with the extension `.dta` is a Stata dataset. Apart from the individual values, datasets also store information on data types and variable labels. To do so, these datasets use specific control characters that make their output unreadable for humans, such as the output of `type popst2.dta`. Unfortunately, every statistical package uses its own control characters for this information. SPSS data files, such as `popst2.sav`, are of little use to Stata. Such files are in a language unknown to us.

The output of `type popst2.raw` is readable for humans. This is a good indication that it is a pure ASCII file, meaning a file without program-specific control characters.

10.2.1 Reading system files from other packages

Many freely accessible datasets are available in SAS, SPSS, or Excel format. Stata has no import filter for reading system files from other statistical packages, except files in SAS export format (SAS XPORT files, `.fda`) and Haver Analytics databases. You can read SAS export files directly into Stata by using the command `fdause` and read Haver databases with the `haver` command. To read all other system files into Stata, you can either use a data conversion program or export the data as an ASCII file from the program in which it was saved. Then you can read them directly into Stata.

We are aware of two programs that convert system files from one statistical package to another: Stat/Transfer by Circle Systems and DBMS/COPY from DataFlux, a subsidiary of SAS Institute. The advantage of using a conversion program is that you can keep variable and value labels that had been assigned to a data file, and in some cases you can even keep missing-value definitions. But to import data into your system, you need not pay for an external program. As we said, you can always save data as an ASCII file and follow the description in the upcoming sections.

If your dataset is in Excel or HTML format, we recommend the following procedure. Once you have read your data into a spreadsheet file, save it as an ASCII dataset in spreadsheet format. In many programs, this is done by choosing **Text (Tab delimited)** from the **Save As...** pulldown menu. Once your file is in text format, you can proceed as shown below.

10.2.2 Reading ASCII text files

Stata has three commands for reading ASCII files: `infile`, `insheet`, and `infix`. The last two commands are simplified special cases of the `infile` command. Once you are at ease using `infile`, you will be able to read in all ASCII files with no problems. We begin with the `insheet` command, which is very easy to use. However, since not all datasets can be read in using `insheet`, we will then explain the use of `infile` in detail.

Reading data in spreadsheet format

In a simple case where data are in an ASCII data file, each observation is written into a separate row, and the columns (variables) are separated by commas or tabs. Spreadsheet programs usually export ASCII files of this type, which is why the format is also known as "spreadsheet format". Windows files are often tab delimited with the file extension `.txt` or comma separated with the extension `.csv`. We have prepared a data file, here called `popst1.raw`:[4]

```
. type popst1.raw
Hessen, 5763, 5837, 5923, 5967, 5981, 6010
Mecklenburg-Vorpommern, 1924,, 1865, 1843, 1832, 1823
Niedersachsen, 7387, 7476, 7578, 7648, 7715, 7780
Nordrhein-Westfalen, 17350, 17510, 17679, 17759, 17816, 17893
Rheinland-Pfalz, 3764, 3821, 3881, 3926, 3952, 3978
Saarland, 1073, 1077, 1084, 1085, 1084, 1084
Sachsen, 4764, 4679, 4641, 4608, 4584, 4567
Sachsen-Anhalt, 2874, 2823, 2797, 2778, 2759, 2739
Schleswig-Holstein, 2626, 2649, 2680, 2695, 2708, 2725
Thueringen, 2611, 2572, 2546, 2533, 2518, 2504
Deutschland, 79753, 80275, 80975, 81338, 81539, 81817
```

The file `popst1.raw` contains the population data for Germany and 10 German states from 1990 to 1995 (Statistisches Bundesamt 1997, 47). The first column contains the state name, the second column the population in 1990, the third column the population in 1991, etc. So, every row begins with the name of the state and continues with the population data for 1990, 1991, 1992, etc., for that state. Every row in this file is a new observation, and for each observation there are seven pieces of information, separated by commas. In the second row there seem to be only six values—the state name (Mecklenburg-Vorpommern) and five numbers. But if you look closely, you will find two commas, one directly after the other. The value between is missing, so we do

4. We use the extension `.raw` because it is the Stata default extension for ASCII files. See section 3.1.8 for more information.

not know what the population was in 1991 for Mecklenburg-Vorpommern, but we do know that the value for 1991 is missing. There are still seven pieces of information in the second row. This is an important point, as the data must be rectangular. Each observation must appear in a single row, and there must be the same number of entries in each row, even if one of the values is missing.

A text file generated by a spreadsheet program may also separate variables by tabs instead of commas. For humans, however, tabs are difficult to distinguish from blanks. The option `showtabs` with the `type` command displays characters representing tabs:

```
. type popst5.raw
. type popst5.raw, showtabs
```

Datasets delimited with commas or tabs can be read into Stata using the command `insheet`. To read in `popst1.raw`, for example, you could type

```
. insheet using popst1.raw, clear
. describe
```

Following `insheet`, the keyword `using` is issued to specify the file to be read in. If the extension of a file is `.raw`, you can omit it. As you see in the output of `describe`, Stata automatically assigned seven variable names, `v1`–`v7`, for the seven columns. If you prefer other variable names, you can indicate these between `insheet` and `using`:

```
. insheet state pop90 pop91 pop92 pop93 pop94 pop95 using popst1, clear
```

When you enter variable names, Stata knows how to deal with numbers, so you can also use

```
. insheet state pop90-pop95 using popst1.raw, clear
```

Many spreadsheet programs store column titles in the first row of spreadsheet files. If this is the case, `insheet` uses the first row as variable names and all other rows as observations. The file `popst2.raw` contains such a row. From

```
. insheet using popst2.raw, clear
```

you obtain the same dataset as the one you created with the previous command.

Although `insheet` is fairly easy to use, checking the state of your data is crucial in doing so. Different spreadsheet programs have different problems exporting ASCII files, so you need to be careful:

- A common problem is that commas are used in numbers as thousands separators, or—depending on the country you work in—as decimal points. In both cases, one numeric value would be split by Stata into two or more variables.
- The cells of spreadsheet programs often contain formulas or references to other cells in the table. If data are carelessly exported as ASCII files, placeholders are easily exported instead of the desired values. If a column contains such a placeholder, this variable is read in as a string variable (see section 5.6). The same

problem occurs if blanks or dots are used for missing values. With `insheet`, the characters separating cells should always be next to each other if the value between them is missing.

- If the individual variables in the ASCII file are separated by blanks, or if characters other than commas or tabs were used to separate cells, you must use the `delimiter` option. See `help insheet` for details.
- You cannot use `insheet` if cells are not separated or if the values of any observations span across multiple rows. Then you need to use another Stata command called `infile`.

You can solve some of these problems by cleaning up your file in a text editor. Here you can replace commas with dots or replace blanks with tabs. However, you should do this only if the process can be automated. And be careful! Some text editors split long lines into several lines, thus destroying the format of your data. Check if blanks are used exclusively to separate cells or if blanks are used within a text cell (e.g., "Mecklenburg Vorpommern"). Furthermore, such processes are not only error prone but also usually do not allow you to document your work. You will have to painstakingly replace everything every time you discover a mistake in your data. Our experience is that, in the long run, the `infile` command is worth learning.

Reading data in free format

Sometimes ASCII files are in free format, meaning that individual variables are separated by blanks, tabs, commas, or line breaks. Take `popst3.raw` as an example:

```
. type popst3.raw
Hessen                  5763
    5837    5923    5967
5981    6010
Mecklenburg-Vorpommern  1924    1892    1865    1843    1832    1823
Niedersachsen           7387    7476    7578    7648    7715    7780
Nordrhein-Westfalen     17350   17510   17679   17759   17816   17893
Rheinland-Pfalz         3764    3821    3881    3926    3952    3978
Saarland                1073    1077    1084    1085    1084    1084
Sachsen                 4764    4679    4641    4608    4584    4567
Sachsen-Anhalt          2874    2823    2797    2778    2759    2739
Schleswig-Holstein      2626    2649    2680    2695    2708    2725
Thueringen               2611    2572    2546    2533    2518    2504
Deutschland             79753   80275   80975   81338   81539   81817
```

Here the information for the state Hessen is for some reason spread over three rows. Unlike in the spreadsheet format, observations in the free format can be spread over several rows. This has an important implication: Stata can no longer automatically identify how many variables the dataset contains. You must enter this information.

ASCII files in free format are read in using the `infile` command. You indicate the number of variables by specifying a variable list. The file `popst3.raw`, for example, can be read with

```
. infile str22 state pop90 pop91 pop92 pop93 pop94 pop95 using popst3.raw, clear
```

or

```
. infile str22 state pop90-pop95 using popst3.raw, clear
```

After specifying `infile`, you enter a variable list, followed by `using` and a filename. From the specified variable list, Stata can infer that there are seven variables to read. Therefore, a new observation starts every seventh entry.[5] One problem reading in `popst3.raw` is the first variable, which contains the name of the state. Since this variable contains text, it must be marked as a string variable (`str` in the variable list). In doing so, you must specify the maximum number of letters the variable may contain. Here, since "Mecklenburg-Vorpommern" is the longest and therefore limiting element, with 22 characters, we use `str22` as the storage type. However, instead of counting the letters, it is usually easier to allow for more space than necessary and later optimize the dataset using `compress` (see section 10.6).

In this form of the `infile` command, everything that is not a blank, tab, comma, or line break is read as the value of a variable, until one of these characters appears. This logic prohibits the use of blanks within string variables and is a common source of error messages. In `popst4.raw`, for example, the hyphen in "Mecklenburg-Vorpommern" has been erased. If you repeat the last command, inserting `popst4.raw` as the filename, you get the following:

```
. infile str22 state pop90 pop91 pop92 pop93 pop94 pop95 using popst4.raw, clear
'Vorpommern' cannot be read as a number for pop90[2]
'Niedersachsen' cannot be read as a number for pop90[3]
'Nordrhein-Westfalen' cannot be read as a number for pop90[4]
'Rheinland-Pfalz' cannot be read as a number for pop90[5]
'Saarland' cannot be read as a number for pop90[6]
'Sachsen' cannot be read as a number for pop90[7]
'Sachsen-Anhalt' cannot be read as a number for pop90[8]
'Schleswig-Holstein' cannot be read as a number for pop90[9]
'Thueringen' cannot be read as a number for pop90[10]
'Deutschland' cannot be read as a number for pop90[11]
(eof not at end of obs)
(12 observations read)
```

What happened? The moment you have a blank in an unexpected place, the allocation of values to variables shifts. "Mecklenburg" is read as a string variable with a maximum of 22 characters and is saved as the variable named `state`. The blank between "Mecklenburg" and "Vorpommern" is understood as the beginning of a new variable. Stata then tries to read "Vorpommern" as a numerical value of the variable `pop90`, fails, and reports this. This mistake is continued through all rows, since `infile` attributes values to one observation until the variable list ends.

To avoid problems with strings containing blanks, you should enclose strings within quotation marks. Missing strings should appear as two consecutive quotes to keep your data rectangular.

5. The same logic applies to the spreadsheet format, which is a special case of the free format, so we can therefore use the above commands to read the files `popst1.raw` and `popst2.raw`.

Reading data in fixed format

Fixed-format data have no specific separating characters between variables. Instead, we know from external information what the numbers at certain positions of the file mean. The file `popst6.raw` is an example of a fixed-format dataset:

```
. type popst6.raw
Hessen                576358375923596759816010
Mecklenburg-Vorpommern192418921865184318321823
Niedersachsen         738774767578764877157780
Rheinland-Pfalz       376438213881392639523978
Saarland              107310771084108510841084
Sachsen               476446794641460845844567
Sachsen-Anhalt        287428232797277827592739
Schleswig-Holstein    262626492680269527082725
Thueringen            261125722546253325182504
```

Here the individual variables cannot be separated unless you know, for example, that the variable `pop90` starts at the 23rd position of each row and ends at the 26th position.

For this type of file, the command `infile` must be used in combination with a dictionary. This technique can also be used for the other data formats. It is the most general way of entering data. The dictionary is an auxiliary file[6] used to define the positions of the variables. Also variable labels and comments can be inserted, and unimportant variables or rows can be omitted. A simple version of such an auxiliary file consists solely of a row with the name of the file containing the data, and a simple list of variables to be read in. Here is an example:

```
                                        top: popst5.dct
 1: dictionary using popst5.raw {
 2:   str22 state
 3:         pop90
 4:         pop91
 5:         pop92
 6:         pop93
 7:         pop94
 8:         pop95
 9: }
10:
                                        end: popst5.dct
```

This dictionary merely lists the variable names and, for string variables, the corresponding specifications. As the positions of the variables are not defined, this dictionary clearly cannot read in `popst6.raw`. But you can use it to read `popst5.raw`. To understand how dictionaries work, try that first—and we will develop the dictionary for `popst6.raw` later.

6. A dictionary can also be written directly before the rows of numbers in the ASCII data. Here, however, the ASCII data would be worthless for other statistical packages. Also writing the dictionary in an editor before rows of numbers is often difficult since many editors cannot read in large datasets. We therefore recommend setting up the dictionary as a external file.

Type the above example in the Stata Do-file Editor (or any other editor), and save it as `popst5.dct`[7] in your working directory. Windows users should make sure that the file is not accidentally saved as `popst5.dct.txt`. Then return to Stata, and type the following command:

```
. infile using popst5.dct, clear
```

Unlike the command on page 303, this command does not contain a variable list. Furthermore, `using` is followed by the reference to the dictionary file instead of to the data file. You can even leave out the extension of the dictionary file, since Stata automatically expects a file with extension `.dct`. The reason is that the `infile` command is entered without a variable list. Here Stata automatically assumes that `using` will be followed by a dictionary.

In this example, this procedure not only seems, but is, unnecessarily complicated. Nonetheless, the use of such a dictionary can be extremely useful. On the one hand, it allows a detailed description of the dataset; on the other hand, it is the only way to read fixed-format datasets.

The following syntax extract shows you some of the more important ways to design a dictionary. Stata allows some further options, but these are rarely necessary and will therefore not be described here. For an overview with examples, type the command `help infile2`.

```
[infile] dictionary [using filename] {
    * comments may be included freely
    _lrecl(#)
    _firstlineoffile(#)
    _lines(#)
    _line(#)
    _newline[(#)]
    _column(#)
    _skip[(#)]
    [type] varname [:lblname] [% infmt] ["variable label"]
}
```

7. You can also find the dictionary among the files you installed at the beginning under the name `popst5kk.dct`.

Let us start with the second-to-last row. Recall that all elements in square brackets are optional. The only required element is therefore a variable name. For every variable, you can also specify the storage type [*type*] (see section 5.6), give a variable label ["*variable label*"], and specify the width of the variable [% *infmt*] and the name of the value label [: *lblname*].

The most important additional element in the syntax for dictionaries is probably `_column(#)`. With `_column(#)`, you can mark at what point in your file a specific variable begins. You can determine the end of the variable by specifying the format. To specify that the variable `pop90` in the file `popst6.raw` begins in the 23rd column and has a width of four columns, you would type the following:

```
_column(23) pop90 %4f
```

We have chosen `%4f` as the format type because the population numbers do not contain more than four characters. With the other variables, proceed correspondingly.

The three line options refer to the rows in your file. Using `_firstlineoffile`, you determine in which row your data begin. Some files might contain a title, references, or comments on the data collection, which you can skip by specifying the first row with real data. Using `_lines`, you can state how many rows constitute an observation. This is necessary for data in fixed format when an observation is spread over several rows. The availability of `_lines` is also helpful for data in free format if you do not want to read all the rows into Stata. You determine the rows from which to read the values using `_line`. The values following this option always refer to the rows within an observation.

Most likely, you will not need a dictionary file right away. However, it is important to know that this possibility exists and how to use it. Let us therefore show you a simple example of a dictionary file in `popst6kk.dct`. We recommend that you copy the dictionary file to `popst6.dct`, use your copy to read the data into Stata, and finally work through some alterations of the dictionary file. You will quickly learn that things are not nearly as complicated as they seem.

top: popst6kk.dct

```
 1: dictionary using popst6.raw {
 2:      * One possibility to read popst6.raw
 3:      * There are others. Just try out!
 4:      _column(1) str22 state  %22s  "State"
 5:      _column(23) pop90 %4f    "Population 1990"
 6:      _column(27) pop91 %4f    "Population 1991"
 7:      _column(31) pop92 %4f    "Population 1992"
 8:      _column(35) pop93 %4f    "Population 1993"
 9:      _column(39) pop94 %4f    "Population 1994"
10:      _column(43) pop95 %4f    "Population 1995"
11: }
```

end: popst6kk.dct

For data in fixed format, the `infix` command offers an easier, although less flexible, alternative to the `infile` command. Such data must be read into Stata using a dictionary, but the `infix` command, however, allows for a simpler structure of the dictionary that you can enter directly at the command line. Since datasets in fixed format without separating characters are less common now, we will not discuss this command further here but instead leave you with an example:

```
. infix str22 state 1-22 pop90 23-26 pop95 43-46 using popst6.raw, clear
(9 observations read)
```

10.3 Inputting data

For one of the examples in discussing graphs, we used temperature data from 1779 to 2004. Parts of these data are listed in table 10.1. The table contains the average temperatures by year, and for July and December, for the small city of Karlsruhe, Germany, from 1984 to 1990. Printed this way, the information is an example of data that are not machine readable.[8] To analyze these data, you first need to input the data by hand using Stata's Data Editor or the `input` command.

Table 10.1. Average temperatures (in °F) in Karlsruhe, Germany, 1984–1990

Time	Yearly	July	December
1984	49.82	65.84	36.86
1985	48.92	68.54	40.28
1986	50.18	67.28	38.66
1987	49.64	67.10	38.12
1988	52.16	66.56	40.82
1989	52.16	68.90	38.48
1990	52.88	68.00	35.24

10.3.1 Input data using the Data Editor

We need to begin this section with no data in memory:

```
. clear
```

You can open the Data Editor by typing the command

```
. edit
```

Typing `edit` opens a window containing an empty data matrix. The upper-left corner of the data matrix is highlighted; this is the currently active cell, and you can

8. We should admit that the data can be found in machine-readable form at http://www.klimadiagramme.de/Europa/special01.htm. Here however, we prefer to stick to the printed table with its Fahrenheit information. The web page contains an HTML table of monthly temperatures in degrees Celsius from 1779 to 2004. If you want to read such data into Stata, just copy the table to a powerful text editor, do some searching and replacing, and import the data into Stata with `insheet` or `infile`.

alter its contents. Above the table, to the left of an input field, you see "var1[1] =". Here you can type the value of the first variable ("var1") for the first observation ("[1]"). If you press the *down arrow* key, the blue highlighting also shifts downward but remains within variable 1; however, the value within the square brackets changes. Every cell therefore corresponds to a new observation.

Now type, say, the value 1984 in the input field to the right of "var1[1]=". Then confirm your entry by pressing *Enter*. The number 1984 is written in the first cell, the second observation becomes the active cell, and the variable is given a name. You can now directly type the value for the second observation, for example, 1985.

Use the mouse to click on the cell in the first row of the second column, and type a value (e.g., 49.82) for "var2[1]" into the input field. Confirm this entry by pressing *Enter*. The dot that appears in the second row of column 2 is a placeholder for a missing value. As soon as a value is entered for an observation, the program adds missing values for the remaining rows that already contain data. This way, the dataset always stays rectangular. If you continue entering data into the second column, the missings are simply overwritten.

Before we end this example, use the mouse to double-click on the gray field with the label `var1` at the top of the first column. A window pops up on the screen, in which you can enter a name and a label for the variable. Type `year` in the first textbox, and click on **OK**. Double-click on the gray field with the label `var2` at the top of the second column. In the pop-up window, type `Year of Measurement` in the first textbox and click on **OK**.

Now close the Editor by pressing *Alt+F4* or clicking on the button that closes windows in your operating system. This will bring you back to the normal Stata window. You can now treat the dataset like any other Stata file.

10.3.2 The input command

Another method to enter data into Stata is the `input` command. The main advantage of this method, compared with entering the data using the Data Editor, is that it can be used in do-files, so you can replicate your steps.

To begin with, we will use the command `input` to enter just one observation of data from table 10.1. Begin by clearing Stata:

```
. clear
```

and then start entering data using `input`. Type the command, and list all variable names for which you want to enter values, e.g.,

```
. input year temp
```

Following this entry, the two variable names appear on the screen, together with the number 1:

```
year temp
  1.
```

Here, `1.` is the entry request of the `input` command. You are asked to enter the values for the first observation. According to table 10.1, these are the numbers 1984 and 49.82. Type these numbers, separated by a blank, in the Command window, and confirm your entry by pressing *Enter*. The entry request for the second observation, `2.`, appears on the screen. Now you could enter the second observation accordingly, but for now, just type

```
. end
```

which brings you back to the normal Stata prompt.

As we said, the main advantage of `input` is that you can easily use the command in do-files. Let us give this a try. Open the Stata Do-file Editor (or any other text editor), and produce the following do-file:

top: crkatemp.do

```
 1: clear
 2: input year mean jul
 3:     1984   49.82   65.84
 4:     1985   48.92   68.54
 5:     1986   50.18   67.28
 6:     1987   49.64    67.1
 7:     1988   52.16   66.56
 8:     1989   52.16    68.9
 9: end
10: exit
```

end: crkatemp.do

After saving the do-file under, say, `crkatemp.do`, you can run the file as usual. Running the do-file will produce the data:

```
. do crkatemp
```

You can also use `input` to add new variables or more observations to an existing dataset. To add observations, type `input` without a variable list. Stata will then automatically let you input further observations. Let us use this feature to add the temperatures for the year 1990, which we have forgotten in the do-file:

```
. input
.  1990 52.88  68
. end
```

To add another variable, type `input` with a new variable name. Here we add the average December temperatures to our dataset:

```
. input dec
.  36.86
.  40.28
.  38.66
.  38.12
.  40.82
.  38.48
.  35.24
```

Here you can omit typing `end`. Stata knows that the dataset contains seven observations and will not let you input more observations if you use `input` with a varlist for an existing dataset. However, if you need to reproduce your steps, a better place to add variables or observation to the dataset would probably be the do-file we produced above.

Which numbers to assign?

So far, we have assumed that the numbers or text you enter are from the printed data. If you collected the data yourself, or if the printed data you enter include text (nonnumerical information) that you would like to use in your analysis, or if data are missing in the sources, you will have to make some coding decisions, which we want to address briefly.

- First of all, make sure that your file has an identification variable. You saw in our file `data1.dta` that the variable `persnr` was an identification variable. For the temperature data we just entered, `year` would serve the same purpose. It is easy to forget to enter an identification variable, especially when the printed data or data you find on the Internet provide this information through sorting. Suppose that you enter grades for a class and, being lazy, you skip the names of the students to avoid typing. That seems fine because the data are ordered alphabetically and you will always know who is who. However, once your data are entered into Stata, you can sort them in various ways, and the relevant mapping information will then be lost. This will become even more problematic when you combine different datasets (see section 10.4).
- An equally important point is how to code missing values. In our population data example, we left the missing information for the population value in Mecklenburg-Vorpommern blank, and once the data are read into Stata, a period will appear indicating the missing value. Other data you enter, however, especially survey data, may contain some information about the missing values, so do not throw away essential information. In an interview, for example, there may be various reasons why data are missing: the respondent was not willing to answer, the respondent does not know the answer, or filter questions had been used, so the respondent was never asked a particular question. If you merely leave this information out and assign a period to a missing value, you will not be able to determine why the value is missing when analyzing the data. For this reason, you should use codes for all variables that explain why a value is missing in a given case. In

many datasets, 97 is used for "don't know", 98 for "answer refused", and 99 for "not applicable". You cannot use this coding if the true answers can contain such numbers. But you can resolve this using by using missing-value codes that always contain one digit more than the answer categories (e.g., 997, 998, 999). This type of coding is then unambiguous but has the disadvantage that it differs across variables. A uniform codification considerably simplifies dealing with missing values. A possible alternative would be to use negative values as codes. Usually, you will not have any negative values as possible answers, so you can use, for example, the values `-1`, `-2`, and `-3` as missing-value codes for all variables. In Stata you can code all these values into 27 different missing-value codes: the usual period, as well as `.a` to `.z`.

- When you enter plain text (such as the name of a state in the population example, names of students, or job titles), you can either use codes when entering them or enter the text directly into Stata. Which of the two options you choose essentially depends on the question asked. If the number of possible answers is limited, you can easily assign a number to every answer. Write down the allocation while entering the data, and merely enter the number in the data file. This procedure is easiest if you are entering the data by yourself. If several people are entering data simultaneously, make sure that the codes have been defined previously (and are known to everyone). On the other hand, if the answer context allows for an unforeseeable number of possibilities, compiling a list before entering the data will be difficult, if not impossible. Here we recommend that you enter the text directly into Stata and read section 5.3.1.
- Sometimes your data will contain dates—not just years, as in the temperature example, but days and months. Then we recommend that you include the century in the year. If you add data later, and if the old and new data are to be joined, the dates must be unambiguous.
- Make sure that every variable contains only one *logical unit*, that is, one piece of information. Sometimes respondents are asked to mark all answer categories that apply to one question. Here each of the possible answers should be stored in a separate variable with a "yes" or "no" indicator. We recommend that you use variable names that display the content of these questions, as in our dataset `data1.dta` the variables `np9501` to `np9507`.
- To avoid errors, we recommend that you enter data twice, merge the datasets afterward, and check for differences (see section 10.4). Sometimes time and financial constraints will often preclude such double entry. Then some other strategies are helpful. For example, when you enter data from a questionnaire that has no codes printed next to the boxes, copy the questionnaire on a transparency, and write the codes behind the boxes on the transparency. When you enter the data, you can then simply place the transparency on top of the questionnaire.[9] Also do not try to modify data or make coding changes while you are entering the data. There is, for example, no point in converting the temperature data we used from Celsius to

9. For details on designing questionnaires, see for example Fowler (1984).

Fahrenheit before you have finished entering the data. You can automate almost everything later within Stata (see chapter 5).

10.4 Combining data

Suppose that you want to analyze how life satisfaction develops over time by using data collected with different surveys at different time points. Or suppose that you want to compare the life satisfaction of German and American citizens by using survey data collected in both countries. Finally, suppose that you want to control for population density in the regression of rents from chapter 8 by using the dataset constructed above. In any of these examples, the information you need is spread over several files, so to perform your analyses, you need to combine the files into a rectangular dataset.

Here we explain how to combine datasets, using as examples several datasets from the German Socio-Economic Panel (GSOEP), which is the same source that we used to produce the dataset `data1.dta` we have been using. The GSOEP is a representative longitudinal study that has collected information annually on more than 12,000 households and nearly 24,000 persons since 1984. At present, the information from the GSOEP is stored in 231 different files and is therefore an excellent resource for demonstrating data-management tasks, both simple and difficult.[10]

To follow our examples, you need to understand the file structure in the GSOEP database. We will provide a brief overview in section 10.4.1.

Stata has three commands for combining different datasets: `merge`, `append`, and `joinby`. `joinby`, however, is needed only in exceptional cases; for more information, see `help joinby` or [D] **joinby**. `merge` and `append` apply more generally. With `merge`, you add variables (columns) to a dataset, and in our examples we will combine data from different points in time, as well as personal information and household information (see section 10.4.2). `append` adds observations (rows) to a dataset. In section 10.4.3, we will show by example how to combine data from the GSOEP with the U.S. Panel Study of Income Dynamics (PSID).

10.4.1 The GSOEP database

In a panel survey, the same people are interviewed several times, so it is not surprising that multiple datasets may contain information from the same persons but are updated every year. Of course, the new information could have been written into the existing data file right away, but usually it is entered into a separate file for storage reasons. Our task later will be to add the variables that are spread over different files to a single file.

10. Two user-written programs exist to ease the process of putting together the files of the GSOEP database. PanelWhiz, by John Haisken-DeNew, is a large-scale project for the data management of various panel datasets, including the GSOEP. PanelWhiz is downloadable from http://www.panelwhiz.eu. The other program, `mkdat`, offers far less functionality than PanelWhiz but is somewhat simpler to use. Read more about user-written programs in chapter 12.

The GSOEP is a household panel, meaning that all persons age 16 and older from randomly selected households are included. Moreover, not only are data on the individuals in households gathered, but also characteristics of the household as a whole (household-level data). However, since the information about the entire household is the same for all persons within the household, these data are stored in separate files: one in which each observation is a household and another in which each observation is a person. The household information can be merged with the personal data when needed. In one of our examples, we will merge household information (e.g., the state in which the household lives) with the personal data.

You will find household panels like the GSOEP in several countries. In fact, one of the first panels of this kind is the PSID, which is a U.S. panel study that has been running since 1964. In recent years, efforts have been made to make the collected data comparable, so that you can now combine PSID data and GSOEP data that are translated into internationally comparable variables. The latter are stored in separate files, as well.

Our data package includes 58 of the original 231 files of the GSOEP database.[11] To get an impression of the data structure, look at the files in the `c:\data\kk2\kksoep` directory by typing

```
. dir kksoep/
  <dir> 10/14/04 08:36 .
  <dir> 10/14/04 08:36 ..
  26.6k 10/14/04 08:36 ahbrutto.dta
  94.7k 10/14/04 08:36 ap.dta
 201.1k 10/14/04 08:36 apequiv.dta
  27.5k 10/14/04 08:36 bhbrutto.dta
  86.4k 10/14/04 08:36 bp.dta
 183.4k 10/14/04 08:36 bpequiv.dta
  26.5k 10/14/04 08:36 chbrutto.dta
  83.4k 10/14/04 08:36 cp.dta
```

The data from the different data collection years (called waves) are stored in different data files. All information gathered in the first year (1984) of the GSOEP is written into files whose names start with the letter `a`, information from 1985 is written to files whose names start with `b`, and subsequent years start with successive alphabetical letters up to the last survey year we have here (year 2002), which is written to files whose names with the letter `s`.

For each year (or, equivalently, letter) you see three file types: `hbrutto`, `p`, and `pequiv`. The observations (rows) of the `hbrutto` files are households, which is indicated by the letter `h` on the second character in the filename. These files contain information about the households that is known prior to the survey, for example, the state in which the household lives. Other types of household-level data of the GSOEP, which are not included among the files you installed in the *Preface*, contain information such as apartment size or monthly rent. More generally, all information that is the same

11. The data files that you downloaded contain only a few of the variables of the original GSOEP files. The number of respondents is reduced, as well: our data are a 50% sample from the observations of the original database (the so-called student sample). In accordance with German regulations for data confidentiality, we randomized parts of the information in the data files.

for all persons within a household is stored in a household-level dataset. Storing this information for each respondent would be redundant and a waste of disk space.

The observations (rows) of the two other file types represent persons, as indicated by the letter `p` in the filename. In general, answers from persons to survey questions are in the `p` files. Other file types at the person level contain variables constructed out of the respondents' answers, such as indices or scales. A special sort of these generated variables are stored in the `pequiv` files, which contain variables that have been made comparable to variables in the PSID.

Our data by no means represent all the information in the original GSOEP database. However, our explanation should be enough to give you an idea of its structure. In panel studies, you cannot collect data at every point in time from all respondents. You may lose contact with some respondents, some may refuse to answer the questionnaires, and some others will die. At the same time, respondents may be added to the study, depending on how the survey is conducted. For example, one policy on follow-ups in the GSOEP is that if spouses get divorced, and one of them moves into the household of another person, all persons of this new household will be interviewed as well. Also panel studies commonly draw refreshment samples from time to time or include more samples. The GSOEP had one additional sample because of the German reunification to collect information from respondents from the former German Democratic Republic and another in 1993–1994 to collect data on immigrants from eastern Europe. A refreshment sample for the entire sample was taken in 2002. Naturally, all these mechanisms lead to changing numbers of observations for each survey year.

We will explain how to combine data from several file types and survey years.

10.4.2 The merge command

The `merge` command is used to add variables to a given dataset. `merge` joins corresponding observations from the dataset that is currently in memory with those from one or more Stata-format datasets stored on the disk. In the simplest case, new variables are added for each observation, as shown in figure 10.2.

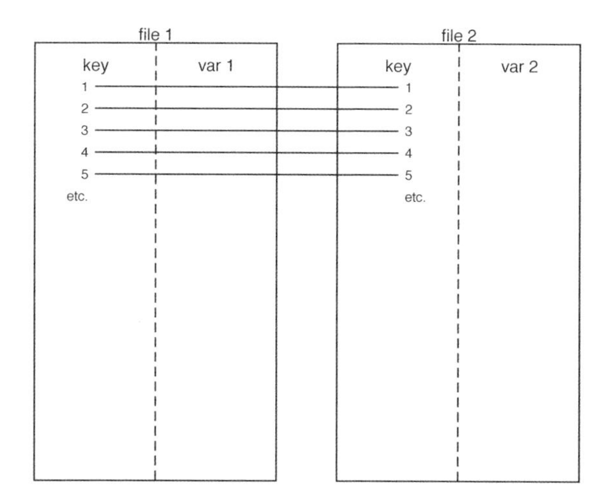

Figure 10.2. Representation of merge for rectangular data

grmerge.do

The basic syntax of the merge command starts with the command itself, an optional variable list, and the specification of the file(s) that you want to merge with the file in memory:

merge [*varlist*] using *filename* ...

The merge command is not complicated, but there are several issues and tricks you should be aware of when merging datasets. We begin with an example for the general merge procedure, followed by an explanation of how to control the result of a merge process and how to merge multiple datasets. Merging individual data with aggregate data and vice versa is explained at the end of this section.

The merge procedure

Suppose that you want to use data from the GSOEP to investigate the relationship between general life satisfaction and hours worked in 2001. The information on life satisfaction is part of the GSOEP personal data files, whereas the number of hours worked is one of the variables that were generated for international comparability and stored in the pequiv files. For the year 2001, the information you want is therefore stored in rp.dta and rpequiv.dta, respectively. To perform your analysis, you need to combine these two datasets so that identical respondents are combined into one row (observation) of the new dataset.

To indicate which observations of different files belong together, you can use a key variable, which is a variable that is part of both datasets and that has a unique value for each observation. In the GSOEP dataset, persnr is such a key variable. Each person who was entered at least once in the GSOEP sample was assigned a personal identification number that does not change over time and is stored in the variable persnr in every

GSOEP file containing personal information. Hence, you can use this variable as a key to match the right observations of the datasets with each other.

To combine two datasets, you begin by loading one of them into the computer's memory; we will use the term "master data" for this file. To use the `merge` command, we then specify the name of the key variable, the word `using`, and the name of the second dataset. We call that dataset the "using" dataset.

Let us try it. Load `rp.dta` into the computer's memory, and merge `rpequiv.dta` with `persnr` as a key. Remember that both datasets are stored in the subdirectory `kksoep` of `c:\data\kk2`. You therefore need to add the directory name to the filename:[12]

```
. use kksoep/rp, clear
. merge persnr using kksoep/rpequiv
```

This seems to have worked. However, we were lucky that it did. To perform a merge with one or more key variables, both datasets have to be sorted by the key variable(s).[13] Stata will issue an error message if one of the two datasets is not sorted. Both of our example datasets are already sorted by `persnr`. If this had not been the case, it would have been necessary to sort the files manually:

```
. use kksoep/rpequiv, clear
. sort persnr
. save myusing, replace
. use kksoep/rp, clear
. sort persnr
. merge persnr using myusing
. erase myusing.dta
```

You could also have used the `sort` option of `merge` to sort the data. We will show examples of this later.

Let us take a look at the merged-data file:

```
. describe
```

The new dataset has again 11,087 observations, and five variables (`d1110402` to `i1111001`) were added to the original file `rp.dta`. The variables `hhnr` and `rhhnr` were part of both datasets, so they are not added from the `using` dataset `rp15501` to the master dataset `rp.dta`. Unless otherwise stated, a variable remains unchanged in the master data when the same variable is also found in the `using` data. Finally, there is a newly added variable, which has not been used in either of the two datasets: `_merge`. This is an indicator variable to check the merge process. A value 3 for `_merge` indicates that all observations from the master data were found in the `using` data. As you can see from

```
. tabulate _merge
```

in our last `merge` command every observation is in both files.

12. Remember from section 3.1.8 that it is not necessary to add the file extension `.dta` after `using`.

13. You will find more information on sorting in section 3.2.1.

You must delete the variable `_merge` before running another `merge` command. After checking the results of your `merge` command, you should therefore

```
. drop _merge
```

Before we proceed, one more word about the key. As mentioned before, the key is used to make sure that each observation of the `using` data is merged to exactly the same observation in the master data. Therefore, the key must uniquely identify the observations of the datasets. However, it is not necessary for the key to be in single variable. Instead you can use a list of variables that jointly identify each observation by specifying a variable list between `merge` and `using`.

Keeping track of observations

The files we merged in the last section had a special property: both files contained exactly the same observations, or, in other words, the data were rectangular. A slightly more complicated case arises if you have to merge nonrectangular data. Then one of the files contains observations that are not part of the other file, and vice versa—a situation that is likely with panel data. Some respondents from an earlier panel might have refused to take part in the later year, whereas in the later year some new respondents appear who had not participated in an earlier year (see figure 10.3 for a schematic).

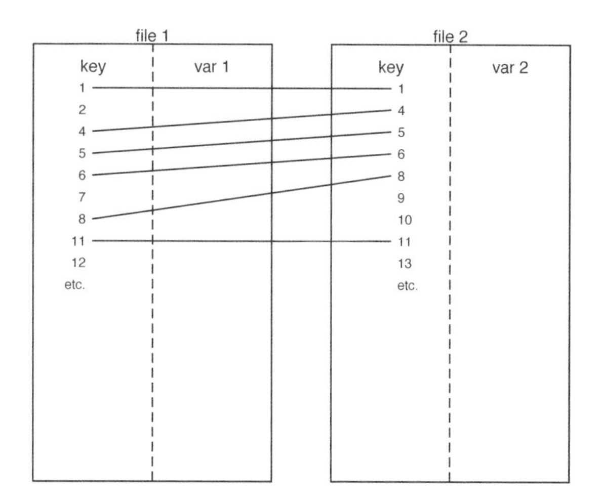

Figure 10.3. Representation of `merge` for nonrectangular data

`grmerge1.do`

If you merge two datasets with different numbers of observations, it is important to keep track of the merging process. Suppose that you want to explore the development of life satisfaction and therefore combine data from 2001 with data collected at the beginning of the panel study. Life satisfaction in 1984 is stored in `ap`. From the command

```
. describe using kksoep/ap, short
```

we learn that this file contains information about 5,796 respondents, whereas the file produced so far contains 11,087 observations. What happens if we merge the two files? Technically, there are no difficulties in doing so. Simply merge the files as before, but this time with the help of the `sort` option of `merge`:

```
. merge persnr using kksoep/ap, sort
```

Now let us note some properties of the merged-data file. As expected, the new file contains variables from both files, and again the variable `_merge` has been added:

```
. ds
. describe, short
```

The combined data file has 14,812 observations, neither just 5,796 from `ap.dta` nor 11,087 from `rp.dta`. What happened? The answer can be seen from `_merge`:

```
. tab _merge
```

_merge	Freq.	Percent	Cum.
1	9,016	60.87	60.87
2	3,725	25.15	86.02
3	2,071	13.98	100.00
Total	14,812	100.00	

The variable `_merge` has three values: 1, 2, and 3; that is, the new dataset contains three different types of observations. Here value 1 indicates that there are 9,016 observations that were only part of the master data. As the file `rp.dta` has been the master data, these are respondents who were interviewed in 2002 but not in 1984. Value 2 indicates observations that are only in the `using` data. This means that for our example, 3,725 respondents were interviewed in 1984 who had not been interviewed in 2002. Finally, value 3 indicates respondents who had been interviewed on both occasions, here 2071. You can find a general summary of the value definitions for `_merge` using `help merge`:

```
_merge ==1      obs. from master data
_merge ==2      obs. from using data
_merge ==3      obs. from both master and using data
```

The variable `_merge` allows us to fine-tune our dataset. Do we want to keep all the respondents, or only those who are interviewed on both occasions? Here we proceed with the latter, which gives us what panel-data analysts call a "balanced" panel dataset:

```
. keep if _merge==3
```

Merging more than two files

Suppose that you want to add data on life satisfaction from every year to your dataset in memory—that is, to merge not just two files, but more than two. Then you can specify more than one using file.

```
. drop _merge
. sort persnr
. merge persnr using kksoep/bp kksoep/cp kksoep/dp kksoep/ep kksoep/fp
```

If you specify multiple `using` files, Stata creates several `_merge` variables: `_merge1` for the first `using` data, `_merge2` for the second, and so forth. Each `_merge` variable contains 0 or 1, indicating whether an observation is present in the corresponding `using` dataset. You can again use these variables to fine-tune the observations to be used in the dataset. For example,

```
. drop if _merge1 + _merge2 + _merge3 + _merge4 + _merge5 <= 2
```

keeps only those observations that were present in at least three of the `using` datasets.

Database-specific merge tools

Many large databases provide extra files to give more information about the observations in the dataset. In the GSOEP, the file `ppfad.dta` contains some general information about all persons for whom the entire GSOEP database has at least one piece of information. Such metadata commonly exist for all nonrectangular data kept in relational databases. For GSOEP, the variables `anetto` to `snetto` of the file `ppfad.dta` indicate whether a specific person has been interviewed in a specific year, and if not, why not. With the help of these variables, you can precisely specify the observations to be kept. So, you could use the information in `ppfad.dta` to first define the observation base, and then merge the other files with the `nokeep` option, which causes `merge` to ignore observations in the `using` data that have no corresponding observation in the master data.

In our example below, we first construct a new variable `nwaves`, which counts how often a person has been interviewed so far. We then drop observations with fewer than five interviews (which is an arbitrary number for our example). Afterward we proceed as before, using the `nokeep` option in the `merge` command.

```
. use kksoep/ppfad, clear
. egen nwaves = anycount(?netto), values(1)
. drop if nwaves < 5
. sort persnr
. save myfile, replace
. foreach stub in a b c d e f g h i j k l m n o p q r s {
.     local files "`files' kksoep/`stub'p"
. }
. merge persnr using `files', nokeep
. drop _merge*
```

We have used a `foreach` loop to construct the list of `using` files. See sections 3.2.2 and 11.2.1 for details.

Merging data on different levels

So far, we have merged datasets where the observations have the same meanings. That is to say, in any file we merged, an observation was a respondent. Now consider a case in which the meanings of the observations in the files are not identical. Such a case arises if you want to merge the person datasets of the GSOEP with the household data. Whereas the observations of the person datasets are individuals, the observations of the household dataset are households. The same case arises if you want to merge the population sizes used in section 10.2.2 with the personal data from the GSOEP. Here the observations are states in one file and persons in the other.

To merge data with different observational levels, you first need to be aware how the observations of the different datasets are linked. This is quite obvious for the personal and household data of the GSOEP. The persons described by the person data live in the households of the household data. The same is true for persons and states: the persons represented in personal data live in the states of the state data. In each case, the observations of any of those datasets somehow belong together. Second, you need to decide for which observational units the analysis should be performed after merging the files. Should the observations of the new data be persons, households, or states? This crucial decision guides the entire remaining process, and the answer depends entirely on your research question. Do you want to make a statement about persons, households, or states? In the following example, we will first merge households and states with the personal data by keeping the data as person data. Afterward, we do the opposite.

The `hbrutto` file of the GSOEP has data on households. If we are to add the household data to personal data, each person belonging to the same household should get the same information from the household data. One observation from the household data has to be added to several observations of the person data. A graphical representation of this logic can be found in figure 10.4.

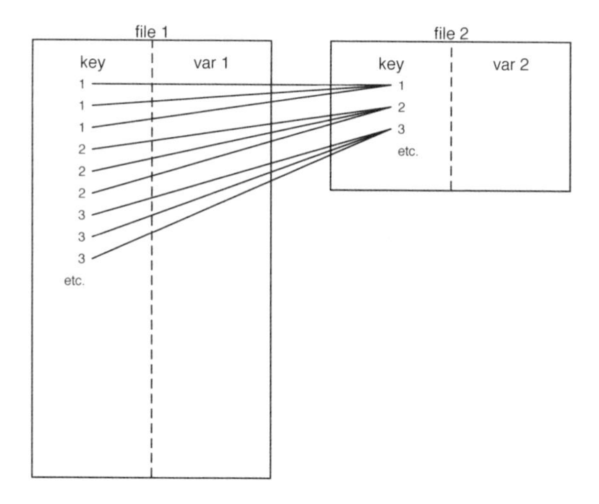

Figure 10.4. Merging aggregate and individual data

grmerge2.do

Again you can merge the different files using a key variable. However, as the household data do not contain the variable **persnr**, this variable cannot be used as a key. Instead, the keys for our merging problem are the household numbers in the **hbrutto** files. Each **hbrutto** file contains the household number in the first interview year and the household number of the current year, and these variables are also part of every person's data. Both numbers together—the first and the current household number—form a key variable; that is, each combination of these two variables occurs in the household data only once. Therefore, you can use both variables together as keys. To merge household information from, say, the 1990 household file gpbrutto to the 2001 person file gp, follow the steps from the second merging example on page 316. You cannot use the sort option when merging aggregate and individual data. The sort option requires that the key in both datasets be unique. Here the key hhnr ghhnr is not unique in the dataset gp, so you have to sort both datasets before running the merge command.

```
. use kksoep/ghbrutto, clear
. sort hhnr ghhnr
. save myusing, replace
. use kksoep/gp, clear
. sort hhnr ghhnr
. merge hhnr ghhnr using myusing, nokeep
. tab _merge
. drop _merge
. erase myusing.dta
```

It does not matter which of the two files you specify as master and using, as long as you know which is which when you interpret the results of _merge:

The same logic applies to merge the population sizes in the file stinfo.dta. This file contains several years of population size data for the German federal states. To assign this information to the GSOEP data, you need to ensure that both datasets contain a key variable with the same name and are sorted by this key variable. Typing

```
. describe using stinfo
Contains data
  obs:            14                          16 Aug 2004 20:36
 vars:             5
 size:           210
-------------------------------------------------------------------------------
              storage  display     value
variable name   type   format      label      variable label
-------------------------------------------------------------------------------
state           byte   %9.0g
pop90           int    %8.0g
pop95           int    %8.0g
pop93           int    %9.0g
sqfeet          float  %9.0g
-------------------------------------------------------------------------------
Sorted by:  state
```

tells us that `stinfo.dta` contains the variable `state` and that the dataset is sorted by this variable. For simplicity, we have prepared this variable with the same codes used in the state variables of the GSOEP database. To merge the population size for 1990, you need to use the variable `gbula`, which holds the state in which the respondent lived in 1990. Key variables must have the same name in both datasets. You can either rename one of the two variables or make a copy of `gbula` with the name `state`; we propose the former. Afterward you can proceed as before. Since you basically want to merge only the variable "population size in 1990", you can use the option `keep()` within the `merge` command. The option `keep()` specifies the variables to be kept from the `using` data:

```
. rename gbula state
. sort state
. merge state using stinfo, keep(pop90)
```

Now for the other way around: suppose that you are to merge person data with household data, or person data with state data, and you want to end up with household or state data, respectively. Here the merging process involves deciding how the values associated with persons can be aggregated. Consider general life satisfaction. Each person in a household has his or her own general life satisfaction. If you want to end up with a household dataset, you need to decide how this different information from one household can be summarized. It might make sense to consider the mean general life satisfaction of all persons from one household. Or you might use the general life satisfaction of the head of the household and merge this information with the household data. The answer once more depends on your research question. But after you decide, you must form a new *aggregated* dataset before you merge your data with the other dataset. The command for forming an aggregated dataset is `collapse`. For example, the command

```
. collapse (mean) gp109, by(ghhnr)
```

calculates the means of the general life satisfaction from 1990 by current household number (ghhnr) and stores the result in a new dataset. This dataset is now no longer person data but household data. You can therefore merge it with other files as easily as before:

```
. save myusing, replace
. use kksoep/ghbrutto, clear
. sort ghhnr
. merge ghhnr using myusing
```

And, of course, the same applies if you want to form state data:

```
. rename gbula state
. collapse (mean) gp109, by(state)
. sort state
. merge state using stinfo
```

10.4.3 The append command

Adding observations to an existing dataset is straightforward. The append command adds a new dataset to the *bottom* of an existing dataset. This means that you extend the data matrix of your current dataset by one or several rows of observations. Identical variables are written one below the other, whereas new variables are added in new columns (see figure 10.5).

(*Continued on next page*)

file 1

var 1	var 2
0	2002
1	1876
0	3000
0	2130
1	1000
etc.	etc.

file 2

var 1	var 2	var 3
0	1238	7
1	1500	9
etc.	etc.	etc.

Figure 10.5. Representation of `append`

`grappend.do`

The structure and function of `append` are simple. The basic syntax is

`append using` *filename* [, `nolabel`]

Let us illustrate this with an example. We have mentioned before that the GSOEP database contains PSID-equivalence files, which are part of the of the Cross-National Equivalent File (CNEF) set assembled by the Cornell College of Human Ecology. The CNEF contains files with equivalently defined variables for panel studies from the United States, Germany, the United Kingdom, and Canada.[14] Like those of the GSOEP, the data of the CNEF are also split into several files: one file for each year and each country. Among the files you installed, we have inserted one of them: a downsized version of the file with data from the American PSID of the year 2001: `pequiv01.dta`. You might take a look at it with

```
. describe using pequiv01
```

14. The panel data distributed in the CNEF are the Panel Study of Income Dynamics (PSID), the German Socio-Economic Panel (GSOEP), the British Household Panel Study (BHPS), and the Canadian Survey of Labour and Income Dynamics (SLID). See http://www.human.cornell.edu/che/PAM/Research/Centers-Programs/German-Panel/upload/b-equiv2.pdf for a more complete description of the CNEF.

This file by and large contains the same variables as the file `rpequiv.dta` from the GSOEP. All variables have equivalent names in both files. The only differences are in the unique person and household numbers, which are called `persnr` and `hhnr` in the GSOEP, but `x1110111` and `x1110201`, respectively, in the PSID.

For a cross-national comparison between Germany and the United States, you need to combine the two files. You begin the process by loading one of the datasets into Stata. Which one you load does not matter, but we begin with the U.S. file `pequiv01`:

```
. use pequiv01, clear
```

Again we will use the term "master file" for the file in the working memory. You can examine that file using `describe`, and you will see that there are 21,396 observations and 11 variables in the data file. One of the variables is `e1110101`, the annual work hours of the respondents in the U.S. dataset.

Before you append another file to the master file, it is a good idea to generate a variable that marks the observations of the master file. This variable will later help you to separate U.S. respondents from German respondents. We will do this by constructing a variable that is 1 for all observations in the current file:

```
. gen country = 1
```

Using `append`, you can now add the respective file of the GSOEP at the bottom of `pequiv01`. The data for Germany are stored in the subdirectory `kksoep`. The command contains the directory name and the variable name (`kksoep/rpequiv`):

```
. append using kksoep/rpequiv
```

The new dataset has now 32,483 observations: 21,396 from the United States and 11,087 from Germany. Variables that are in both datasets can have valid values for all observations. On the other hand, variables that are part of only one of the two datasets get a missing value for any observation from the dataset without that variable. Therefore, the variables `persnr` and `hhnr` have a missing value for all observations from the United States, whereas the variables `x1110111` and `x1110201` are missing for all German respondents:

```
. describe, short
. summarize x1110111 x1110201 hhnr persnr rhhnr country
```

The same applies for the newly generated variable `country`, which has the value 1 for all U.S. respondents and missing for German respondents. After replacing the missing value in `country` with zero and doing some labeling, you can use `country` to make the intended comparison:

```
. replace country=0 if country >= .
. label define countrylb 0 "Germany" 1 "US"
. label values country countrylb
. histogram e1110101 if e1110101 > 0, by(country)
```

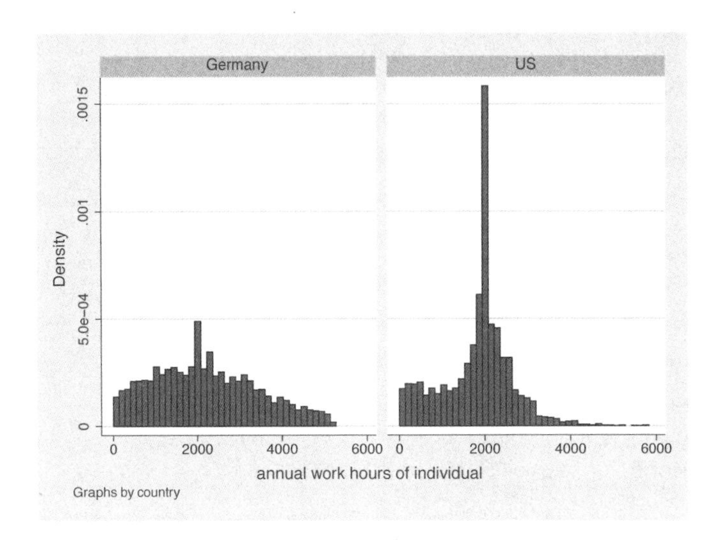

10.5 Saving and exporting data

You have taken a lot of trouble to construct a dataset with which you and your colleagues can easily work. Saving this dataset, on the other hand, is straightforward: with the command `save filename` you save the dataset in memory with the name `filename.dta` as a Stata system file in the current working directory. With the option `replace` you can overwrite a previous version of the same dataset.

Saving the data is therefore fairly easy. Just before saving, however, you should consider three issues:

1. If no data type is specified, variables are by default created as "floats" (see section 5.6). These variables often take up more storage space than is necessary. The command `compress` optimizes all variables with regard to their storage space, with no loss of information. You should therefore always use `compress` before saving.
2. Working with a dataset is often easier if the variables are in a certain order. The commands `order` and `aorder` can be used to order the variables in the dataset.
3. Like variables and values, the dataset can also be given a label. The corresponding command is `label data`. The data label is shown in the output of `describe` on the upper right. In `data1.dta`, for example, it is "SOEP´97 (Kohler/Kreuter)".

So far, we have focused on the different ways of reading data from other formats into Stata. At times, however, you will need to read a Stata dataset into another program. Of course, you can do this using the software packages described in section 10.2.1. However, since practically all statistical packages can read ASCII files, it is just as easy to save your Stata data to an ASCII file.

In Stata, you can save ASCII files in spreadsheet format and in free format. To save in spreadsheet format, you can use `outsheet`; to save in free or fixed format, you can use `outfile`. Both commands basically work the same way: after the command, you can specify a variable list followed by the code word `using` and the filename.

For both commands, the contents of string variables are written between quotes. If you do not want to use quotes, specify the `noquote` option. *Important:* Variables with value labels are saved as string variables, and you can use the `nolabel` option to prevent this.

10.6 Handling large datasets

Stata always loads the entire dataset into the computer's working memory. The memory therefore sets certain limits on the size of workable datasets, which if exceeded, will result in two problems:

1. very slow processing speed and
2. the display of the error message "no room to add more variables".[15]

Unfortunately, the first problem can also occur when you are dealing with smaller datasets. To help you deal with oversized datasets, we will first explain some basics about working memory in modern operating systems. Then we will make some recommendations for using large datasets.

10.6.1 Rules for handling the working memory

The working memory is part of the hardware of your computer. It consists of two parts, the physical and the virtual working memory. The physical working memory consists of special chips (RAM), which quickly remember and retrieve pieces of information. The virtual working memory is a file on the hard disk ("swap file"), which in principle works like physical memory but much more slowly.

Stata loads its data into the working memory. Whether the data then end up in physical memory or the swap file is determined by the operating system. If the data end up in physical memory, the processor can access them quickly; if they end up in the swap file, it takes longer. You can even hear the access to the swap file, since it will involve a lot of reading and writing on the hard disk.

For Stata to work efficiently, the datasets should be kept in physical memory if possible. Generally, the swap file is used only when the physical memory is full. The memory does not, however, contain only Stata data. The operating system needs part of the memory, and every program that is running uses memory. So, to prevent the first problem, close all applications that you do not really need.

15. The error message "matsize too small" does not indicate an oversized dataset. This error message can usually be prevented easily. See `help matsize` for details.

Programs likely to use a lot of memory are Internet browsers, various office packages, and spell checkers.

Closing applications, unfortunately, does not eliminate the error message "no room to add more variables". This message appears if the working memory is too small to store another variable. Many Stata procedures create temporary variables, so that error message may appear even when you are not trying to create or load any new variables.

Stata merely uses the working memory according to your specifications. You can find out how much working memory you have from

```
. memory
                                              bytes
-----------------------------------------------------------------------
Details of set memory usage
    overhead (pointers)                     129,932          0.62%
    data                                  1,494,218          7.12%
                                         ------------------------------
    data + overhead                       1,624,150          7.74%
    free                                 19,347,362         92.26%
                                         ------------------------------
    Total allocated                      20,971,512        100.00%
-----------------------------------------------------------------------
Other memory usage
    set maxvar usage                      2,001,666
    set matsize usage                        88,800
    programs, saved results, etc.            17,185
                                         ---------------
    Total                                 2,107,651
-----------------------------------------------------
Grand total                              23,079,163
```

The row "Total allocated" reports how much working memory is reserved for Stata, here 20,971,512 bytes, or 20 megabytes (MB). The remaining rows indicate how the working memory is being used by Stata. You can adjust the amount of working memory reserved for Stata, for example, by using

```
. clear
. set memory 20m
Current memory allocation

                    current                                 memory usage
    settable          value     description                 (1M = 1024k)
    --------------------------------------------------------------------
    set maxvar         5000     max. variables allowed            1.909M
    set memory          20M     max. data space                  20.000M
    set matsize         100     max. RHS vars in models           0.085M
                                                            -----------
                                                                 21.994M
```

A total of 20 MB, or 20,480,000 bytes, is reserved. However, this command can be entered only when the working memory is empty.

If the error message "no room to add more variables" appears, the working memory reserved for Stata is too small. To prevent the second problem, remember the following rule: the amount of working memory reserved for Stata should be somewhat larger than the dataset. How much larger the reserved working memory should be depends on what you intend to do. To create new variables, you need more space than you do to analyze data. Should you be working with categorical variables, you will probably create many dummy variables, which also take up a lot of space. A rule of thumb is that you need $(4 \times n)/1024$ MB of working memory for every new variable.[16]

To be sure you have sufficient working memory, many users reserve a lot of working memory for Stata. This is, however, not necessarily a good idea, since it may force Stata to use virtual memory. The swap file is used whenever the reserved working memory exceeds the available memory.

10.6.2 Using oversized datasets

Ideally, the amount of working memory reserved would be large enough to easily fit the dataset and small enough to prevent the use of the swap file. For some datasets, it is impossible to comply with this rule. This is the case for datasets that are larger than the RAM. If you often work with such datasets, we have only one piece of advice: buy more RAM. If this is an exceptional situation, we suggest the following:

- Do you need all variables in the dataset? If not, import only the variables you need:

```
. use persnr gender room using data1, clear
(SOEP'97 (Kohler/Kreuter))
```

- Do you need all observations in the dataset? If not, import only the observations you need:

```
. use data1 if gender==1 in 1/1000, clear
(SOEP'97 (Kohler/Kreuter))
```

- Does your dataset take up more storage space than necessary? Try reading in your dataset a little at a time and optimizing it. To do so, you first import only specific observations or variables, optimize the storage space used by this partial dataset with `compress`, and save it under a new name. Repeat this procedure for the remaining variables or observations, and join the different partial datasets using `merge` or `append` (section 10.4).[17]
- Does your dataset contain many identical observations? You should transform the dataset into a frequency-weighted dataset (page 63).

16. This formula assumes that you create new variables according to the default setting as "floats". You can find detailed related information on the Internet at http://www.stata.com/support/faqs/data/howbig.html.
17. If you do not know beforehand which variables or how many observations your dataset contains, see the dataset with `describe using data1.dta`. You can use the `describe` command even without loading the data into Stata's working memory. The only condition is that you add `using` to the command.

Only if none of this helps should you use the swap file. However, never do so interactively. Always write a do-file (you do that anyway) and test your do-file on a small partial dataset. If the do-file runs correctly, you can apply it to the big dataset. Before embarking on this you should type

```
. set virtual on
```

The Stata dataset is then handled in a way that increases the speed of saving in the swap file. In any case, the best thing is to start the do-file just before closing time, the weekend, or your annual holiday. And do not forget to

```
. set virtual off
```

when you return. `set virtual on` causes Stata to optimize memory in such a way that accessing the swap file is quicker, but it forces Stata to spend time doing more optimizations than are necessary when the dataset fits in RAM.

10.7 Exercises

- Obesity

  1. Point your browser to the interactive database on European Quality of Life of the European Foundation for the Improvement of Working and Living Conditions[18]. On this web page, find the tables for the percentage of people who are very overweight. Create a Stata dataset holding this table.
  2. Merge to this file the information on the share of people who have strong underweight from the same web site. Save this file for later use.
  3. Download data from the National Health and Nutrition Examination Study (NHANES) using the following command:

     ```
     . webuse nhanes2, clear
     ```

     (If you get the error message "`no room to add more observations`", please type `clear`, followed by `set memory 30m` before using the above command; see section 10.6 for details.)
  4. Produce an aggregated dataset for the share of people who have strong overweight and strong underweight by region. Thereby using the definition of the European Foundation for the Improvement of Working and Living Conditions (ignoring the differences in the age of the observations).
  5. Produce an integrated dataset of the American figures and European figures from 2005.

- Income

  1. Merge the files `ap` and `apequiv` of the subdirectory `kksoep` of the file package of this book.

18. http://www.eurofound.europa.eu/areas/qualityoflife/eurlife/index.php

2. Produce a dataset holding the "HH Post-Government Income" of all the GSOEP equivalence files (`kksoep/*pequiv`). Use only those respondents that have records in each of the data files ("balanced panel design").
3. Produce the same dataset as above, but this time use those respondents for whom there are records in at least 10 data files.
4. Merge together the files `ahbrutto` and `ap`. Explore whether there is a difference when you reverse the order in which you merge the two data files?
5. Create a household file out of `ap`. That is a data file where each household is one observational unit (row). Add `ahbrutto` to this file.

11 Do-files for advanced users and user-written programs

When working on a substantial data analysis, you will often want to reuse a previous command. For example, you might want to repeat your analysis or replicate it with different data, or your dataset might contain a series of similar variables that should be treated in the same way. Perhaps during your analysis, you are recalculating certain measured values or creating certain graphs for which there are no existing commands in Stata. In all of these situations you can save work by learning the programming tools that we will introduce in this chapter.

We will distinguish among four programming tools: macros, do-files, programs, and ado-files. We will cover these in detail in section 11.2. In section 11.3, we will show you how you can use these tools to create your own Stata commands. However, before we start, we will give two examples of possible ways to use these tools.

11.1 Two examples of usage

Example 1

Imagine that you are working with one of our example datasets, the panel dataset `progex.dta`. The dataset contains the life satisfaction of 1,761 interviewees for every survey wave of the German Socio-Economic Panel (GSOEP). The dataset is organized so that answers from every survey wave are in a different variable. You can view the structure of the dataset through the following commands:[1]

```
. use progex
. describe
. list in 1
```

The variable for the respondent's life satisfaction appears 19 times in this file. You will find the data from 1984 in `ap6801`, those from 1985 in `bp9301`, those from 1986 in `cp9601` and so on (see section 10.4.1 for an explanation of the structure of variable names in the GSOEP).

Imagine that you want to generate new variables, which differ between respondents with a life satisfaction below and above the yearly average life satisfaction. Then you would have to enter commands like the following for each year:

1. Please make sure that your working directory is `c:\data\kk2`. More information on this can be found on page 3.

```
. label define lsat .a "No answer" 1 "below avarage" 2 "above avarage"
. summarize ap6801 if ap6801 > -1
. generate lsat1984 = 1 if ap6801 <= r(mean) & ap6801 > -1
. replace lsat1984 = 2 if ap6801 > r(mean) & ap6801 < .
. replace lsat1984 = .a if ap6801 == -1
. label value lsat1984 lsat
. label variable lsat1984 "Life Satisfaction High/Low (1984)"
```

Repeat these commands (except the first) for the answers from 1985, 1986, and so on. After 114 commands, you will have generated and labeled your variables. Naturally, you will not have created all of these commands interactively, but instead you will have entered the above block into a do-file. Once you have done this, you can simply copy the entire block 19 times and change the years in the variable names. Nevertheless, this still requires a lot of work and is comparatively error prone.

What would happen if you wanted to use the median as the differentiation point instead of average, and you wanted to put the missing category −1 into a category 3 for "other"? Or what if you wanted to separate those responses with a life satisfaction of 0 and 10, respectively?

You can probably guess what we are driving at. Overall, it seems slightly unwieldy to have a do-file that repeats only one set of commands. One solution for this problem can be found in the do-file `crlsat.do`, which you downloaded already. In this file, we used a `foreach`-loop (see section 3.2.2) together with more tricks that we will discuss in this chapter.

Example 2

Suppose that when carrying out logistic regressions, you would rather specify Aldrich and Nelson's Pseudo-R^2 (Aldrich and Nelson 1984, 57) than McFadden's, which is the one calculated by Stata (McFadden 1973). To do this, you can calculate the value with the `display` command after a logistic regression:

```
. gen satisfied1984 = ap6801==10
. logistic satisfied1984 gebjahr
. display "p2 = " e(chi2)/(e(chi2) + e(N))
```

However, instead of writing the equation each time, you might write a small ado-file to accomplish this task. If you save the ado-file on your hard drive, in the future you can enter the name of the ado-file after the logistic regression to get Aldrich and Nelson's Pseudo-R^2. Our version of this ado-file can be installed using `net install p2` from http://www.stata-press.com/data/kk2/.[2]

2. Notes on installing ado-files can be found in chapter 12.

Now please load our dataset `data1.dta`.

```
. use data1, clear
```

11.2 Four programming tools

Here we discuss several features of Stata that you will find yourself using repeatedly. We begin with an introduction to local macros, which allow you to store results, strings, and other items. We then show you how to write your own commands and parse syntax. We end with discussions of do-files and ado-files.

11.2.1 Local macros

We have already dealt with local macros in chapter 4 and covered the long-term storage of internal results in various parts of the book. Here we want to introduce local macros as a more general concept.

A local macro contains strings of characters. When you *define* a local macro, a name is given to these characters. Once it has been defined, you can enter the name of the macro instead of the characters.

You can define a local macro with the command `local`. Let us use this command to define the local macro `a`:

```
. local a "inc yedu ybirth"
```

This command assigns the string of characters `inc yedu ybirth` to the macro name `a`. To distinguish between the macro name and character string in this example, you have to put the character string in quotation marks. You need quotation marks only if you wish to include spaces in front of the first character or behind the last character.

If you wish to use the content of local macro `a`, you have to inform Stata that "a" is the name of a local macro and not the name of a variable. To do this, you place a single opening quotation mark (`` ` ``) before the name of the local macro and a single closing quotation mark (`'`) after it. On many American keyboards, the opening quote is found toward the top left (near the *Esc* key), whereas the closing quote is found more right, near the *Enter* key. On European keyboards, the position of both characters changes substantially from country to country. Often the opening quote is used to produce the French *accent grave*, which forces you to press the space key before the sign appears on the screen.

Whenever Stata comes across a local macro, the macro is immediately replaced by its contents. Whenever you have to type `inc yedu ybirth`, you can instead simply type the macro name, e.g.,

```
. summarize `a'

    Variable |       Obs        Mean    Std. Dev.       Min        Max
-------------+--------------------------------------------------------
      income |      3034    1349.207    1245.701          0      12438
        yedu |      3292    11.38548    2.408573          7         18
      ybirth |      3340     1951.72    18.33337       1902       1981
```

With this command, Stata sees `summarize inc yedu ybirth` once the macro has been expanded. If it does not, then you have probably used the wrong characters for identifying the local macro. Make sure that you have used the correct quotation marks around the macro name `a`.

Local macros enable you to save a lot of typing. However, remember three key points if you are using local macros for this purpose:

- The name of a local macro must be no more than 31 characters long.
- A local macro can contain up to 165,200 characters. This figure is for Stata/IC. For Small Stata, the figure is 8,681. For Stata/MP and Stata/SE, the figure is 1,081,511 characters.
- Local macros apply only within the *environment* in which they are defined. If you define a local macro in a do-file, the macro is accessible only in that do-file. If the do-file has finished, the local macro is no longer defined. If you interactively define a local macro, Stata will recognize it as long as you are working interactively. As soon as you start a do-file, you cannot use the interactively defined macro. You will be able to use the interactively defined macro again only once the do-file has finished and you have begun working interactively again. One advantage of this is that you can use the same macro names in different contexts without having to worry about mixing them up.

Calculating with local macros

Besides saving typing, you can also use local macros to carry out calculations. There are two options available for this, which we show in the next example.

```
. local m1 2+2
. local m2 = 2+2
```

The commands differ by the use of equal-sign. Both commands work and apparently lead to the same result:

```
. display `m1'
4
. display `m2'
4
```

However, there is one difference, which you will see if you embed the macro name in quotation marks. With quotation marks, `display` displays text, whereas without quotation marks, display shows the value of an expression (section 3.1.6). For this reason, the results above show the evaluation of the expression saved in the local macro. In both instances, the result was 4. If, however, you display the contents of the local macros as text, then it becomes clear that the first macro contains the expression, whereas the second contains the result.

```
. display "`m1'"
2+2

. display "`m2'"
4
```

Note a crucial difference when multiplying:

```
. display 2*`m1'
6
```

because $2 \times 2 + 2 = 6$ and

```
. display 2*`m2'
8
```

because $2 \times 4 = 8$.

Combining local macros

You can combine several local macros. Here are a few examples. Please type these commands to get some practice and understanding of local macros. However, before typing these command lines, remember to check that your working directory (page 3) contains our do-files and datasets:

```
. local a dir *.
. local b dta
. local c do
. `a'`b'
. `a'`c'
. local b `a'`b'
. display "`b'"
. `b'
```

Changing local macros

Imagine that you wanted to add the number 1 to an existing macro, in this case, i, which contains the number 5. The command is as follows:

```
. local i 5

. local i = `i' + 1
```

In the second command, the macro `i` on the right-hand side of the equation must be embedded in the characters ‘ and ’, whereas the macro `i` on the left-hand side must not. The reason behind this is that `‘i’` is replaced by the number 5; the command is therefore converted into `local i = 5 + 1`. If you had enclosed the `i` on the left side in quotation marks, the command would have become `local 5 = 5+1`. This, in fact, would have created a new local macro with the name “5”.

Besides this minor glitch, it is not difficult to modify local macros. Here are some examples:

```
. local i 1
. local i = ‘i’ + 10
. di ‘i’
11
. local i = ‘i’ + ‘i’
. di ‘i’
22
. local i "i is ‘i’"
. di "‘i’"
i is 22
```

You can redefine a local macro whenever you want with the contents of an expression, a string, or an extended macro function (see section 11.3.6). Two commonly used macro-extension operators add or subtract the number 1 from a macro, either immediately before Stata expands the macro or immediately thereafter. Consider, for example

```
. local i 1
. di ‘i++’
1
. di ‘i’
2
```

Here we first define the local macro `i`. Afterward we display `i` with an extension operator (`++`), meaning that the macro is expanded, and then one is added to the macro; that is, the content of the macro is changed to 2. If you put the extension operator in front of the macro name, 1 is added to the macro before the macro is expanded. The same applies to the extension operator `--` for subtracting 1:

```
. di ‘++i’
3
. di ‘i--’
3
. di ‘--i’
1
```

11.2.2 Do-files

Say that you want your computer to display the words "hello, world". To achieve this, you can type the following command:

```
. display "hello, world"
hello, world
```

You would have to repeatedly type this command every time you wanted to see "hello, world" displayed. Admittedly, this is not too much of a problem, as the command is not very long. Nevertheless, the command could be longer, so you might want to know how to save yourself some work.

Please enter the following do-file and save it under the name `hello.do` in your working directory. As always, remember to write only the Stata commands. The solid line with the filename and the line number indicate that the commands are placed into a file.

```
——————————————————————— top: hello.do ———
1: display "hello, again"
2: exit
——————————————————————— end: hello.do ———
```

Once you have done this, you can execute `hello.do` by typing

```
. do hello
```

We get the same result that we would receive by entering `display "hello, world"`, but now we just have to type `do hello` to see the message again. Of course, a do-file will typically have many more lines of code.

11.2.3 Programs

Besides do-files, the command `program` offers a second option for accessing many commands at the same time. The way `program` works can be better illustrated through an example. Please type the commands below. The first line begins the definition of a program named `hello`; you may use any other name that it is not longer than 31 characters. When you are finished with the first line and press the *Enter* key, you will see that Stata displays `1.` at the beginning of the next line. This is Stata's way of prompting you to type the first command of the program. Please type the command without typing the number.

```
. program hello
  1. display "Hello, world"
  2. end
```

When you confirm this entry by pressing *Enter*, the prompt for the second command that the program should carry out appears on the screen. Here we want to end the program, so we type `end` in the Command window. This will now return you to the normal prompt.

Now that the program is defined, you can type `hello` in the Command window:

```
. hello
Hello, world
```

You entered `hello`, and Stata replied with "Hello, world".

The `program` command defines programs. However, in contrast to do-files, these programs are stored in the computer's memory and not in a file on the computer's hard disk. If you type some word into the Stata Command window, Stata will look in the memory (see section 10.6) for something that is called the same as the word you just entered into the command line. Therefore, if you enter `hello` in the window, Stata will search the memory for a program called `hello` and will find the hello program you previously saved there with the `program` command. The Stata commands between `program` and `end` are then carried out.

In many respects, these programs are similar to the do-files you have already seen. However, there are some differences:

- Do-files are saved in a file on the computer's hard drive, whereas programs are stored in memory.
- Do-files are not deleted when Stata is closed or the computer is shut down. Programs are lost when Stata is closed.
- A do-file is accessed by entering `do` *filename*, whereas a program is accessed by entering the program name without any command in front.
- Since Stata searches for programs in memory, programs must therefore be loaded before they can be accessed. Do-files must be saved to the hard drive.
- Do-files display the results of Stata commands that have been carried out, *as well as* the commands themselves. Programs only display the results.

The most important difference between do-files and programs is that do-files remain available for access on a long-term basis. Programs, on the other hand, are available only during a Stata session. In the following section, we will therefore concern ourselves with the options available to us for storing programs over a long period. Before we do this, we need to look at several typical problems that may arise when saving and accessing programs.

The problem of redefinition

Imagine that you want your computer to display "Hi, back" instead of "Hello, world" when you type `hello`. So, you try reentering the program:

```
. program hello
hello already defined
r(110);
```

Stata knows that a program in memory is already named "hello" and does not allow you to overwrite it. First, you must delete the old version from the RAM before you can create the new version:

```
. program drop hello
. program hello
  1. display "Hi, back"
  2. end
. hello
Hi, back
```

The problem of naming

Imagine that you wanted the program to be called `q` instead of `hello`:

```
. program q
  1. display "Hello, world"
  2. end
. q
```

Surprisingly, this displays the settings of various Stata parameters. The reason is that the letter `q` is a shortcut for the Stata command `query`, which is a "built-in" Stata command. Stata searches for programs only when it has not found a built-in Stata command with the specified name. For this reason, you should never define a program with the same name as a built-in command.

To ascertain whether a command with a given name already exists, enter `which` *commandname*. If Stata replies with "Command *commandname* not found as either built-in or ado-file", you may use the name for your program.

The problem of error checking

Stata checks the syntax of a program only once it has been executed:

```
. program hello2
  1. displai "Hello, world"
  2. end
```

Here `displai` was entered instead of `display`, so Stata will detect an error when executing the program:

```
. hello2
unrecognized command:  displai
r(199);
```

Because the individual commands are not repeated on screen as they are in do-files, it is often hard to find an incorrectly entered command in lengthy programs. By using the command `set trace on`, you can instruct Stata to display a program's commands. This enables you to follow the program command by command while it is being executed, and thus find the incorrect command. Lines that are to be executed begin with a hyphen. If a line in the program contains a macro, another line beginning with an equal-sign shows the line with expanded macros. Unfortunately, the trace creates a large volume of output when it is dealing with lengthy programs. So do not forget to switch off the trace once you have found the error in a program: `set trace off`. See [P] **trace** or `help trace` for more information.

11.2.4 Programs in do-files and ado-files

Earlier we interactively stored a program by entering the respective commands in the Command window. One of the disadvantages of this process is that the stored programs are lost when the Stata session ends (at the latest). Another disadvantage is that you are unlikely to enter a lengthy program without typos, which will become apparent only when the program is executed. To correct the typo, you will have to reenter the entire program. If you want to store a program on a long-term basis, then you have to enter the definition of the program in a do-file.

To help you learn how to define programs in do-files, you should rewrite the file `hello.do`, which you created on page 339, as follows:

top: hello.do

```
1: program hello
2:     display "hello, again"
3: end
4: exit
```

end: hello.do

This do-file contains a new version of the `hello` program from the previous section. The difference is that our program should now display "hello, again". More importantly, the definition of the program should now be written in a do-file. We have slightly indented the list of commands between `program` and `end`. This helps us to find the beginning and end of the program definition but does not affect Stata in any way. Now that the program definition is written into the do-file we do not have to rewrite the definition of the program every time we close Stata; executing the do-file is sufficient. Let us try it. Please save the do-file and execute it. If you have followed all of our steps in section 11.2.3, then you should receive an error message:

```
. do hello
. program hello
hello already defined
r(110);
```

The reason for the error message is that Stata recognizes the names of the programs already loaded in memory and will not allow them to be overwritten. As we have already interactively defined `hello`, it is currently being stored in memory. Therefore, you must delete this older version before you can create the new version. This is best done directly in the do-file through `capture program drop hello`. By using `program drop hello`, the `hello` program is deleted from memory, while `capture` ensures that no error messages will follow if there is no such program (also see section 2.2.3). Please change `hello.do` to

```
top: hello.do
1: capture program drop hello
2: program hello
3:     display "hello, again"
4: end
5: exit
end: hello.do
```

Save these changes, and try it again:

```
. do hello
```

The program `hello` has not been executed, as it has been stored in memory only through the commands in the do-file. However, we can now execute the program interactively:

```
. hello
hello, again
```

We can also call up the program directly in the do-file as follows:

```
top: hello.do
1: capture program drop hello
2: program hello
3:     display "hello, again"
4: end
5: hello  // <- Here we call the execution of the program
6: exit
end: hello.do
```

Here the program is first defined and then executed. Give it a try:

```
. do hello
```

A further option for saving programs is offered by ado-files, which are also known as "ados". How ados work becomes clearer if we go back one step. First, delete the program `hello` from memory:

```
. program drop hello
```

Then reload `hello.do` in your Editor. Remove the commands in `hello.do` related to deleting and calling up the program. The file `hello.do` should now look like this:

top: hello.do

```
1: program hello
2:     display "hello, again"
3: end
4: exit
```

end: hello.do

Once you have saved it, the do-file can be executed with

```
. run hello
```

The command `run` is the same as the `do` command, the only difference being that after you enter `run`, the individual command lines do not appear on the screen. If you enter `run hello`, `hello.do` is *quietly* executed and at the same time the program `hello` is loaded in memory. After this, you can interactively execute the program:

```
. hello
hello, again
```

Now to the ado-files: if we save a do-file with the extension `.ado` instead of `.do`, the above-mentioned steps are both automatically carried out. Entering the name of the ado-file will suffice. Go ahead and try it. Save your do-file with the extension `.ado` under the name `hello.ado` and then enter the following:

```
. program drop hello
. hello
hello, again
```

It works. To understand why, let us take a closer look at the steps that take place after `hello`. In general, after a command has been entered, Stata takes the following steps:

1. Stata checks if `hello` is an internal command. If so, then Stata will have executed it. As `hello` is not an internal command, it moves on to the next step.
2. Stata checks if `hello` is stored in memory. If so, the program would be executed. As we deleted the `hello` program from memory shortly before entering the command `hello`, Stata would not find a program called `hello` in memory and would move to the next step.
3. Stata searches for the file `hello.ado`. Here Stata searches in various places on the hard drive, including the working directory. When Stata finds the file `hello.ado`, it essentially gives itself the command `run hello.ado` and subsequently checks if the program `hello` is now in memory. If so, as in this example, then the program is executed. If not, then Stata will display the error message "unrecognized command". If the ado-file contains subprograms, Stata will also load any subprograms of `hello.ado` and makes them available solely to `hello`.

Because of the second step, programs that are defined by ado-files must be deleted from memory before any potential changes to the program take effect. If you rewrite `hello.ado` as follows

top: hello.ado

```
1: program hello
2:     display "hi, back again"
3: end
4: exit
```

end: hello.ado

and then enter `hello`, at first nothing will have changed:

```
. hello
hello, again
```

You must first delete the old program from memory before the ado-file can be reimported. To do so, you can use the `program drop` command or the `discard` command. `discard` forces Stata to drop all automatically loaded programs from memory and reload them from disk the next time they are called; see `help discard` for details.

```
. discard
. hello
hi, back again
```

Let us summarize what we have learned so far: if we enter the command `hello` at the command prompt, Stata will find the program `hello` in memory and then execute it, or it will not find it. In the latter case, Stata searches for `hello.ado`, loads the program in memory, and executes it. The command `hello` therefore works just like a normal Stata command. In fact, the command `hello` is a normal Stata command: an "external" Stata command.

Stata generally differentiates between external and internal commands. We have already mentioned the idea of internal commands a couple of times, so we will now define them in more detail. Internal commands are commands programmed in the C programming language that are compiled for various hardware platforms and operating systems. External commands, on the other hand, are the ado-files that we have just introduced you to: programs that are saved in files with the extension `.ado`. Nearly all Stata commands are ado-files. If you enter the command `adopath`, you will receive a list of directories that Stata searches for ado-files.[3] You can view the contents of ado-files with any editor.

11.3 User-written Stata commands

As we said at the beginning, ado-files comprise nothing more than the definition of a program. If you enter the name of an ado-file in Stata, the program is stored in memory

3. The directory structure of ado-files is explained in detail in section 12.3.3.

and then executed. In effect, an ado-file therefore behaves just like every other Stata command; or, as we stated above, the ado-file is a normal Stata command.

To execute an ado-file, you must save it on the hard drive where Stata can find it, such as the personal ado-directory. The command `adopath` will indicate where the personal ado-directory can be found on your computer. In Windows, the personal ado-directory will generally be found at `c:\ado\personal` (see section 12.3.3).

Here, with the help of an example, we will briefly demonstrate how ados are programmed. We will not cover all the programming options that are available to you. Instead consider this section as a sort of *ladder* that allows you to climb a fruit tree. To pick the best fruit, you will have to choose the best branches by yourself. All the Stata commands and the tools described in the previous section are available for your use. The Stata Internet courses NetCourse 151 and NetCourse 152 will give you a detailed introduction to programming in Stata (section 12.1). Before you take the time to program your own Stata command, you should thoroughly check to see whether someone has already done the work. The Statalist archive and the sources of information listed in section 12.3.3 should help you do this. Nevertheless, it is certainly worthwhile learning a little bit about Stata programming. If you are using ados written by other users, it is especially useful to be able to "read" what is actually happening in Stata code. This is also true for the commands supplied with the program.

In section 7.3.3, we showed you how you can generate a graph with superimposed kernel density estimates. For your benefit, we have saved the commands that we used in a file called `denscomp.do`, which you downloaded.

top: denscomp.do

```
1: use data1, clear
2: gen linc = log(income)
3: gen xhelp = log(_n*250) in 1/50
4: kdensity linc if gender == 1, generate(xvals1 fmen) at(xhelp) nograph
5: kdensity linc if gender == 2, generate(xvals2 fwomen) at(xhelp) nograph
6: graph twoway connected fmen fwomen xhelp ///
7:   , title(Income by Gender) ytitle(Density) xtitle(Log (Income)) ///
8:     legend(label(1 "Men") label(2 "Women"))
9: exit
```

end: denscomp.do

If you run the do-file, you can view the graph on your screen again. The following section will describe how to generate such graphs with one command for arbitrary datasets. The command should have roughly the following syntax:

`denscomp` *varname* [*if*] [*in*] , `by(`*varname*`)` `at(`*varname*`)` [*twoway_options*]

That is, to produce our example graph, *varname* is substituted by `linc`, `by(`*varname*`)` becomes `by(gender)`, and `at(`*varname*`)` becomes `at(xhelp)`, respectively.[4] Ideally,

4. If you find this program useful, Thomas Steichen's highly efficient package in STB-46, gr33 should interest you. Notes on installing the package can be found in section 12.3.1. You can install the version developed here with `net from http://www.stata-press.com/data/kk2` and `net install denscomp`. Detailed explanations of these commands can be found in section 12.3.1.

your command should be available for all datasets and variables. However, for that you will have to implement a few generalizations. Look more closely at denscomp.do, as every command line currently contains an element related to our specific question on the income comparison for men and women.

Before you begin to program the generalization of the do-file, you should create the first version of the ado-files in an editor and then save it under denscomp.ado. This first version is almost identical to the above denscomp.do. In fact, you can just save denscomp.do with the name denscomp.ado. Once you have done this, you can make the following changes:

1. Delete the use command, as well as both generate commands used to generate the variables linc and xhelp.
2. Type the command program denscomp in the first line of the file.
3. Type the command version 10 in the second line of the file. You already know this command from the chapter on do-files. It indicates which Stata version this program has been written for. This guarantees that your program will work error free in later Stata versions.
4. Type the command end to indicate the end of the program in the next-to-last line of the file. The command exit ends the ado-file, not the program.

The preliminary version of denscomp.ado should now look like this:

top: denscomp.ado

```
1: program denscomp
2: version 10
3: kdensity linc if gender == 1, generate(xvals1 fmen) at(xhelp) nograph
4: kdensity linc if gender == 2, generate(xvals2 fwomen) at(xhelp) nograph
5: graph twoway connected fmen fwomen xhelp ///
6:    , title(Income by Gender) ytitle(Density) xtitle(Log (Income)) ///
7:      legend(label(1 "Men") label(2 "Women"))
8: end
9: exit
```

end: denscomp.ado

You should then check to see that everything still works. Checks such as these should be made after every major step. So please save this ado-file in your current working directory, and then type denscomp.

If you run denscomp, you will now receive an error message:

```
. denscomp
fmen already defined
r(110);
```

The reason for this is that the variables fmen and fwomen were both created in denscomp.do and remain in the dataset. denscomp.ado then tries to create them; however, as they already exist, you will receive an error message. Therefore, you have to first delete fmen and fwomen before you run denscomp.ado.

```
. drop fmen fwomen
. denscomp
```

If you did not run `denscomp.do` above, you will probably receive the error message

```
. denscomp
. variable linc not found
r(111);
```

If you receive instead the error message "no variables defined" then you have not loaded the dataset. Please load the `data1.dta` dataset, and type `denscomp` again. The variable `linc` is used in `denscomp.ado`, but this will not be in your dataset if you haven't run `denscomp.do`. Here please type the following commands (see section 7.3.3):

```
. use data1, clear
. generate linc = log(inc)
. generate xhelp = log(_n * 250) in 1/50
```

You can try again after these steps:

```
. denscomp
```

The graph should now be displayed on your screen. We created the graph with one command; however, `denscomp` will only create the graph with the same variable and options. Now we will generalize `denscomp`.

11.3.1 Parsing variable lists

The first step in generalizing `denscomp.ado` is to allow the graph to be displayed for arbitrary variables.

This can be achieved with the `syntax` command. The `syntax` command is used to define the structure with which a user is to call our program. Here we assume that the program is accessed through the name of a command and a variable name. This setup is parsed to the program in line 3 of the following program snippet:

```
——————————————————————————— top: denscomp.ado ————
1: program denscomp
2: version 10
3: syntax varname(numeric)
4: kdensity `varlist' if gender == 1, generate(xvals1 fmen) at(xhelp) nograph
5: kdensity `varlist' if gender == 2, generate(xvals2 fwomen) at(xhelp) nograph
✂
9: end
——————————————————————————— end: denscomp.ado ————
```

We will print only those parts of the ado-file that we talk about. However, we have printed the final version of `denscomp.ado` on page 359.

The `syntax` command has two functions: verifying input and parsing program arguments. That is, `syntax` checks whether the user's input makes sense and either produces an error message or parses the input so the program can use it.

Let us clarify. As we code `syntax varname`, Stata tries to interpret everything written behind `denscomp` as a variable name. If `denscomp` is called up without a variable name, Stata will issue an error message, indicating that a variable name is required. Moreover, Stata will check whether the user gave one or more variable names. As we specified `syntax varname`, Stata will produce an error message if the user accidentally specified more than one variable name. If we had decided to allow the user to specify more than one variable name, we would have coded `syntax varlist`, with the effect that the user could have specified a list of variable names. Finally, as we specified `syntax varname(numeric)`, Stata will check whether the variable input by the user is a numeric variable or a string variable. If it is a string variable, Stata will issue an error message and exit the program immediately.

The second function of `syntax` is to parse program arguments given by the user. The `syntax` command in our example causes Stata to save the entered variables name in the local macro `varlist` when the program is called up. With this in mind, we can use `` `varlist' `` in our program instead of the hard-coded variable name `linc`—see lines 4 and 5 of the program code.

We now call this program using `denscomp linc`. Go ahead and try it. You will first have to save `denscomp.ado` and then delete the old version of the program from memory with `discard` and delete the variables `fmen` and `fwomen`. Afterward, your program should run free of errors.

```
. discard
. drop fmen fwomen
. denscomp linc
```

11.3.2 Parsing options

These changes have made the program more generalized. However, the distributions of men and women are still being compared with each other. We should be able to have the option of using the graph to compare other subgroups. We would like to have an option we could use to define the variable whose values are to be compared. For example, this option could be called `by(`*varname*`)`. The question that remains is how to transfer these options to the program.

For this, we again turn to the `syntax` command. As indicated above, by using `syntax` we are informing a program about the structure of the command. To enable the option `by(`*varname*`)`, we amend the `syntax` command as shown in line 3 below:

top: denscomp.ado

```
1: program denscomp
2: version 10.0
3: syntax varname(numeric), by(varname)
4: kdensity `varlist' if `by'==1, gen(xvals1 fmen) at(xhelp) nograph
5: kdensity `varlist' if `by'==2, gen(xvals2 fwomen) at(xhelp) nograph
```

end: denscomp.ado

With these changes, the program `denscomp` now expects to be called with an option after the name of a numeric variable. We add the option by placing a comma in the `syntax` command. As we did not place the list of options (we refer to a list of options even though currently only one option is allowed) in square brackets, we must specify the option. If we had used `syntax varname(numeric) [, by(varname)]` the option would not have been required. Square brackets in `syntax` commands work just as described in the syntax diagram in the online help.

As it stands, we only allow one option: `by()`. Only one variable name is allowed inside the brackets of `by()`. There is no restriction on numeric variables; hence in this case, string variables are also allowed. As above, if we mistakenly type the name of a variable into `by()` that does not exist in the dataset, `syntax` will display an error message. Also `syntax` will place the variable name given in `by()` in the local macro called `by`. You can see how we use this local macro in lines 4 and 5 of our program snippet above.

You can apply the same logic for the `at()` option in the two `kdensity` commands in lines 4 and 5. The `kdensity` commands are both called with `at(xhelp)`. However, this will work only for datasets that contain the variable `xhelp`. What we need instead is an option for `denscomp` to declare the name of the numerical `at()` variable. Therefore, we type

top: denscomp.ado

```
✂
3: syntax varname(numeric), by(varname) at(varname numeric)
4: kdensity `varlist' if `by'==1, gen(xvals1 fmen) at(`at') nograph
5: kdensity `varlist' if `by'==2, gen(xvals2 fwomen) at(`at') nograph
6: graph twoway connected fmen fwomen `at' ///
```

end: denscomp.ado

We used the keyword `numeric` in the `at()` option because `kdensity` can compute densities only at numerical points. Stata will issue an error message if the user specifies a string variable. To try out these changes, save your file, and type the following commands in Stata. You need to use the `discard` command to remove the current ado-file from memory. Only then will the changes become effective.

```
. discard
. drop fmen fwomen
. denscomp linc, by(gender) at(xhelp)
```

You should now see the familiar graph. If not, carefully check your file for any errors. To determine which of the program lines is faulty, you can use the command `set trace on`, which was described on page 342. A further help in locating errors is to display suspect program lines with display. To do so, you must modify the program with lines such as these:

```
                                              top: denscomp.ado
✂
3: syntax varname(numeric), by(varname) at(varname numeric)
4: display "kdensity `varlist' if `by'==1, gen(xvals1 fmen) at(`at') nograph"
                                              end: denscomp.ado
```

The program is then called up, and the output of display shows you the command line after macro expansion. This often helps to find typos. Once the program is error free, delete the display commands. During your search for errors, do not forget to type the commands discard and drop xvals1 xvals2 fmen fwomen after every change made to the ado-file.

11.3.3 Parsing if and in qualifiers

Before we refine our ado-file further, we would like to show you how to include if and in qualifiers in the program. Once again, our starting point is syntax:

```
                                              top: denscomp.ado
✂
3: syntax varname(numeric) [if] [in], by(varname) at(varname numeric)
4: marksample touse
5: kdensity `varlist' if `by'==1 & `touse', gen(xvals1 fmen) at(`at') nograph
6: kdensity `varlist' if `by'==2 & `touse', gen(xvals2 fwomen) at(`at') nograph
                                              end: denscomp.ado
```

Entering [if] and [in] lets you use if and in qualifiers. As both of them are in square brackets, neither element is required. Again syntax saves the qualifiers that have been entered by the user in local macros—here, in if and in, respectively. In doing so, the whole expression, including the keywords if and in, is saved in the macro.

So far, we have enabled if and in qualifiers. That is, Stata will not print an error message when the user invokes denscomp with an if qualifier. Now we must make our program exclude observations that do not satisfy those conditions. One solution is to use the keep command to delete all the observations for which the preconditions are not applicable and only execute the following commands for the remaining observations. A better solution would be to use marksample (see line 4). marksample generates a temporary variable, and in our example the name of this temporary variable is touse (short for: to use).[5] In this variable, all cases that are to be used are assigned the value 1, and all other cases are assigned the value 0. marksample receives the information about which cases to use from the if and in qualifiers.

If the temporary variable touse has been defined by marksample, this variable can be used to restrict the statistical commands in the program to the observations marked with 1. You see examples of how to achieve this in lines 5 and 6 of the program snippet

5. We discuss temporary variables in section 11.3.7 in more detail. Like local macros, they are cursory and are available only in a given environment.

above. Because touse is a temporary variable, it is placed between single quotation marks just like a local macro. In fact, touse *is* a local macro containing the name of the temporary variable. And you do not need to specify if 'touse'==1, as touse contains only 0 and 1, which is Stata's way to express false and true (see section 3.1.6).

11.3.4 Generating an unknown number of variables

Until now, we have only specified the basic characteristics of the ado-file. Now we will generalize our command. Accomplishing this is often the most time-consuming part of programming Stata commands.

Before we carry on, let us contemplate our situation. Please save your current ado-file and start your ado with:

```
. discard
. drop fmen fwomen
. denscomp linc, by(emp) at(xhelp)
```

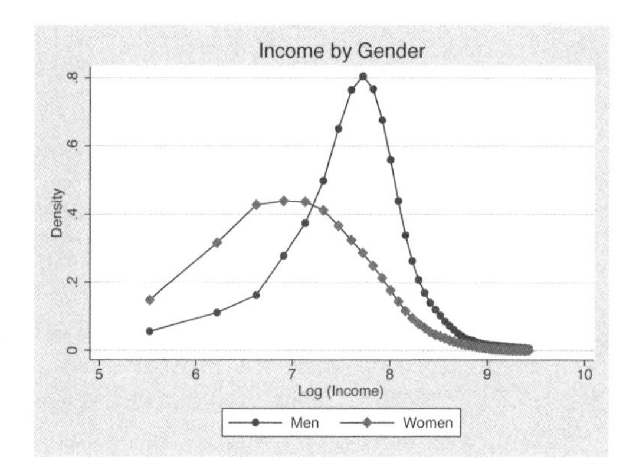

This time, we have used employment status as the variable in the option by(), In principle, this should work.[6] Nevertheless, something is not quite right here. The graph shows only two density curves, even though we were expecting one for every type of employment status, which would mean a total of six curves. According to the labeling, the graph still shows the income of men and women, instead of the income of individuals with varying employment status.

Apparently, there is still some work to be done. We should start with the first problem: why are only two density curves being displayed, instead of six? The cause lies in lines 5 and 6 of the program:

6. If your program still will not work despite a thorough check for errors, then you can fall back on our file denscomp2.txt. This contains the version of our program with which we generated the graph. You can use it by simply replacing your ado-file with the program definition in denscomp2.txt.

top: denscomp.ado

```
✂
3: syntax varname(numeric) [if] [in], by(varname) at(varname numeric)
4: marksample touse
5: kdensity `varlist' if `by'==1 & `touse', gen(xvals1 fmen) at(`at') nograph
6: kdensity `varlist' if `by'==2 & `touse', gen(xvals2 fwomen) at(`at') nograph
```

end: denscomp.ado

These lines contain the calculations for the density of the values 1 and 2 for the variable given in by(). If we type the command with the option by(emp), the local macro by is replaced by emp and the densities for the first two values of employment status are displayed. There are no commands for any further categories.

The problem, however, is that the number of categories can vary with the variable specified in by(). This in turn means that we do not know how often we want the program to execute the kdensity command.

If we knew the number of categories, we could create a loop in our program that would execute kdensity for every variable in by() and stop when there are no remaining categories. Therefore, we must somehow find out how many categories the variable in by() has.The easiest way to get this information is to use the command levelsof.

In general, levelsof is used to look into a specific variable and to store the existing distinct values of this variable in a local macro. The variable emp, for example, contains values of 1–7, although the value 5 is not used. The following command will store these values into the local macro K:

```
. levelsof emp, local(K)
. di "`K'"
```

Now let us see how you can use this in our program. You find the levelsof command in line 5 of the program excerpted below. We put quietly in front of the command. quietly can be used in front of any Stata command to suppress all its output.

top: denscomp.ado

```
✂
5: quietly levelsof `by' if `touse', local(K)
6: foreach k of local K {
7:     kdensity `varname' if `by'==`k' & `touse', gen(xvals`k' fx`k') ///
8:             at(`at') nograph
9: }
```

end: denscomp.ado

As you can see in line 5, we used levelsof to store the distinct values of the by() variable in the local macro K. Then we set up a foreach loop over all elements of the local macro K. Within the loop, kdensity is executed for each employment category. We changed the option gen() of the kdensity command in line 7. Instead of generating hard-coded variable names xvals1, xvals2, fmen, and fwomen, we generate variables with names according to the element name of the loop. That is, if we specify denscomp with the option by(emp), we generate variables xvals1, xvals2, xvals3, xvals4, xvals5, xvals6, and xvals7; and fx1, fx2, fx3, fx4, fx6, and fx7 since there are no observations for which emp is equal to 5.

Next, because the number of fx variables is no longer fixed, we have to adapt the graph command accordingly. For this reason, we create a list with the names of the generated fx variables within the loop and then save it in a local macro called yvars (line 9):

top: denscomp.ado

```
✂
 6: foreach k of local K {
 7:     kdensity `varlist' if `by'==`k' & `touse', gen(xvals`k' fx`k') ///
 8:            at(`at') nograph
 9:     local yvars "`yvars' fx`k'"
10: }
```

end: denscomp.ado

We can now include the local macro yvars in the graph command instead of fmen and fwomen:

top: denscomp.ado

```
✂
11: graph twoway connected `yvars' `at' ///
```

end: denscomp.ado

Now it's time for another search for any errors:

```
. discard
. drop fmen fwomen
. denscomp linc, by(emp) at(xhelp)
```

It now looks a lot better. Nevertheless, it is far from perfect. In particular, the labeling of the graph remains a problem. We still have a bit of work ahead of us.

11.3.5 Default values

The first question is how to label the graph. We suggest that you make the default title "variable by by variable". "Density" would be suitable as the title of the y axis, and the title of the x axis should contain the variable name. The meaning of the data area would be explained in the legend by the categories of the by variable.

Let us start with the title of the graph. A simple way of achieving the desired result is to change the graph command:

top: denscomp.ado

```
✂
11: graph twoway connected `yvars' `at' ///
12:    , title("`varlist' by `by'") ytitle("Density") xtitle("Log (Income)") ///
13:      legend(label(1 "Men") label(2 "Women"))
```

end: denscomp.ado

This, however, has the disadvantage of being tied to our example. To be able to change the title when calling up the program, we first have to include the corresponding option in the syntax statement at the beginning of the program:

top: denscomp.ado

```
3:  syntax varname(numeric) [if] [in], by(varname) at(varname numeric) ///
4:    [ TItle(string asis) ]
```

end: denscomp.ado

With line 4, we allow the user to specify an option called `title()`, where an arbitrary string can be specified, which is parsed inside the program `asis`; this means that quotation marks are part of the macro contents. The user now can enter a title when calling up the program. As we do not want to be obliged to enter a title, we have written the option in square brackets. The two capital letters indicate the minimum abbreviation, so the user can specify `ti()` instead of `title()`.

As was the case before, the `syntax` command saves everything within the parentheses in the local macro `title`. We can use this local macro in the title option of the `graph` command further down:

top: denscomp.ado

```
12:  graph twoway connected `yvars' `at' ///
13:     , title(`title') ytitle(Density) xtitle(Log (Income)) ///
```

end: denscomp.ado

Now we no longer have a default title, so we will add one. Add the following line to the program just before the `graph` command:

top: denscomp.ado

```
11:  local title = cond(`"`title'"' == `""', `"`varlist' by `by'"', `"`title'"')
```

end: denscomp.ado

Two things are new in this line: the function `cond()` and the compound quotes. Let us first explain the `cond()` function. The function requires three arguments, each separated by a comma (`cond(`x`,`a`,`b`)`). The first argument contains a condition that will be evaluated. If this condition is true, the expression in the second argument is returned, whereas if the condition is false, the expression in the third argument is returned. Consider the following simplified version of the `cond()` function in line 11: `cond("`title'" == "", "`varlist' by `by'", "`title'")`. Now think of a user who specified no title option. Then the first argument of the condition would become `"" == ""`. This condition is true, so the second argument is returned: `local title = "`varlist' by `by'"`. Hence, we will have a default title. If the user specifies `title(My title)`, the first argument becomes `"My title" == ""`. This condition is false. Here Stata returns the third argument (`local title = "My title"`), which sets the title to the user-specified string.

Now for the compound quotes. Consider a user who includes `title("My" "title")` when invoking `denscomp`. That is, this user wants the word "My" as the first line of the title, and "title" as the second. Again the first argument in the `cond()` function will be false. However, this time, the simplified version of line 11 will become `local title = ""My" "title""`. The problem here is that Stata cannot distinguish the opening quotation in front of "My" from a closing quotation mark. In fact, Stata cannot distinguish opening and closing quotation marks at all. To differentiate between the beginning of a quotation and the end of a quotation, we use compound quotes. To start a quotation, we use `‘"` (a single opening quotation, followed by a double quote) and to end a quotation we use `"’` (a double quote, followed by a single closing quotation). We recommend that you always use compound quotes inside programs. They never cause problems but are often necessary.

The same procedure may be used for the other characteristics of the graph, e.g., the title of the y and x axes:

```
                                                      top: denscomp.ado
✂
 3: syntax varname(numeric) [if] [in], by(varname) at(varname numeric) ///
 4:   [TItle(string asis) YTItle(string asis) XTItle(string asis) * ]
✂
12: local ytitle = cond(‘"‘ytitle’"’ == ‘""’, ‘"Density"’, ‘"‘ytitle’"’)
13: local xtitle = cond(‘"‘xtitle’"’ == ‘""’, ‘"‘varname’"’, ‘"‘xtitle’"’)
14: graph twoway connected ‘yvars’ ‘at’ ///
15:   , title(‘title’) ytitle(‘ytitle’) xtitle(‘xtitle’) ‘options’
16: end
                                                      end: denscomp.ado
```

In our `graph` command, we have not specified any other options, such as colors or marker symbols. Therefore, apart from the explicitly defined defaults, the graph resembles a standard Stata graph. At the same time, we want to be able to define all of graph's characteristics whenever we call up the program. To do so, we have also specified the wildcard element in the `syntax` command (line 4). The star (`*`) means that the user can enter further options in addition to those that have been explicitly allowed. These options are saved by `syntax` in the local macro `options` and can then be included in the `graph` command as shown in line 16.

You should now try out these changes. Please type

```
. drop fx*
. discard
. denscomp linc, by(emp) at(xhelp) ti(My Graph) xtitle(My X-Title)
> ytitle(My Y-Title)
```

11.3.6 Extended macro functions

Now all that is missing is an informative legend. This is a little tricky. By default, the variable labels of the data being plotted are used as the text in the legend. We can use this default setting by labeling the individual variables generated by `kdensity` with the appropriate caption. Ideally, the legend should inform us of the category of the variable in `by()`. We will therefore use the value label of the variable in `by()` as the text for the legend.

We can achieve this by using the extended macro function `label`. The general syntax for extended macro functions is

`local` *macroname*: *extended macro function*

That is, using a colon instead of an equal-sign informs Stata that an extended macro function is to follow.

Here we add lines 14, 19, and 20 to the program, which concerns the `foreach` loop calculating the various density estimations. Also we slightly changed the command for displaying the graph (lines 26, 27).

```
                                                        top: denscomp.ado
✂
14: local i 1
15: foreach k of local K {
16:     kdensity `varlist' if `by'==`k' & `touse', gen(xvals`k' fx`k') ///
17:          at(`at') nograph
18:     local yvars "`yvars' fx`k'"
19:     local label: label (`by') `k'
20:     local leglab `" `leglab' label(`i++' `"`label'"') "'
21: }
✂
25: graph twoway connected `yvars' `at' ///
26:   , title(`title') ytitle(`ytitle') xtitle(`xtitle') legend(`leglab') ///
27:   `options'
                                                        end: denscomp.ado
```

To customize the legend on a graph, we use the `legend()` option. If our graph had two data series, one for men and one for women, we could then specify `legend(label(1 "Men") label(2 "Women"))`. Here we do not know the number of categories in the `by` variable, so we use the local macro `i` to keep a count as we loop through them.

Before the loop begins, the macros `label` and `leglab` are empty. During the first iteration of the loop, the macro `label` is filled with the value label corresponding to the first level of the `by()` variable. The `leglab` macro is then set to `label(1 "`*text*`")`, where *text* is the text contained in `label`. The macro `i` is also incremented so that next time it will be equal to 2. In later iterations of the loop, the `leglab` macro is modified by adding elements of the form `label(`*x* `"`*text*`")`, where *x* is the value of the `by()` variable and *text* is the corresponding value label.

The local macro `leglab` can then easily be used for the definition of the legend in the `graph` command (line 26).

With this change, the first raw version of the ado-file would now be ready. To check the situation so far, you should try the program again:

```
. discard
. drop fx*
. denscomp linc, by(gender) at(xhelp)
. drop fx*
. denscomp linc, by(edu) at(xhelp)
```

Before you work through the next section, you should make sure your program will run without errors, as locating errors will become more difficult in the next section. So please try out as many combinations of `graph` options, `if` and `in` qualifiers, and variables as you can.

11.3.7 Avoiding changes in the dataset

A program that generates variables is usually inconvenient. No one really enjoys deleting the variables generated by the previous command every time they use it. Therefore, an obvious solution would be to delete the newly generated variables at the end of every program; but this is not much better either, as it assumes that the program will actually run until it finishes. If the user aborts the program or if it crashes and displays an error message, the variables generated up until then will remain in the dataset.

The solution is to use temporary variables. You have already come across a temporary variable when you were using `marksample`. Temporary variables are automatically removed when a program exits and are defined in two steps:

1. Declaring the variables' names
2. Generating the temporary variables

You declare temporary variables using the `tempvar` command, and you generate them using `generate`. However, the variable names are enclosed in single quotation marks, as is the case with local macros.

If, for example, you wanted to include the temporary variables `x1` and `x2` in a program, you could code

```
. tempvar x1 x2
. gen `x1' = 1
. gen `x2' = `x1'/_n
```

Here we generated variables within the `kdensity` command to hold the density estimations. To generate these variables as temporary variables, you declare them within the `foreach` loop and call these temporary variables in option `gen()` of the `kdensity` command:

```
——————————————————————————————— top: denscomp.ado ———
✂
12:     foreach k of local K {
13:         tempvar g`k'
14:         quietly gen `g`k'' = `by'==`k' if `by' < . &  `touse'
15:         local G "`G' `g`k''"
16:     }
——————————————————————————————— end: denscomp.ado ———
```

Note the two single quotation marks. The first indicates the end of the local macro `k`, whereas the second indicates the end of the temporary variable. You also need to replace every other instance of `fx`k'` in the program with ``fx`k''`. After this step, a first beta version of the ado-file is ready, and figure 11.1 shows the entire code.

top: denscomp.ado

```
 1: program denscomp
 2: version 10.0
 3: syntax varname(numeric) [if] [in], by(varname) at(varname numeric) ///
 4:   [TItle(string asis) YTItle(string asis) XTItle(string asis) * ]
 5: marksample touse
 6:
 7: // Store levels of By-Variable
 8: quietly levelsof `by' if `touse', local(K)
 9:
10: // Check Graph-Options
11: if "`options'" ~= "" {
12:     foreach k of local K {
13:         tempvar g`k'
14:         quietly gen `g`k'' = `by'==`k' if `by' < . &  `touse'
15:         local G "`G' `g`k''"
16:     }
17:     graph twoway scatter `G' `at', `options' nodraw
18: }
19:
20: // Calculate Kernel-Densities
21: local i 1
22: foreach k of local K {
23:     tempvar fx`k' xvals`k'
24:     kdensity `varlist' if `by'==`k' & `touse', generate(`xvals`k'' `fx`k'') ///
25:          at(`at') nograph
26:     local yvars "`yvars' `fx`k''"
27:     local label: label (`by') `k'
28:     local leglab `" `leglab' lab(`i++' `"`label'"') "'
29: }
30:
31: // Graph-Defaults
32: local title = cond(`"`title'"' == `""', `"`varlist' by `by'"', `"`title'"')
33: local ytitle = cond(`"`ytitle'"' == `""', `"Density"', `"`ytitle'"')
34: local xtitle = cond(`"`xtitle'"' == `""', `"`varname'"', `"`xtitle'"')
35:
36: // Display the Graph
37: graph twoway connected `yvars' `at' ///
38:   , title(`title') ytitle(`ytitle') xtitle(`xtitle') ///
39:     legend(`leglab') `options'
40: end
```

end: denscomp.ado

Figure 11.1. Beta version of `denscomp.ado`

11.3.8 Help files

Once you are satisfied with your ado, you should write a help file. Help files are straightforward ASCII files that are displayed on screen by Stata when one types

```
. help command
```

For Stata to find the help file for a command, the help file must have the same name as the ado-file, albeit with the file extension `.sthlp`. Here the help file would therefore be called `denscomp.sthlp`.

For your help file to have the same look and feel as common Stata help files, you can use the Stata Markup and Control Language (SMCL). In Stata, all output is in SMCL format, be it normal command output or output from invoking a help file. In SMCL, directives are used to affect how output appears. If, for example, your help text includes

```
                                              top: denscomp.sthlp
✂
15: {p 4 4 2}
16: {cmd:denscomp} uses {help kdensity} to estimite densities of varname
17: for categories of by().
                                              end: denscomp.sthlp
```

when you call up help with `help denscomp`, the line "denscomp uses kdensity to estimate densities of varname for categories of by()" will be displayed, with "denscomp" displayed in white and "kdensity" displayed in blue and clickable. That is, by clicking on **kdensity**, you will call up the help text for `kdensity`.

In [P] **smcl**, you will find further information on help, as well as recommendations for organizing your help.

11.4 Exercises

1. Create a local macro that holds the name of the directory in which you have stored the data package of this book.
2. In Stata, change to an arbitrary directory and load the `data1.dta` into memory using your local macro.
3. Change to the directory in which you have stored our data package using your local macro.
4. Create a new dataset of all the GSOEP person files (`kksoep/*p`). Save this dataset for later use.
5. Rename the variables `anetto`, `bnetto`, ..., `snetto` into `netto1984`, `netto1985`, ..., `netto2002` with a loop.
6. Rename all variables for life-satisfaction into `lsat1984`, `lsat1985`, ..., `lsat2002` with a loop.
7. Add to your last loop a command that keeps track of the original variable name using `note`. After reloading your dataset, rerun the loop for life satisfaction.
8. Save your loop as a program `mysoepren`. Run the program after having reloaded your dataset.
9. Store your program `mysoepren` in the file `mysoepren.ado` in the current working directory of your computer. Exit and relaunch Stata and rerun `mysoepren`.
10. Change `mysoepren.ado` such that you can parse an arbitrary list of variables to be renamed. Rerun the program for the life-satisfaction variables and the "Netto"-Variables.

11. Change `mysoepren.ado` such that you can use arbitrary numbers for the new variable names. Rerun the program for the life-satisfaction variables and the "Netto"-Variables using numbers 1–19 instead of 1984–2002.
12. Change `mysoepren.ado` such that you will get a confirmation if the number of variable names to be renamed and the number of new variable names match. Use the command to rename only the first 5 occurrences of the household number.

12 Around Stata

12.1 Resources and information

In addition to the manuals, there are several resources available to help you learn Stata. The Stata web page http://www.stata.com has the most complete and up-to-date versions of official resources, as well as links to a wealth of user-written resources. Of particular interest are

- The *Stata Journal* (SJ):[1] A printed peer-reviewed quarterly journal, containing articles about data analysis with Stata. The *Stata Journal* replaced the *Stata Technical Bulletin* (STB)[2]. The *Stata Journal* also publishes Stata software for general use (see section 12.3.1).
- FAQs:[3] A web page with answers to frequently asked questions.
- The Statalist:[4] An open Internet discussion forum where you can post questions about tricky Stata problems or general statistical techniques. Even simple questions will be posted and answered, so do not be shy. However, we recommend that you consult Stata's excellent documentation and check out the Statalist archive[5] before you post a question to Statalist. To participate, send an email to *majordomo@hsphsun2.harvard.edu* with "subscribe statalist" in the body of the message. The subject line remains empty. To unsubscribe, send an email with "unsubscribe statalist" to the same address. Participating in Statalist is free.
- Stata Bookstore:[6] A collection of books about Stata, as well as an excellent selection of books regarding particular statistical techniques; many of these books use Stata for illustration. You can order the books online.
- Stata Press:[7] The publisher of the Stata manuals, the *Stata Journal*, and many books on Stata. The Stata Press offers a web page that also contains all datasets used in the manuals.

---

1. http://www.stata-journal.com
2. http://www.stata.com/support/stb/faq/
3. http://www.stata.com/support/faqs/
4. http://www.stata.com/statalist/
5. http://www.stata.com/statalist/archive/
6. http://www.stata.com/bookstore/
7. http://www.stata-press.com

- Stata NetCourses:[8] Internet courses offered and run by StataCorp. Once a week the instructor will post lecture notes and assignments on a password-protected web site. The discussion between participants and instructors takes place on a message board at the NetCourse web site. Instructors will respond to questions and comments usually on the same day as they are posted. NetCourses are available on beginner and advanced levels. There is a course fee for the NetCourses. Participation requires a valid license number.
- Additional links:[9] Links to many other resources, the impressive collection of Stata textbook and paper examples from UCLA Academic Technology Service (ATS).[10]

12.2 Taking care of Stata

In the past, a new Stata version has been released roughly every 2 years. However, this does not mean that there is no development between new releases. On the contrary, Stata updates are available on average every couple of months. These updates include improvements to existing commands, as well as more commands that have been added since the last Stata version. Even if you are not interested in more commands, some of the changes may, for example, affect the overall speed of the program. We therefore recommend that you regularly update your Stata installation. Updating Stata is free.

The changes provided in the updates may affect internal and external commands. Internal commands are part of the file `wstata.exe`.[11] If internal commands have changed since the last update, you need to replace the old executable with the new file.

External commands are called ado-files or ados. The functionality of ados is the same as for internal Stata commands. As a user, you will not notice the difference. In fact, many Stata commands are ados. You will get new ados with almost every Stata update.

We distinguish between official and unofficial ados. Official ados are distributed by StataCorp with the installation CD-ROM and with the updates. Everyone who installed Stata will have them. Unofficial ados will be only on those machines where the user has installed them. Here we will show you how to update the internal commands and the official external commands. The next section will discuss how to install unofficial external commands.

Taking care of the official Stata installation takes place with the command `update`. To begin, please type

8. http://www.stata.com/netcourse/
9. http://www.stata.com/links/resources3.html
10. http://www.ats.ucla.edu/stat/stata/
11. This is true for Stata for Windows. Other versions of Stata use different names for the internal commands, for example, `xstata` for the Linux version of Stata.

```
. update query
(contacting http://www.stata.com)
Stata executable
    folder:              C:\Program Files\Stata10\
    name of file:        wstata.exe
    currently installed: 15 Oct 2007
    latest available:    27 May 2008
Ado-file updates
    folder:              C:\Program Files\Stata10\ado\updates\
    names of files:      (various)
    currently installed: 13 Nov 2007
    latest available:    27 May 2008
Recommendation
    Type -update all-
Click to edit automatic update checking preferences
```

The command `update query` compares information of the installed files to the latest available version of these files. In our example, internal (Stata executable) and external commands (ado-files) are outdated. Stata therefore recommends to update both of them. To do so, type

```
. update all
```

If you want to update only external commands, you can enter `update ado`, or to update internal commands, enter `update executable`. Stata will then connect to the Internet, search for the required files, and save them at the right place on your machine. That's all that is to be done for the ados. The new executable with the internal commands will, however, be downloaded and saved as `wstata.bin` in your Stata directory. This file needs to be renamed to `wstata.exe`. This can be done automatically with the Stata command

```
. update swap
```

In case `update swap` does not work for you (e.g., because you do not have write permission in this particular directory), you can always rename the files manually. You need to close Stata, rename the old executable `wstata.exe` as e.g., `wstata.old`, and rename `wstata.bin` as `wstata.exe`.

In any case, after successfully updating, you get a list of changes by typing `help whatsnew`.

The procedure described above assumes that you have a direct connection to the Internet. If this is not the case for your computer, you can download the official update as a compressed zip archive.[12] Once you are at your machine, you unzip the archive with the appropriate program. However, you need to make sure that you keep the archive directory structure. Now you can proceed as described above, using the option `from()` of the `update` command. Just type the name of the directory that you used to unpack the zip archive name within the parentheses (e.g., `update ado, from(c:\temp)`).

12. http://www.stata.com/support/updates/

12.3 Additional procedures

Any Stata user can write ados (see chapter 11), so many statisticians and others all over the world have written ados for statistical procedures that do not exist in Stata. Once an ado is written, any user can install it and use it just like any other command. Here we will show you how to install such user-written programs.

12.3.1 SJ and STB ado-files

We call ados published with the *Stata Journal* or the *Stata Technical Bulletin* SJ-ados and STB-ados, respectively. Stata itself knows about SJ-ados or STB-ados; that is, the command `search` will provide information about their existence. SJ- and STB-ados can be downloaded for free from the Internet.

Say that you want to present the results of all the many regressions you run in a format appropriate for publication. You want to display the t statistic or the standard error in parentheses behind each coefficient, you want to add stars to indicate their significance level, and you want to print summary measures like R^2 below the coefficients. To do so, you could of course type all numbers by hand in your word-processing program, but this would be time consuming, error prone, and not easily reproduced. If you are lucky, you might find a Stata command that does the work for you. One way to find such a command is to use the Stata command `search`. Try, for example

```
. search regression output
```

which will give you several hits. Among them are

```
  (output omitted)
SJ-7-2  st0085_1  . . . . . . . . . . . . .  Making regression tables simplified
        (help estadd, estout, _eststo, eststo, esttab if installed) .  B. Jann
        Q2/07   SJ 7(2):227--244
        introduces the eststo and esttab commands (stemming from
        estout) that simplify making regression tables from stored
        estimates
  (output omitted)
```

The entry you see refers to issue 2 in volume 7 of the *Stata Journal*. This means that there is an article with the title "Making regression tables simplified" written by Ben Jann. There is an entry number attached to the article, `st0085_1`, that we will discuss a little later. More important is the description "(help estadd, estout ... if installed)". This line indicates that you are dealing with a program (or several programs), because only programs can be installed. From the short description of the program, we learn that in this version some small bugs had been fixed. Our example introduces additions (`eststo` and `esttabs` to an already existing program (`estout`). The program itself simplifies the creation of regression tables from stored estimates. Sounds good! That is what we were looking for. You will learn more if you read the respective SJ article or if you install the program and read the online help.

The easiest way to install SJ- or STB-ados is to click on the entry number. This will open a window that guides you through the installation. However, if you want to install a program within your do-files, you can use the `net` commands. Then the installation happens in two steps:

1. The command `net sj` *volume-issue* will connect you to a web address of the *Stata Journal* holding all ados published in the specified issue. Likewise, the command `net stb` *issue* will do the same for STB-ados.
2. The command `net install` *pkgname* will install the program *pkgname*.

For example, to install the program `estout`, you would type

```
. net sj 7-2
________________________________________________________________________________
http://www.stata-journal.com/software/sj7-2/
(no title)
________________________________________________________________________________

DIRECTORIES you could -net cd- to:
    ..                Other Stata Journals

PACKAGES you could -net describe-:
    gr0001_3          Update:  Generalized Lorenz curves and related graphs
    st0085_1          Update:  Making regression tables simplified
    st0097_1          Update: Generalized ordered logit/partial
                      proportional odds models for ordinal dependent variables
    st0122            Fit population-averaged panel-data models using
                      quasileast squares
    st0123            Maximum likelihood and two-step estimation of an
                      ordered-probit selection model
    st0124            Two postestimation commands for assessing confounding
                      effects in epidemiological studies
    st0125            Estimation of Nonstationary Heterogeneous Panels
    st0126            QIC criterion for model selection in GEE analyses
________________________________________________________________________________
```

which will show a list of all available programs in volume 7 issue 2 of the SJ. Among others, you find the package "Making regression tables simplified" with the entry "st0085_1". This is the same entry number you have seen with `search`, and it is the package name that you use to install the program:

```
. net install st0085_1
```

If you want to see a description of the package before you install it, you can type

```
. net describe st0085_1
```

12.3.2 SSC ado-files

Another source of user-written Stata commands is the Statistical Software Components (SSC) archive maintained by Boston College.[13] Many of the commands stored there

13. http://ideas.repec.org/s/boc/bocode.html

are discussed on Statalist. In Stata, the `ssc install` command can be used to install them; for more information, type `help ssc`.

For example, in the SSC archive is a command called `listtex` written by the British biostatistician Roger Newson. The program writes the output of Stata's `list` command in a file that can be processed by word processing systems (MS Word, Open Office Writer) or presentation software (Power Point, Open Office Impress).

```
. ssc install listtex
```

12.3.3 Other ado-files

In addition to SJ-, STB-, and SSC-ados, there are many other user-written commands. You can get an idea of the variety of ados by following one of the links displayed when you type

```
. net from http://www.stata.com
```

Most of these ados can be installed with `net install`, so the only challenge is to actually find the right ado for your problem. However, before we discuss how to find the right ado, you should know how to install ado-files in case Stata's `net` facilities do not work. This can happen if

- the computer you work on is not connected to the Internet,
- you programmed your own ado,
- the author of the ado did not prepare the file appropriately. To detect an ado package with `net from` and to install it with `net install`, there must be a table of contents file (`stata.toc`) and a package description file stored under the URL. If an author does not provide these files, you will not be able to use the `net` commands to install the ado,
- the `stata.toc` refers to packages on another computer. Then the particular file must be downloaded with `net get` and installed by hand, as explained below.

Usually you can get the desired ado packages outside Stata and install them by hand. Every ado package should have at least two files: the ado-file itself and an accompanying help file. The ado-file has the extension `.ado`, the help file the extension `.sthlp` or `.hlp`. To install both of them, you copy the files in your personal ado-directory.

You can determine the location of your personal ado-directory with

```
. adopath
```

Official ados are saved in the first two directories (`UPDATES` and `BASE`). The third directory, called `SITE`, is meant for network administrators to install ados that should be accessible for all network users. The fourth directory, here indicated with "`.`", refers to your current working directory. Finally, you find the address of your personal ado-directory (`PERSONAL`), in which you can copy ados "by hand". Ados you wrote yourself should be stored here. The most common place for the personal ado-directory under Windows is `c:\ado\personal`; however, this depends on the local installation. You must create `c:\ado\personal` by hand; this can be done using the Stata command `mkdir`. The `c:\ado\plus` directory is created by Stata when you first use `net install` or `ssc install`.

Finally, the `PLUS` directory contains ados you have downloaded with Stata's Internet facilities (e.g., `net install` and `ssc install`), and `OLDPLACE` is the place where ados before Stata 7.0 were installed. The order of the path matches the order with which ados will be searched on your machine when you type a command (see section 11.2.4).

Now back to the interesting question: where do you find the "right" ado? We have already discussed this above with respect to the SJ-ados. For SSC-ados, you can also browse the SSC archive with any Internet browser.[14] Also you will find several links on the Stata web site.[15] The best way to find Stata ados, however, is by adding the option `all` to `search`. This searches in the online help, *as well as* the FAQs at the Stata web site, the *Stata Journal*, and many other Stata-related Internet sources. You can try `search, all` with arbitrary keywords, known package names, or just author names. `search` displays results with the addresses of entries matching your search term. If you click on the address, you will be led through the installation process.

12.4 Exercises

1. Find out whether your Stata is up to date.
2. If necessary, update your Stata to the most recent version *or* explain to your local network administrator how he or she can update Stata.
3. Explore the changes of the most recent Stata update.
4. Install the following frequently used SJ ado-files:

    a. The value label utilities `labeldup` and `labelrename` from SJ 5-2.
    b. The package `dm75` of STB-53 for safe and easy matched merging.
    c. The lean graphic schemes from SJ 4-3.
    d. The enhanced graphing model diagnostic tools described in SJ 4-4.

14. http://ideas.repec.org/search.html
15. http://www.stata.com/links/resources.html

5. Install the following ado-packages from the SSC archive:

   a. `egenmore`
   b. `fitstat`
   c. `tostring`
   d. `vclose`
   e. `fre`

6. Find and install user-written packages for the following tasks:

   a. Center metric variables
   b. Draw thematic maps
   c. Sequence analysis

7. Starting from http://www.stata.com/users/, explore the ado-directories of the following users or institutions:

   a. Nicholas J. Cox
   b. Bill Rising
   c. University of California in Los Angeles (UCLA)
   d. Jeroen Weesie
   e. Scott Long and Jeremy Freese
   f. Ben Jann

References

Agresti, A. 1984. *Analysis of Ordinal Categorical Data.* New York: Wiley.

Aiken, L. S., and S. G. West. 1991. *Multiple Regression: Testing and Interpreting Interactions.* Newbury Park, CA: Sage.

Aldrich, J., and F. Nelson. 1984. *Linear Probability, Logit, and Probit Models.* Beverly Hills: Sage.

Anderson, J. A. 1984. Regression and ordered categorical variables (with discussion). *Journal of the Royal Statistical Society, Series B* 46: 1–30.

Andreß, H.-J., J. A. Hagenaars, and S. Kühnel. 1997. *Analyse von Tabellen und kategorialen Daten. Log-lineare Modelle, latente Klassenanalyse, logistische Regression und GSK-Ansatz.* Berlin: Springer.

Anscombe, F. J. 1973. Graphs in statistical analysis. *American Statistcian* 27: 17–21.

Baltagi, B. 2008. *Econometric Analysis of Panel Data.* 4th ed. New York: Wiley.

Baum, C. F. 2006. *An Introduction to Modern Econometrics Using Stata.* College Station: Stata Press.

Belsley, D. A., E. Kuh, and R. E. Welsch. 1980. *Regression Diagnostics: Identifying Influential Data and Sources of Collinearity.* New York: Wiley.

Berk, K. N., and D. E. Both. 1995. Seeing a curve in multiple regression. *Technometrics* 37: 385–398.

Berk, R. A. 2004. *Regression Analysis: A Constructive Critique.* Thousand Oaks: Sage.

Bollen, K. A., and R. W. Jackman. 1990. Regression Diagnostics: An Expository Treatment of Outliers and Influential Cases. In *Modern Methods of Data Analysis,* ed. J. Fox and J. S. Long, 257–291. Newbury Park, CA: Sage.

Chambers, J. M., W. S. Cleveland, B. Kleiner, and P. A. Tukey. 1983. *Graphical Methods for Data Analysis.* New York: Chapman and Hall.

Cleveland, W. S. 1979. Robust locally weighted regression and smoothing scatterplots. *Journal of the American Statistical Association* 74: 829–836.

———. 1984. Graphical methods for data presentation: Full scale breaks, dot charts, and multibased logging. *American Statistician* 38: 261–269.

———. 1993. *Visualizing Data*. Summit, NJ: Hobart Press.

———. 1994. *The Elements of Graphing Data*. 2nd ed. Summit: Hobart Press.

Conover, W. J. 1999. *Practical Nonparametric Statistics*. Wiley Series in Probability & Mathematical Statistics, New York: Wiley.

Cook, R. D. 1977. Detection of influential observations in linear regression. *Technometrics* 19: 15–18.

Cook, R. D., and S. Weisberg. 1994. *Introduction to Regression Graphics*. New York: Wiley.

———. 1999. *Applied Regression Including Computing and Graphics*. New York: Wiley.

Cox, N. J. 2002a. Speaking Stata: How to face lists with fortitude. *Stata Journal* 2: 202–222.

———. 2002b. Speaking Stata: How to move step by: step. *Stata Journal* 2: 86–102.

———. 2004. Speaking Stata: Graphing distributions. *Stata Journal* 4: 66–88.

Diekmann, A. 1998. Die Bedeutung der Sekundäranalyse zur Kontrolle von Forschungsergebnissen. In *Statistik im Dienste der Öffentlichkeit*, ed. W. Haug, K. Armingeon, P. Farago, and M. Zürcher. Bern: Bundesamt für Statistik.

Diggle, P. J., K.-Y. Liang, and S. L. Zeger. 1994. *Analysis of Longitudinal Data*. New York: Oxford Univeristy Press.

Efron, B., and R. J. Tibshirani. 1993. *An Introduction to the Bootstrap*. New York: Chapman and Hall.

Emerson, J. D., and D. C. Hoaglin. 1983. Stem-and-Leaf Displays. In *Understanding Robust and Exploratory Data Analysis*, ed. D. C. Hoaglin, M. Frederick, and J. W. Tukey, 7–30. New York: Wiley.

Fahrmeir, L., R. Künstler, I. Pigeot, and G. Tutz. 1997. *Statistik. Der Weg zur Datenanalyse*. Berlin: Springer.

Fienberg, S. E. 1980. *The Analysis of Cross-Classified Categorical Data*. Cambridge, MA: MIT Press.

Fisher, R. A. 1935. The logic of inductive inference. *Journal of the Royal Statistical Society, Series A* 98: 39–54.

Fowler, F. J. 1984. *Survey Research Methods*. Beverly Hills: Sage.

Fox, J. 1991. *Regression Diagnostics: An Introduction*. Newbury Park, CA: Sage.

———. 1997. *Applied Regression Analysis, Linear Models, and Related Methods*. Thousand Oaks: Sage.

———. 2000. *Nonparametric Simple Regression: Smoothing Scatterplots.* Thousand Oaks: Sage.

Freedman, D. A. 2004. The Ecological Fallacy. In *Encyclopedia of Social Science Research Methods*, ed. M. Lewis-Beck, A. Bryman, and T. Liao. Thousand Oaks: Sage.

Gould, W., J. Pitblado, and W. Sribney. 2003. *Maximum Likelihood Estimation with Stata.* 3rd ed. College Station: Stata Press.

Greene, W. H. 2008. *Econometric Analysis.* 6th ed. New York: Prentice Hall.

Gujarati, D. N. 1995. *Basic Econometrics.* New York: McGraw-Hill.

Hagle, T. M. 1996. *Basic Math for Social Scientists: Problems and Solutions*, vol. 109. Thousand Oaks: Sage.

Hair, J. F., R. E. Anderson, R. L. Tatham, and W. C. Black. 1995. *Multivariate Data-Analysis with Readings.* 4th ed. London: Prentice-Hall International.

Hamilton, L. C. 1992. *Regression with Graphics: A Second Course in Applied Statistics.* Belmont: Duxburry Press.

Hardin, J. W., and J. M. Hilbe. 2003. *Generalized Estimating Equations.* New York: Chapman & Hall/CRC.

Hosmer, D. W., Jr., and S. Lemeshow. 2000. *Applied Logistic Regression.* 2nd ed. New York: Wiley.

Howell, D. C. 1997. *Statistical Methods for Psychology.* Belmont: Duxbury Press.

Huff, D. 1954. *How to Lie with Statistics.* New York: Norton.

Kennedy, P. 1997. *A Guide to Econometrics.* 3rd ed. Cambridge: MIT Press.

King, G., R. O. Keohane, and S. Verba. 1994. *Designing Social Inquiry.* Princeton: Princeton University Press.

Kish, L. 1965. *Survey Sampling.* New York: Wiley.

Kreuter, F., and R. Vailliant. 2007. A survey on survey statistics: What is done and can be done in Stata. *Stata Journal* 7: 1–21.

Lee, E. S., R. N. Forthofer, and R. J. Lorimor. 1989. *Analyzing Complex Survey Data.* Newbury Park, CA: Sage.

Lehtonen, R., and E. J. Pahkinen. 1995. *Practical Methods for Design and Analysis of Complex Surveys.* New York: Wiley.

Levy, P. S., and S. Lemeshow. 2008. *Sampling of Populations: Methods and Applications.* 4th ed. New York: Wiley.

Long, J. S. 1997. *Regression Models for Categorical and Limited Dependent Variables.* Thousand Oaks: Sage.

Long, J. S., and J. Freese. 2006. *Regression Models for Categorical Dependent Variables Using Stata.* 2nd ed. College Station: Stata Press.

Mallows, C. L. 1986. Augmented partial residuals. *Technometrics* 28: 313–319.

Marron, J. S. 1988. Automatic smoothing parameter selection: A survey. *Empirical Economics* 13: 187–208.

McFadden, D. 1973. Conditional Logit Analysis of Qualitative Choice Behavior. In *Frontiers in Econometrics*, ed. P. Zarempka, 105–142. New York: Academic Press.

Mitchell, M. N. 2008. *A Visual Guide to Stata Graphics.* 2nd ed. College Station: Stata Press.

Mooney, C. Z., and R. D. Duval. 1993. *Bootstrapping: A Nonparametric Approach to Statistical Inference.* Newbury Park, CA: Sage.

Moore, D. S. M., and G. P. M. McCabe. 2005. *Introduction to the Practice of Statistics.* 5th ed. San Francisco: Freeman.

Mosteller, F., and J. W. Tukey. 1977. *Data Analysis and Regression: A Second Course in Statistics.* Reading, MA: Addison–Wesley.

Pannenberg, M., R. Pischner, U. Rendtel, and G. Wagner. 1998. Sampling and Weighting. In *Desktop Companion to the German Socio-Economic Panel Study (GSOEP). Version 2.0*, ed. J. Haisken-De New and J. R. Frick, 97–118. Berlin: DIW.

Pearson, K. 1900. On the criterion that a given system of deviations from the probable in the case of a correlated system of variables is such that it can be resonably supposed to have arisen from random sampling. *Philosophical Magazine, Series 5* 50: 157–175.

Popper, K. R. 1994. *Logik der Forschung. 10. Auflage.* Tübingen: J.C.B. Mohr.

Rabe-Hesketh, S., and B. Everitt. 2007. *A Handbook of Statistical Analyses Using Stata.* 4th ed. Boca Raton: Chapman & Hall/CRC.

Rabe-Hesketh, S., and A. Skrondal. 2008. *Multilevel and Longitudinal Modeling Using Stata.* College Stata, TX: Stata Press.

Raftery, A. E. 1995. Bayesian Model Selection in Social Research. In *Sociological Methodology Volume 25*, ed. P. V. Marsden, 111–163. Oxford: Blackwell.

Schnell, R. 1994. *Graphisch gestützte Datenanalyse.* München u. Wien: Oldenbourg.

Schnell, R., and F. Kreuter. 2005. Separating interviewer and sampling-point effects. *Journal of Official Statistics* 3: 389–410.

Skinner, C. J., D. Holt, and T. M. F. Smith. 1989. *Analysis of Complex Surveys.* Chichester, UK: Wiley.

Statistisches Bundesamt, ed. 1997. *Statistisches Jahrbuch 1997.* Stuttgart: Metzler-Poeschel.

Stine, R. 1990. An Introduction to Bootstrap Methods. In *Modern Methods of Data Analysis*, ed. J. Fox and J. S. Long, 325–372. Newbury Park, CA: Sage.

Tufte, E. R. 2001. *The Visual Display of Quantitative Information.* 2nd ed. Cheshire, CT: Graphics Press.

Veall, M. R., and K. F. Zimmermann. 1994. Evaluating pseudo-R^2's for binary probit models. *Quality and Quantity* 28: 151–164.

Wainer, H. 1984. How to display data badly. *American Statistican* 38: 137–147.

Wooldridge, J. M. 2002. *Econometric Analysis of Cross Section and Panel Data.* Cambridge, MA: MIT Press.

Author index

Subject index

E